Imaging Floods and Glacier Geohazards with Remote Sensing

Imaging Floods and Glacier Geohazards with Remote Sensing

Editors

Francesca Cigna
Hongjie Xie
Karem Chokmani

MDPI • Basel • Beijing • Wuhan • Barcelona • Belgrade • Manchester • Tokyo • Cluj • Tianjin

Editors
Francesca Cigna
Italian Space Agency (ASI)
Italy

Hongjie Xie
University of Texas at San Antonio
USA

Karem Chokmani
Institut National de la Recherche Scientifique (INRS)
Canada

Editorial Office
MDPI
St. Alban-Anlage 66
4052 Basel, Switzerland

This is a reprint of articles from the Special Issue published online in the open access journal *Remote Sensing* (ISSN 2072-4292) (available at: https://www.mdpi.com/journal/remotesensing/special_issues/floods_glacier_geohazards).

For citation purposes, cite each article independently as indicated on the article page online and as indicated below:

LastName, A.A.; LastName, B.B.; LastName, C.C. Article Title. *Journal Name* **Year**, *Volume Number*, Page Range.

ISBN 978-3-0365-0066-9 (Hbk)
ISBN 978-3-0365-0067-6 (PDF)

Cover image courtesy of NASA/Christy Hansen.

Contents

About the Editors

Francesca Cigna (Ph.D.), Researcher in Earth observation and data analytics at the Italian Space Agency (ASI), holds a Ph.D. in Earth sciences and remote sensing and MSc and BSc in environment and territory engineering. Her key areas of interest and specialty are satellite radar, InSAR, natural and man-made hazards and risks, shallow geological processes, ground instability and land changes, landscape archaeology and cultural heritage. She leads research on Earth observation from space-borne platforms, image processing and time series analysis, mainly focused on radar through past, current and future SAR missions such as the COSMO-SkyMed constellation, ESA's ERS-1/2 and ENVISAT and the Copernicus Sentinel-1 mission. She collaborates with the Working Group on Disasters of the Committee for Earth Observation Satellites (CEOS).

Hongjie Xie (Ph.D.), Professor and Department Chair, Department of Geological Sciences, University of Texas at San Antonio (UTSA), USA. He is also the founding director for the NASA MIRO Center for Advanced Measurements in Extreme Environments at UTSA. His primary research interests are remote sensing and GIS technologies, theories, and applications. He has strong experience in image processing, computer software development and GIS programming, and is particularly interested in interdisciplinary research by integrating remote sensing, GIS, field measurements to geology (the Earth and the Mars), surface hydrology, cryosphere, terrestrial ecosystems, urban development, and environmental studies.

Karem Chokmani (Ph.D.) is Professor in remote sensing and statistical hydrology at the National Institute of Scientific Research (INRS) in Canada and Scientific Leader of the Laboratory for Environmental Remote Sensing by Drone. He holds a Ph.D. in geomatics and MSc and BSc in agricultural engineering. His fields of interest focus on the estimation of water resources at the local and regional level through statistical hydrology, remote sensing and geomatics. He is interested in the study of spatial and temporal variability of water resources in a changing climate. He is engaged in the development of stochastic modeling approaches for monitoring and assessment of changes in the cryosphere. He also works on the development of algorithms for the monitoring of water quality of rivers and lakes using remote sensing.

Preface to "Imaging Floods and Glacier Geohazards with Remote Sensing"

Remote sensing plays a pivotal role in understanding where and how floods and glacier geohazards occur; their severity, causes and types; and the risk that they may pose to populations, activities and properties. By providing a spectrum of imaging capabilities, resolutions and temporal and spatial coverage, remote sensing data acquired from satellite, aerial and ground-based platforms provide key geo-information to characterize and model these processes.

This book includes research papers published in the Special Issue "Imaging Floods and Glacier Geohazards with Remote Sensing" of the journal Remote Sensing.

Launched in mid-2018, the Special Issue gathered 1 editorial and 11 research articles on novel technologies (e.g., sensors, platforms), data (e.g., multi-spectral, radar, laser scanning, GPS, gravity) and analysis methods (e.g., change detection, offset tracking, structure from motion, 3D modeling, radar interferometry, automated classification, machine learning, spectral indices, probabilistic approaches) for flood and glacier imaging.

Through target applications and case studies distributed globally, these articles contribute to the discussion on the current potential and limitations of remote sensing in this specialist research field, as well as the identification of trends and future perspectives.

Francesca Cigna, Hongjie Xie, Karem Chokmani
Editors

 *remote sensing*

Editorial

Imaging Floods and Glacier Geohazards with Remote Sensing

Francesca Cigna [1],[*] and Hongjie Xie [2]

[1]   Italian Space Agency, Via del Politecnico snc, 00133 Rome, Italy
[2]   Center for Advanced Measurements in Extreme Environments and Department of Geological Sciences,
     University of Texas at San Antonio, San Antonio, TX 78249, USA; hongjie.xie@utsa.edu
[*]   Correspondence: francesca.cigna@asi.it

Received: 23 November 2020; Accepted: 24 November 2020; Published: 26 November 2020

Geohazards associated with the dynamics of the liquid and solid water of the Earth's hydrosphere, such as floods and glacial processes, may pose significant risks to populations, activities and properties. Adverse weather, tsunamis, storm surges, sea level rise or even changes in land use (e.g., infrastructure projects and resource exploitation) may cause coastal, fluvial and surface-water inundations. Heavy snowmelt, ice jams and dam failure can lead to catastrophic flooding. Rock, snow and ice avalanches impacting glacial lakes can trigger outburst floods. Sea ice and icebergs may disrupt ship circulation along sea lanes worldwide.

Understanding how these geohazards occur, their severity, causes and types and the damage they cause helps to design and improve forecasting methods and risk mitigation approaches. By providing a spectrum of imaging capabilities, resolutions, temporal and spatial coverage, remote sensing plays a pivotal role in achieving these objectives.

Developed within the "Remote Sensing in Geology, Geomorphology and Hydrology" section of the journal *Remote Sensing* as part of a growing series of thematic volumes (e.g., [1–5]), the Special Issue "Imaging Floods and Glacier Geohazards with Remote Sensing" [6] was launched in mid-2018 with the aim to gather research articles and reviews on the use of satellite, aerial and ground-based remote sensing to image floods and glacier geohazards. One of the key goals was to collect research studies on novel technologies (e.g., new sensors and platforms), data (e.g., multi-spectral, radar, laser scanning, GPS and gravity) and analysis methods (e.g., change detection, offset tracking, structure from motion, 3D modeling, radar interferometry, automated classification, machine learning, spectral indices and probabilistic approaches), as well as case studies distributed globally and discussions of current trends and future perspectives in this research field.

The Special Issue project was collaboratively led by an international team of three Guest Editors: Dr Francesca Cigna from the Italian Space Agency in Italy, Prof Hongjie Xie from the University of Texas at San Antonio in the USA and Prof Karem Chokmani from the National Institute of Scientific Research in Canada. The three Guest Editors handled a total of 19 manuscripts over the 21 months when the call for papers was disseminated and the system was open for submissions, namely from May 2018 until the end of February 2020 [6]. The average time from submission to acceptance was 63 days, while the time from acceptance to online publication was 4 days. The first paper was published on 24 March 2019, and the last on 8 May 2020.

In total, 66 authors contributed to the submitted manuscripts, and a team of 35 anonymous international experts in the field of flood and glacier remote sensing was involved in the peer-review process to help the Guest Editors ensure a rigorous assessment of the submissions during the course of the Special Issue project. On average, 2–3 reviewers provided feedback on each manuscript, and some reviewers were involved in the assessment of more than one submission.

In the following paragraphs, this editorial paper provides an overview of the research articles composing the Special Issue (Table 1), via a summary of the remote sensing data and methods used and the initial scientific impact achieved in the first few months after publication of the last paper.

Table 1. Remote sensing data, methods and areas of interest discussed in the 11 research papers composing the Special Issue (sorted in ascending order, according to the publication date). Notation: BSI, Bare Soil Index; DEM, Digital Elevation Model; DInSAR, Differential Interferometric SAR; GEE, Google Earth Engine; GEOBIA, GEographic Object-Based Image Analysis; GPS, Global Positioning System; GRACE, Gravity Recovery and Climate Experiment; LWM, Land and Water Mask; NDMI, Normalized Difference Moisture Index; NDVI, Normalized Difference Vegetation Index; NDWI, Normalized Difference Water Index; NPCRI, Normalized Pigment Chlorophyll Ratio Index; RTK, Real-Time Kinematic; SAR, Synthetic Aperture Radar; SfM, Structure from Motion; SRTM, Shuttle Radar Topography Mission; SWIR, Short Wave InfraRed; UAV, Unmanned Aerial Vehicle.

Article	Remote Sensing Data and Methods	Event and/or Area of Interest
Paul 2019 [7]	Corona KH4-A/B declassified, Landsat-5/7/8, Sentinel-2, QuickBird and WorldView; SRTM and High Mountain Asia DEMs; red/SWIR band ratios, contrast-enhancement, DEM differencing	2004, 2007 and 2016 glacier collapses and surges at Amney Machen mountain range (China)
Benoudjit and Guida 2019 [8]	TerraSAR-X and Sentinel-1 SAR; Landsat-5 and Sentinel-2; iterative optimization through stochastic gradient descent, NDWI, supervised classifier, automated flood extent mapping	2007 flood in Tewkesbury (UK) and 2015 flood in Mawlamyine (Myanmar)
Amitrano et al. 2019 [9]	Sentinel-1 SAR; frequency-domain offset tracking	2017 monitoring of Petermann, Nioghalvfjerdsfjorden and Jackobshavn Isbræ (Greenland) and Thwaites (Antarctica) glaciers
Uddin et al. 2019 [10]	Sentinel-1 SAR; Landsat-8; SRTM; DEM; land use/land cover mapping in GEE; NDVI, NDWI, NDMI, BSI and NPCRI indices; LWM, GEOBIA, supervised image classification, machine learning	2017 floods in whole country of Bangladesh
Lin et al. 2019 [11]	Sentinel-1 SAR; King Air 350ER aerial photos; SPOT-6, WorldView and QuickBird; distribution normalization, Bayesian probability, probabilistic thresholding, classification	2016 flood in Lumberton, North Carolina (USA)
Idowu and Zhou 2019 [12]	GRACE; terrestrial water storage anomaly; precipitation data; flood potential index	2012 flood in Lower Niger River Basin (Nigeria)
Ai et al. 2019 [13]	GPS RTK; ArcticDEM, UAV photogrammetric DEMs; elevation changes estimation, interpolation methods, DEM generation	2013–2015 mass changes at Austre Lovénbreen and Pedersenbreen glaciers (Svalbard)
Sebastiá-Frasquet et al. 2019 [14]	Sentinel-2; precipitation and wind data; Secchi disk depth, suspended particulated matter; chlorophyll *a* concentration, turbidity mapping	2017–2018 turbidity of Albufera de Valencia lagoon (Spain)
Sajjad et al. 2020 [15]	Landsat-8; GPS data; Google Earth; supervised classification, land use/cover change detection, modified NDWI	2014 flood in Lower Chenab plain (Pakistan)
Avian et al. 2020 [16]	Terrestrial laser scanning; Sentinel-1 SAR; Sentinel-2; UAV photos; automatic cameras; DInSAR; SfM	2001–2019 monitoring of Pasterze glacier (Austria)
Liang and Liu 2020 [17]	Sentinel-1 SAR; Planet and DMCii imagery; riverine flood depth, storm surge water height, land cover; National Map 3D Elevation Program DEM; NDWI, data fusion, daily inundation probability, weight of evidence	2017 inundations in Harris, Texas (USA)

The published Special Issue comprises 11 research articles. The pictorial word cloud in Figure 1 combines the thematic keywords used in their main metadata (namely, their titles, abstracts and keywords), while Table 1 summarizes the remote sensing data and methods used, and the areas of interest investigated in each article.

Figure 1. Thematic keywords of the 11 research papers composing the Special Issue "Imaging Floods and Glacier Geohazards with Remote Sensing" [6] of *Remote Sensing* (created with wordclouds.com).

The cloud shows that among the most frequently used keywords, there are not only terms relating to the geophysical processes involved (e.g., surge, mass, deficit and storm) or the approaches employed for flood and glacier imaging (e.g., supervised classification and comparison) but also specific data types (e.g., digital elevation models), sensors and missions (e.g., Sentinel-1). The latter data, in particular, were exploited in more than one article (see Table 1) and reflect a higher-level trend that can be observed in the recent specialist literature in this field which increasingly exploits satellite Synthetic Aperture Radar (SAR) imagery. A significantly high number of articles also focused on image classifiers, probabilistic approaches and elevation and change detection methods.

Multi-sensor and multi-platform approaches were also quite common across the papers on specific events or sites, and so were studies focused on algorithm development and testing. Most contributions focused on flood events, hazard and risk, while only four on glacier monitoring.

MDPI's article metrics powered by TrendMD were exploited with the aim to gather a flavor of the visibility of the Special Issue across the journal readership in the first 6 months after the publication of the last paper. TrendMD uses technologies such as Google Analytics by Google Inc. to track the use of and interaction with webpages made by visitors (e.g., abstract and full-paper views and downloads). The metrics for the 11 articles of the Special Issue show that since the publication of the first article at the end of March 2019, the Special Issue received more than 15,500 views in total over the 20-month-long time span between March 2019 and November 2020. This reflected an average number of 100 views/month for each paper. A positive exception is represented by the boosted performance of the article by Uddin et al. 2019 [10], which attracted over 4400 views since its publication in July 2019 and as of mid-November 2020, i.e., approximately 260 views/month.

The overall 64 citations in the indexed literature received as of mid-November 2020 also provide an indication of the good scientific impact that the Special Issue is building across the scientific community in the first few months after publication. A portion of these citations were made by articles published in MDPI open-access journals, including *Remote Sensing, Sustainability, Water, Hydrology* and *Applied Sciences*, while many others were received from articles in scientific journals of other publishers, focused on the fields of hydrology, remote sensing and environmental and Earth sciences. Looking at the scale of single articles, while generally, most of the papers of the Special Issue received 1 to 6 citations so far, two apparent positive outliers are the research articles by Benoudjit and Guida 2019 [8] and Uddin et al. 2019 [10], with outstanding achievements of 16 and 30 citations already attracted, respectively.

Overall, the body of literature collected in the Special Issue provides a good representation of the current state of the art and trends in this topical research field, showcasing remote sensing tools currently used for imaging, characterizing and modeling floods and glacier processes. A wide range of platforms, data sources, processing and analysis methods and models have been presented and discussed, with several cases studies distributed globally. The Special Issue, thus, contributes, together with other thematic volumes published in *Remote Sensing*, to the technical and scientific discussion on the use of remote sensing data in geology, geomorphology and hydrology.

Author Contributions: Conceptualization, F.C. and H.X.; formal analysis, F.C.; data curation, F.C.; visualization, F.C. and H.X.; writing—original draft preparation, F.C.; writing—review and editing, H.X. All authors have read and agreed to the published version of the manuscript.

Funding: This research received no external funding.

Acknowledgments: This Special Issue was developed following on from the invitation sent to Francesca Cigna by Kevin Zhang, Managing Editor at MDPI. The Guest Editors would like to acknowledge the authors who contributed to this Special Issue with their papers and express their gratitude to the anonymous reviewers for providing their feedback on the submitted manuscripts and helping the authors to enhance the scientific quality of their papers. Karem Chokmani from the National Institute of Scientific Research (Quebec City, QC, Canada) is sincerely acknowledged for his work and commitment in handling manuscripts submitted to the Special Issue. The *Remote Sensing* Editorial Office and the team of Assistant Editors are greatly acknowledged for their support and assistance to setup, kick-off, manage and publish this Special Issue.

Conflicts of Interest: The authors declare no conflict of interest.

References

1. Alcântara, E.; Park, E. Editorial for the Special Issue "Remote Sensing of Large Rivers". *Remote Sens.* **2020**, *12*, 1244. [CrossRef]
2. Topouzelis, K.; Papakonstantinou, A.; Singha, S.; Li, X.M.; Poursanidis, D. Editorial on Special Issue "Applications of Remote Sensing in Coastal Areas". *Remote Sens.* **2020**, *12*, 974. [CrossRef]
3. Domeneghetti, A.; Schumann, G.J.-P.; Tarpanelli, A. Preface: Remote Sensing for Flood Mapping and Monitoring of Flood Dynamics. *Remote Sens.* **2019**, *11*, 943. [CrossRef]
4. Rodríguez-Fernández, N.; Al Bitar, A.; Colliander, A.; Zhao, T. Soil moisture remote sensing across scales. *Remote Sens.* **2019**, *11*, 190. [CrossRef]
5. Cigna, F.; Tapete, D.; Lu, Z. Remote sensing of volcanic processes and risk. *Remote Sens.* **2020**, *12*, 2567. [CrossRef]
6. Cigna, F.; Xie, H.; Chokmani, K. MDPI Remote Sensing: Special Issue "Imaging Floods and Glacier Geohazards with Remote Sensing". Available online: https://www.mdpi.com/journal/remotesensing/special_issues/floods_glacier_geohazards (accessed on 12 November 2020).
7. Paul, F. Repeat Glacier Collapses and Surges in the Amney Machen Mountain Range, Tibet, Possibly Triggered by a Developing Rock-Slope Instability. *Remote Sens.* **2019**, *11*, 708. [CrossRef]
8. Benoudjit, A.; Guida, R. A novel fully automated mapping of the flood extent on sar images using a supervised classifier. *Remote Sens.* **2019**, *11*, 779. [CrossRef]
9. Amitrano, D.; Guida, R.; Di Martino, G.; Iodice, A. Glacier monitoring using frequency domain offset tracking applied to sentinel-1 images: A product performance comparison. *Remote Sens.* **2019**, *11*, 1322. [CrossRef]

10. Uddin, K.; Matin, M.A.; Meyer, F.J. Operational flood mapping using multi-temporal Sentinel-1 SAR images: A case study from Bangladesh. *Remote Sens.* **2019**, *11*, 1581. [CrossRef]

11. Lin, Y.N.; Yun, S.H.; Bhardwaj, A.; Hill, E.M. Urban flood detection with Sentinel-1Multi-Temporal Synthetic Aperture Radar (SAR) observations in a Bayesian framework: A case study for Hurricane Matthew. *Remote Sens.* **2019**, *11*, 1778. [CrossRef]

12. Idowu, D.; Zhou, W. Performance evaluation of a potential component of an early flood warning system-a case study of the 2012 flood, lower Niger River Basin, Nigeria. *Remote Sens.* **2019**, *11*, 1970. [CrossRef]

13. Ai, S.; Ding, X.; Tolle, F.; Wang, Z.; Zhao, X. Latest geodetic changes of Austre Lovénbreen and Pedersenbreen, Svalbard. *Remote Sens.* **2019**, *11*, 2890. [CrossRef]

14. Sebastiá-Frasquet, M.T.; Aguilar-Maldonado, J.A.; Santamaría-Del-ángel, E.; Estornell, J. Sentinel 2 analysis of turbidity patterns in a coastal lagoon. *Remote Sens.* **2019**, *11*, 2926. [CrossRef]

15. Sajjad, A.; Lu, J.; Chen, X.; Chisenga, C.; Saleem, N.; Hassan, H. Operational monitoring and damage assessment of riverine flood-2014 in the lower chenab plain, Punjab, Pakistan, using remote sensing and gis techniques. *Remote Sens.* **2020**, *12*, 714. [CrossRef]

16. Avian, M.; Bauer, C.; Schlögl, M.; Widhalm, B.; Gutjahr, K.H.; Paster, M.; Hauer, C.; Frießenbichler, M.; Neureiter, A.; Weyss, G.; et al. The status of earth observation techniques in monitoring high mountain environments at the example of pasterze glacier, austria: Data, methods, accuracies, processes, and scales. *Remote Sens.* **2020**, *12*, 1251. [CrossRef]

17. Liang, J.; Liu, D. Estimating daily inundation probability using remote sensing, riverine flood, and storm surge models: A case of hurricane harvey. *Remote Sens.* **2020**, *12*, 1495. [CrossRef]

Publisher's Note: MDPI stays neutral with regard to jurisdictional claims in published maps and institutional affiliations.

remote sensing

Article

Repeat Glacier Collapses and Surges in the Amney Machen Mountain Range, Tibet, Possibly Triggered by a Developing Rock-Slope Instability

Frank Paul

Department of Geography, University of Zurich, 8057 Zurich, Switzerland; frank.paul@geo.uzh.ch; Tel.: +41-44-6355175

Received: 19 February 2019; Accepted: 19 March 2019; Published: 24 March 2019

Abstract: Collapsing valley glaciers leaving their bed to rush down a flat hill slope at the speed of a racing car are so far rare events. They have only been reported for the Kolkaglacier (Caucasus) in 2002 and the two glaciers in the Aru mountain range (Tibet) that failed in 2016. Both events have been studied in detail using satellite data and modeling to learn more about the reasons for and processes related to such events. This study reports about a series of so far undocumented glacier collapses that occurred in the Amney Machen mountain range (eastern Tibet) in 2004, 2007, and 2016. All three collapses were associated with a glacier surge, but from 1987 to 1995, the glacier surged without collapsing. The later surges and collapses were likely triggered by a progressing slope instability that released large amounts of ice and rock to the lower glacier tongue, distorting its dynamic stability. The surges and collapses might continue in the future as more ice and rock is available to fall on the glacier. It has been speculated that the development is a direct response to regional temperature increase that destabilized the surrounding hanging glaciers. However, the specific properties of the steep rock slopes and the glacier bed might also have played a role.

Keywords: glacier surge; glacier collapse; rock-slope instability; hazard; Landsat; Sentinel 2; Tibet

---

1. Introduction

Glacier surges have recently been in the focus of several scientific studies [1], largely because remote sensing data with a sufficiently high spatial and temporal resolution allow accurate tracking of surface and morphological changes as well as creation of dense time-series of flow velocities [2–12]. During a surge, large amounts of ice are transported at comparably high velocities (several m/day) from an upper reservoir area downward to a receiving zone, possibly creating a strong advance of the terminus. After a surge, the ice becomes stagnant and melts away while in the reservoir zone new ice accumulates [8]. Whereas an improved understanding of possible surge mechanisms slowly emerges [13–15], collapses of glaciers (where the ice is removed from a flat bed) have so far only rarely been reported. The two most prominent and best-documented examples are the 2002 collapse of Kolkaglacier in the Caucasus [16–19] and the 2016 collapse of two small valley glaciers in the Aru mountain range of Tibet [20–23]. In both regions the collapses caused long-distance (several km) mass movements with very high velocities (200 to 290 km/h) [19,22] that have not been thought possible before they occurred. In contrast to the frequent ice break-offs at steep hanging glaciers [24], the Kolka and Aru glaciers rested on beds of a comparably low slope and should thus have been resistant to mechanical failure.

The history of events leading to the failure was different in both cases and they are thus unique on their own. Kolkaglacier is a well-known surge-type glacier [17] that is mostly nourished by avalanche snow and has been removed from its bed during its 2002 collapse, possibly due to a massive loading with debris from rock fall of its surrounding slopes [25]. Kolkaglacier had a regular surge in 1969/70 and might have also collapsed in 1835 and 1902 [17]. The Aru glaciers were not known as surging before [23], but showed a typical pattern of surge-related elevation changes before they collapsed [21]. The twin-glacier collapse removed the lower parts of both glaciers, likely due to a change in thermal/hydrologic conditions at the glacier bed [22].

The collapses reported here for a glacier (GLIMS ID G099443E34824N, Inventory ID CN5J352E0017) located on the north-western slope of the Amney Machen (A'nyê Maqên) mountain range (34.822° N, 99.44° E) took already place in 2004, and again in 2007 and 2016. Figure 1a shows the location of the mountain range and the glacier and Figure 1b is an oblique perspective view up-glacier showing the study region after the second collapse in 2008. The surges and collapses of this glacier have so far not been reported [26] or analyzed in the scientific literature, although they repeatedly buried a country road that has been reconstructed after each event (Figure A1a). Hence, local authorities are aware of the collapses and an information board describing the 2004 avalanche had been installed. The board can be seen in a picture from a tourist [27] and shows the advancing glacier a few weeks before its third collapse in 2016 (Figure A1b). The small valley glacier (size about 1.5 km^2) has a homogenous, gently sloping tongue (slope about 13°) with a length of about 2 km and an elevation range of 450 m (from 4800 to 5250 m). The tongue is nourished by ice from a steep (mean slope 36°) and its once connected upper part, that reaches as high as 5900 m.

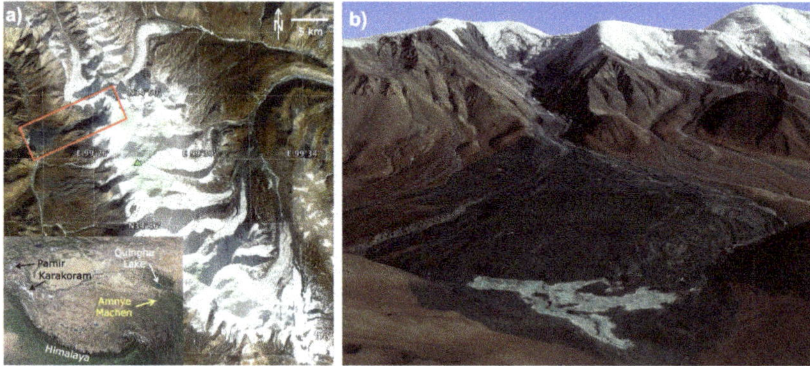

Figure 1. (**a**) Location of the Amney Machen mountain range in north-eastern Tibet, China. The collapsed glacier is marked with a red square. The location of the study region is marked in the inset with a yellow arrow. Image sources: Screenshots from Google Earth. (**b**) Oblique perspective view of the collapsed glacier as seen on a Quickbird image acquired on 13.1.2008, a few months after the second collapse. The steep and partly already glacier-free rock walls can be seen in the background. The extent of the 2004 collapse is still visible in the foreground, the deposit from the 2007 collapse can be seen on top of it. Image source: Screenshot from Google Earth.

The Amney Machen mountain range is covered by numerous glaciers (covering about 80 km^2), of which several have been classified in a previous study as surge type and strongly advancing between 1966 and 1981 [28]. Time series of Landsat images (available since 1987) reveal that several of the larger glaciers in this mountain range have surged again during the past three decades. Without temperature measurements, any assumptions about thermal conditions of the ice and rock walls are speculative. However, thermokarst features that can be found in the valley floor to the southwest of the mountain range and the permafrost zonation map by Gruber (2012) [29] are providing evidence that the glaciers higher up could be poly-thermal or cold-based. The purpose of this study is providing an overview on the glacier surges and collapses between 1987 and 2017 based on the analysis of freely available satellite image time series. A further analysis of the events using numerical modeling and field data is worthwhile but beyond the scope of this study.

2. Datasets

2.1. Satellite Data

The analysis of the surges and collapses of the glacier is based on optical satellite images from different sources (Table A1). Declassified reconnaissance imagery from the Corona KH4-A and B missions acquired in 1964 and 1969 are among the earliest sources of information about the region. With a spatial resolution of about 3 to 5 m they reveal several details of the glacier tongue and its forefield. However, sun-lit snow is over exposed. Landsat imagery covering the period 1987 to 2016 are mostly taken from Landsat 5 (21 scenes) at 30 m resolution (red-band), and Landsat 7 (14 scenes) at 15 m resolution (panchromatic band). The striping of Landsat scenes after 2003 due to the failure of the scan-line corrector had little impact on this, as the study region is located close to the scene center. Four Landsat 8 scenes acquired from 2013 to 2016 (15 m resolution) and four Sentinel 2 scenes acquired in 2016 and 2017 (10 m resolution) have been used for the more recent analysis. Collectively, satellite data provide an image in about every year for the full period (none in 1992/98), allowing a continuous interpretation of events. Several further scenes have been used for a detailed analysis of the collapses, partly also including scenes with snow cover and clouds. Key images of the surges and collapses are provided as a separate dataset in the Supplemental Material.

Additional analysis was performed using very high-resolution imagery acquired by Quickbird on 13 January 2008 and likely one of the Worldview satellites in winter 2016/17. These images were directly analyzed at maximum resolution in Google Earth and maps from bing.com. For the latter, the image source is not provided. The Corona and Landsat 5/7 scenes have been downloaded from earthexplorer.usgs.gov. The Landsat 8 and Sentinel 2 images were obtained from remotepixel.ca.

2.2. Digital Elevation Models (DEMs)

For topographic analysis and calculation of elevation changes, the 30 m version of the SRTM DEM (acquired in February 2000) has been used in combination with the 8 m High Mountain Asia (HMA) DEM acquired in 2015 [30]. Hillshade versions of both DEMs are shown for the study region in Figure 2. The latter suffers locally from data voids but is otherwise of very high quality. The SRTM DEM has a more rough or 'bumpy' surface but is otherwise looking reasonable. The SRTM DEM was downloaded from earthexplorer.usgs.gov and the HMA DEM from nsidc.org.

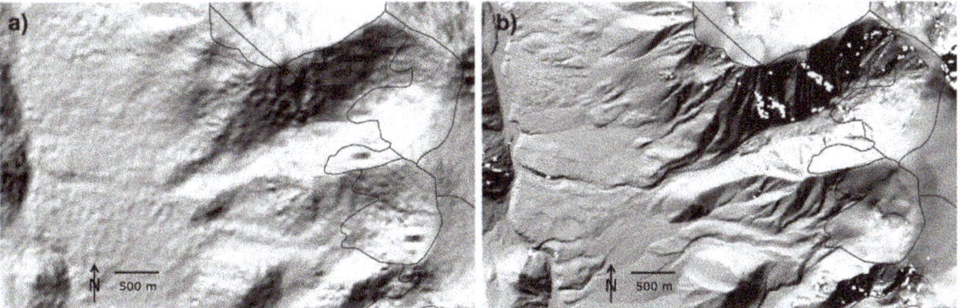

Figure 2. Visual comparison of the two DEMs for the region around the collapsed glacier (outlines from RGI6.0 in black). (**a**) The 30 m SRTM DEM from 2000 (source: earthexplorer.usgs.gov), (**b**) the 8 m HMA DEM from 2015 (source: nsidc.org/data/HMA_DEM8m_CT).

3. Methods

3.1. Satellite Data

As a base for quantitative assessments, glacier outlines were mapped automatically from the Sentinel 2 scene acquired on 4 August 2017 using the red/SWIR band ratio method [31]. The various extents of the glacier, the three debris fans, and the lake were created by on-screen digitizing using

ArcGIS from ESRI. Contrast-enhanced versions of the highest resolution dataset (panchromatic bands from Landsat 7 and 8) are used for this purpose. To identify the chronology of the surge and collapse events, all images were displayed in chronological order in a Geographic Information System (GIS) and flicker-images (going back and forth between two scenes) are used to follow the changes described below. The changes are difficult to see in side-by-side comparisons of static images, so some of the multi-panel images have been arranged to facilitate top-bottom comparison. For improved visibility of the changes the reader is referred to the Supplemental Material.

3.2. Topographic Analysis

The HMA DEM provided elevation and slope values of the glacier and the surrounding mountain range. In combination with the digitized glacier extents and deposit regions, elevation ranges have been calculated to determine mean slope values. The related horizontal distances were directly measured in the GIS using the Sentinel 2 image from 2017 in the background, i.e., the geometric reference has UTM zone 47N with WGS 1984 datum. Elevation changes were calculated by subtracting the SRTM DEM from the HMA DEM, after both were resampled bilinearly to 16 m resolution and co-registered. Horizontal shifts were smaller than the cell size, but the different vertical datum caused a bias of about 30 m that was subtracted. Due to artifacts in the difference DEM, elevation changes were only analyzed locally. A manually digitized centerline through the glacier and deposit area was used to derive elevation values from the co-registered DEMs.

4. Results: The History of the Surge and Collapse Events

4.1. The 1960s Situation

The Corona images from 1964 (Figure 3) and 1969 (not shown) revealed that the glacier was connected to its eastern and southern tributary during that time and had a well developed tongue with some debris cover along its northern margin. In the 1969 image, a crevasse across its entire width is visible in its flat upper part (indicating a steep slope in the bedrock) and a small rock outcrop (in the following RO-A) already existed in the northern part of its accumulation region. The terminus was flat and retreating from 1964 to 1969 by about 50 m. The river leaving the glacier has eroded an increasingly deeper trench in the foothill section of the mountain slope before reaching the main river (Qu'ngoin He) draining the region into the Yellow River (Huang He). In its lower part, the outflow is thus spatially well constrained, whereas in its upper part, erosion was less intense and a potential flooding would have been able to leave the riverbed and spread out over a larger region. Apart from this, nothing unusual (or a previous surge) can be detected on the images. The hill-slope section below the glacier seems to be intact, i.e., not impacted by a previous collapse.

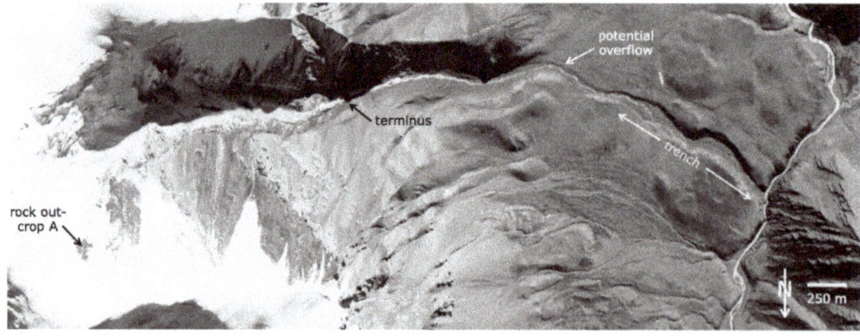

Figure 3. Corona image from 1964 showing the glacier being connected to its accumulation area and a glacier forefield that is free of any avalanche deposits. Note: South is up to highlight the relief. Image source: earthexplorer.usgs.gov.

4.2. The 1987–1995 Surge

The few available Landsat Multispectral Scanner images covering the period 1972 to 1982 are too sparse and too coarse to reveal any substantial variability in terminus position. It is thus assumed here that the glacier had been about stagnant during the 1970s until 1985. The following Landsat TM time series reveals a surging glacier tongue that advanced about 700 m from 1987 to 1995 (Figure 4). The size of RO-A increased during this time and by 1997 it connected to the ice-free terrain further down and was thus no longer an outcrop. As it is unlikely that all the ice has been removed by melting (other glaciers at lower elevations are stable), it can be assumed that a larger part of it was deposited on the lower glacier in form of ice avalanches. From 1996 to 2000, the terminus had been stationary and the lower part of the glacier tongue showed the typical post-surge down-wasting.

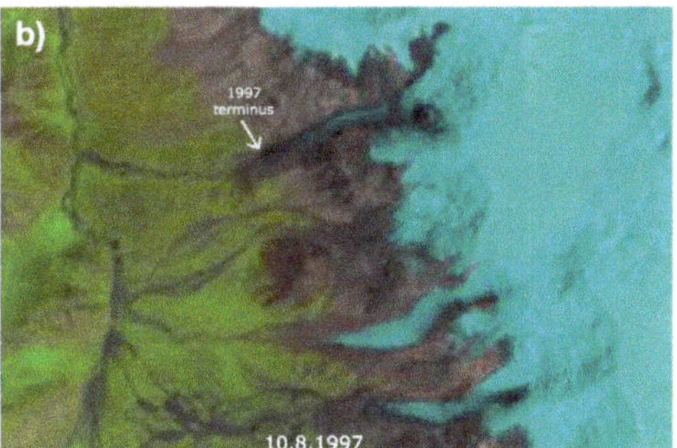

Figure 4. Landsat images from (**a**) 15 August 1987 and (**b**) 10 August 1997 showing the first surge of the glacier. The false colour images are showing (clean) ice and snow in cyan, rocks and gravel in pink to purple, and vegetation in light green. Image source: earthexplorer.usgs.gov.

4.3. The Developing Slope Instability

From August 2000 to August 2002, the area of the ice-free terrain around the former RO-A further increased, particularly in its upper parts (Figure 5a,b). The 2002 image also shows substantial darkening of the northern part of the glacier (in its flat section), indicating that substantial rock fall occurred before, maybe as a part of ice avalanches, maybe independently. This debris-covered part remained until 2003. From 2002 to 2003, the lower part of the glacier did not change much but a further rock-outcrop (RO-B) developed in the steep part of its eastern accumulation region. When also considering the later very high-resolution satellite images, it seems that between August and September 2003, a larger part of the main eastern tributary broke off and slid down onto the lower part of the glacier. Afterwards, the terminus started surging.

4.4. The 2003/4 Surge and Collapse

The glacier advanced by about 230 m and collapsed at some point between 26.1.2004 (Figure 5c) and 3.2.2004 as revealed by a largely snow-covered Landsat TM image. The ETM+ pan image form 11.2.2004 shown in Figure 5d has already much less snow so that the debris fan becomes visible. The bright trace in the panchromatic band of the ETM+ image and the still snow covered part visible on the TM image is indicating that a larger part of the collapsing glacier tongue has been deflected to the south by a double moraine wall at the end of the short valley and left the river bed at the 'potential overflow' point marked in Figure 3. However, the related ice/rock/water mixture must have been sufficiently fast/mobile to also overflow the c. 40 m high moraine wall. The marks of this first collapse can still be identified in the very high-resolution image acquired by Quickbird on 13.1.2008 (cf. Figure 1b).

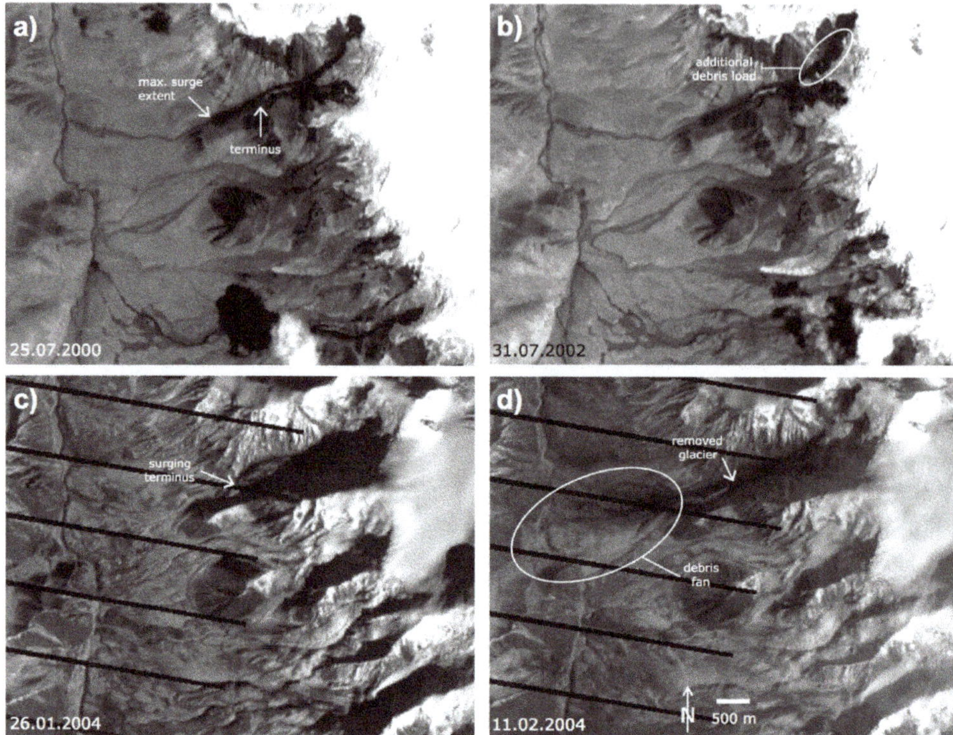

Figure 5. Landsat ETM+ pan images from (**a**) 25 July 2000, (**b**) 31 July 2002, (**c**) 26 January 2004, and (**d**) 11 February 2004 showing in (**a**) the active terminus in 2000 compared to the maximum surge extent, (**b**) the loading with debris by 2002, in (**c**) the maximum surge extent in 2004 before the collapse, and in (**d**) the debris fan after the collapse. Image source: earthexplorer.usgs.gov.

The full extent of the deposit has been mapped using the 15 m panchromatic band of the Landsat ETM+ scene acquired on 5 August 2004 (Figure A2a). With a measured area of 2.16 km^2 and an arbitrarily assumed mean thickness of maybe 10 m, the volume of the avalanche would be 20–25 million m^3, which is much less than both avalanches in the Aru mountain range [21]. The Landsat image also shows a lake (area 0.353 km^2) that likely formed as a result of the blocking of the main river by the avalanche deposit (Figure A2a). Until 14 September 2004 the lake grew to 0.534 km^2 and drained sometime between 29 June and 15 July 2005 at an even slightly larger size of 0.71 km^2.

4.5. The 2007 Surge and Collapse

From 2004 to September 2007, the glacier advanced by about 300 m before it collapsed again at some time between 23 September and 2 November 2007 (Figure 6). Its maximum extent before the collapse was much smaller than in 2004, close to its 'normal' minimum extent (cf. Figure A2a,b). This indicates that the surge/collapse mechanism might have been different, that lubrication by melt water might have played a role, or that the glacier has reached a different point of dynamic stability. It is well possible that after three years, the collapsed material was more a loose ice/rock mélange rather than a homogenous glacier. The contrast-stretched close-up of the Quickbird scene acquired on 13 January 2008 (Figure 7) reveals that a small part of the flat, lower glacier remained in its bed. This part is separated from highly crevassed dirty ice higher up that is still connected to the southern tributary. In other words, the rupture did not follow a potential failure zone, but occurred somewhere across the glacier.

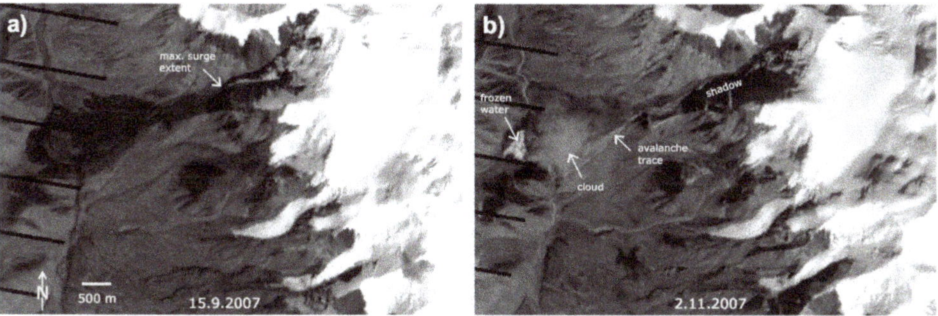

Figure 6. Two Landsat 7 ETM+ pan images showing the surging glacier in 2007: (**a**) before and (**b**) after the collapse. Image source: earthexplorer.usgs.gov.

The resulting avalanche deposit is well visible on a Landsat ETM+ image from 16 August 2008 (Figure A2b) that has been used for mapping its extent (1.46 km^2). The very low reflectance in the ETM+ panchromatic band (that stretches well into the near infrared) indicates that the deposit has a high water content, likely from the melting ice inside. The area covered by the deposit is considerably (32%) smaller than in 2004, but a part of the ice–debris mixture flowed again over the double moraine, indicating high flow velocities. The oblique perspective view in Figure 1b shows the fresh deposit on top of the older, more extended deposit from the 2004 collapse. The nadir-view of the new deposit (Figure 8) reveals several further details. It shows the larger extent of the 2004 collapse (that has flattened out) with the new 2007 deposit on top of it. The latter has still considerable relief, indicating high ice content. At the southern margin of the deposit, even crevasses can be seen, indicating that most of the former glacier might be located here. The deposit is also filling a large part of the trench and after a sharp turn to the south some material has been moved out of the trench and deposited higher up. An additional stream on top of the main deposit (with a different surface pattern) indicates that possibly a second wave of material came down after the main collapse. The image also shows the buried country road and the overtopped moraine wall.

Figure 7. The large image shows a contrast-enhanced close-up of the glacier remnant (between the arrows and the dashed line) after its 2007 collapse as seen on the 13 January 2008 Quickbird image (image source: Screenshot from Google Earth). The smaller inset shows about the same region after the 2016 collapse on an undated satellite image. Image source: Screenshot from bing.com.

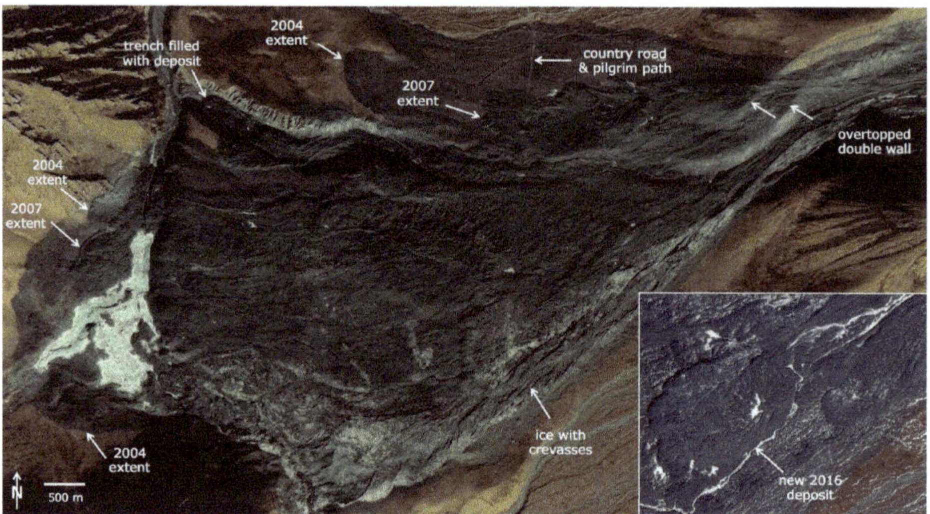

Figure 8. Close-up showing the avalanche deposit zone as seen on a Quickbird scene acquired on 13 January 2008, i.e., about 3 months after the 2007 collapse. Remnants from the 2004 collapse can be seen as well. Image source: Screenshot Google Earth. The inset shows a close-up of the third deposit from the 2016 collapse. Image source: Screenshot from bing.com.

4.6. Further Head Wall Degradation after 2009

From 2008 to 2009, another larger part of the steep glacier around the second rock outcrop (RO-B) started separating. The beginning of it is already visible on the Quickbird image from 13 January 2008 (Figure 1b). At some point between August 2010 and April 2011, this glacier part collapsed as well and likely ended on the lower glacier tongue. Interestingly, the failure zone did not follow the major crevasses but was located higher up. This means that the further development of the glaciers on this rock wall is difficult to predict. In the 15 years from 1997 to 2012, the original rock outcrop RO-A developed into a large, heart-shaped region of steep glacier-free terrain with only a small remnant of the former eastern glacier tributary left.

4.7. The Surge and Collapse in 2016

From 2011 to 2016, the glacier surged again by about 700 m (Figure 9) and the ice-free region around RO-B extended further upwards. The last phase of the surge and following collapse can be followed on Sentinel-2 images at 10 m spatial resolution (Figure 10). The glacier was still surging on 30 July 2016 (Figure 10a), reaching its maximum extent shortly after 28 September 2016 (Figure 10b), and collapsing until 18 October 2016 (Figure 10c). The inset in Figure 7 shows that approximately the same part of the glacier remained in its bed as after the 2007 collapse. Whereas the big gap to its upper eastern part is still there, this part is much less crevassed than in 2008.

A photograph that was taken on 12 September 2016 by G. Butler (Figure A1b) shows the advancing tongue about three weeks before its collapse. As one can see from the image, the terminus looks dual-layered with an upper layer of brighter ice over the 'normal' tongue. The image also shows some vegetation growth on top of the 2007 avalanche deposit (in the foreground), the fine grained material covering the mountain flanks to the north and south of the glacier valley (middle ground), and the steep and dark, now nearly ice-free rock wall in the background. The path through the avalanche deposit shown in Figure A1a reveals that the rocky material is rather small-grained and that the ice underneath the debris layer melted away in the meantime.

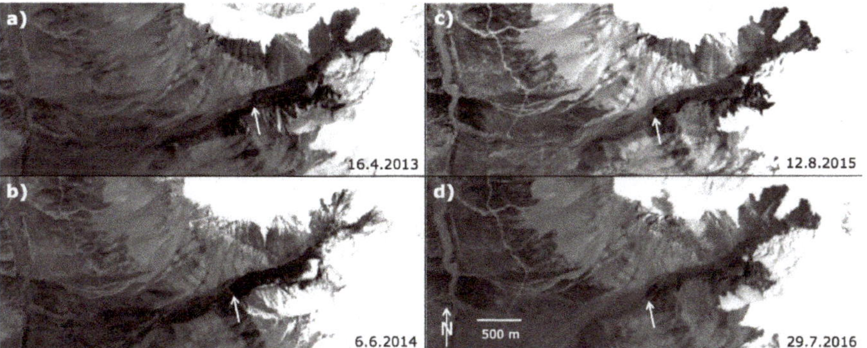

Figure 9. Time series of the 2013–2016 surge as seen with the Landsat 8 OLI panchromatic band. The white arrow marks the position of the terminus. Image acquisition dates are: (**a**) 16 April 2013, (**b**) 6 June 2014, (**c**) 12 August 2015, and (**d**) 29 July 2016. Image source: earthexplorer.usgs.gov.

The Sentinel-2 image from 4 August 2017 shows the deposit in nadir-view (Figure 10d) and was used to digitize its extent. It is smaller (area 1.25 km^2) than the one from 2007 and the avalanche has this time seemingly not overtopped the moraine walls at the end of the short valley. Hence, all material was diverted to the south-west and the deposit covers a larger region here than in 2007; very similar to 2004 (Figure 11). However, it reached not that far down, i.e. did likely not cross the main river. The inset in Figure 8 shows a subset of the fresh deposits on top of the former ones.

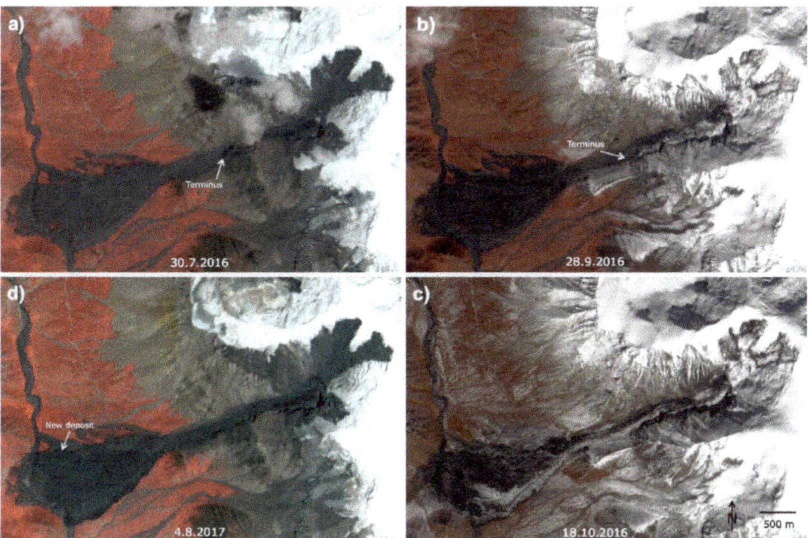

Figure 10. Time series of the 2016 glacier collapse as seen on false-colour composite images (near infrared, red, and green as RGB) acquired by Sentinel-2. (**a**) The image from 30 July 2016 shows the glacier during its surge. (**b**) Maximum extent on 28 September 2016 about 1 week before the collapse. (**c**) After the collapse on 18 October 2016. (**d**) The deposit from the last collapse (in darker shades of grey) in summer 2017. The destroyed country road (that is also used as a pilgrim path) has been re-established over the fresh deposit. Image source: Copernicus Sentinel data 2016 and 2017.

In Figure 11, the Sentinel-2 image from 4 August 2017 is shown along with overlays of all three deposit extents and the various glacier extents (minimum and maximum) for comparison. Remarkable are the large maximum surge extents from 2004 and 2016 compared to the extent before the 2007 collapse. The growth of rock outcrop RO-B from 2007 to 2017 and the re-established path crossing the fresh avalanche deposit (Figure A1a) can also be seen.

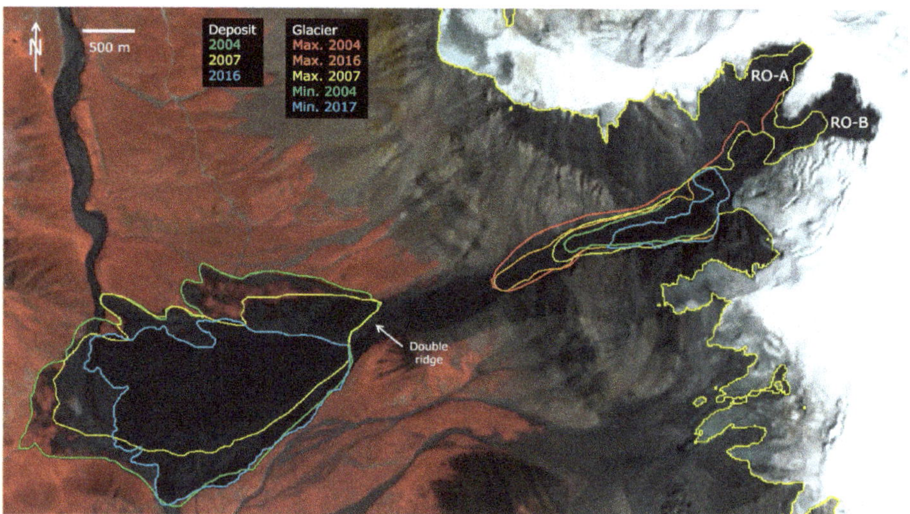

Figure 11. Extents of the deposits and glaciers for the three surges shown on the background of Figure 6d: Deposits from the 2004, 2007, and 2016 collapses are marked green, yellow, and blue, respectively. Maximum glacier extents (before the collapse) from 2004, 2007, and 2016 are shown in red, yellow, and orange, whereas minimum extents (after the collapse) from 2004 and 2017 are shown in green and blue, respectively. Image source: Copernicus Sentinel data 2017.

The close-up from the very high-resolution satellite image in Figure 12 (acquired after the third collapse) reveals that the bedrock is covered with fine-grained material and that the rock cliffs point downwards. There is thus not much resistance offered by the terrain and a hanging glacier that might get lubricated at its bed will have a severe stability problem. Furthermore, larger parts of the rock outcrops seem to be covered by thin ice and the rocks as well as the fine-grained material are comparably dark (i.e. they absorb much energy). It is thus well possible that the remaining glacier on this steep slope will degrade further and maybe slide down, thereby providing additional material to the lower glacier tongue. This could trigger further glacier surges and collapses in the future. Further Landsat 8 and Sentinel-2 images from 2017 and 2018 indicate that the glacier started advancing again in autumn 2017 but was in a stable position over most of 2018 at about the maximum 2007 extent shown in Figure 11.

4.8. Elevation Changes and Topographic Analysis

Elevation changes between the SRTM DEM and the HMA DEM over the 2000–2015 period are depicted in Figure 13 showing the difference grid and elevation profiles along a centerline through the deposit and the glacier tongue for both DEMs. Apart from regions with strong elevation gain (blue) and loss (red) that are related to DEM artifacts in regions of steep terrain and low contrast (snow, shadow), elevation changes over glaciers are also well recognizable. When starting at the top, the strong ice loss in the region of the former tributaries and later rock outcrops A and B (cf. Figure 12) is clearly visible (values are from about −40 to −100 m). Despite artifacts being close, the heart-shaped regions of the two rock outcrops match very well to the area of elevation loss.

Figure 12. Close-up of the instable slope with the two large, heart-shaped rock outcrops (RO-A and RO-B) that have merged in the meantime. Some remaining debris-covered ice is visible (partly snow covered) in the outcrops that seem to be covered with dark and fine-grained sediment. The former eastern accumulation area is now limited to a small remaining tongue (cf. Figures 4 and A1). A larger piece of ice in the valley floor is still connected to the southern tributary but a large crevasse is separating it from the rest of the glacier. Image source: Screenshot from bing.com.

Going further down, the (former) glacier tongue is also sharply depicted in the difference image. At the acquisition date of the first DEM (SRTM), the tongue was still close to its maximum extent from the first surge (Figure 4b) that was never reached again afterwards (Figure 11). Accordingly, from 2000 to 2015, this lower region showed a pronounced thickness loss (about 40 m). When the second DEM was acquired in 2015, the glacier was again surging (Figure 9c) and the bluish region just in front of the maximum 2007 extent in Figure 13 is depicting the slightly higher (about 5–10 m) elevation of the terminus at this position compared to the year 2000 surface. The advancing tongue can also be seen in the black profile shown in the inset of Figure 13. Higher up, the glacier surface in 2015 is generally lower than in 2000 (also up to 40 m), and in the zone of the steep cliff, elevations are about the same.

Much more subtle and within the DEM uncertainty are the elevation changes for the deposit region. However, closer inspection reveals a majority of bluish cells (elevations about 5–10 m higher) within the limits of the 2007 deposit and more variable changes outside. This would confirm that this is indeed a region of mass deposit, but the uncertainties are too high for a quantitative determination of the deposited volume.

When the study region is separated in four units A: deposition, B: transition, C: flat glacier, D: steep glacier (see bottom of Figure 13), the elevation, length, and slope values presented in Table 1 are found from the HMA DEM. The mean slope values indicate that the area of deposition is indeed comparably flat (<6°) and that the flat tongue of the glacier is in the typical range for valley glaciers (about 13°). The transition zone is somewhat steeper and the mean slope of the rock walls is 36°, However, locally, values exceeding 50° are found. The overall slope or Fahrböschung (measured from the upper point of the collapsed glacier to the lowest point of the deposit) is 10.9° and thus within the range of similar events [18].

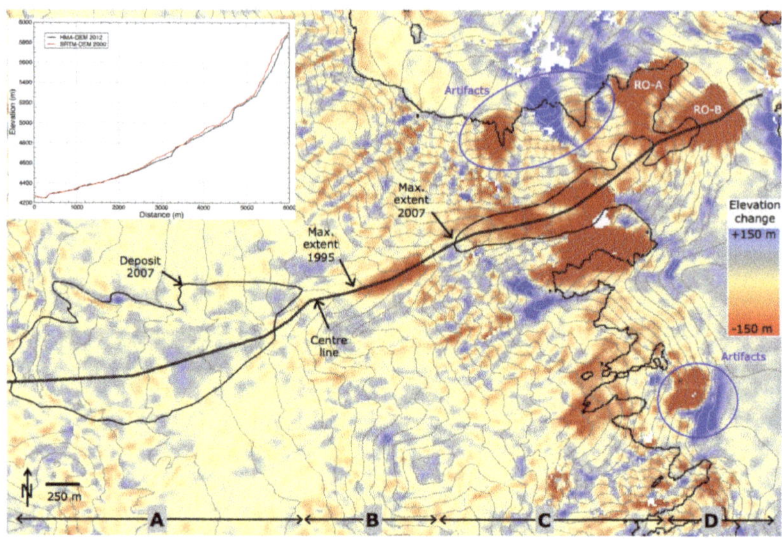

Figure 13. Elevation difference grid (SRTM-HMA DEM) over the 2000 to 2015 period showing volume loss (red) of the glacier but also artefacts to the north and south of it. Glacier outlines and the extent of the avalanche deposit from 2007 are shown in black, the elevation values for the inset are sampled along the dotted line. Elevation contour lines (grey) have 50 m equidistance, sections A, B, C, and D are described in the text. Elevation differences have been limited to a range of ±150 m. The inset shows elevation profiles extracted from both DEMs along the center line.

Table 1. Topographic characteristics of the different zones as derived from the HMA-DEM.

	Zone Name	Elevation (m)			Length (m)	Mean Slope (°)
		Minimum	Maximum	Range		
A	Deposit	4250	4450	200	2000	5.7
B	Transition	4450	4800	350	1200	16.3
C	Flat glacier	4800	5250	450	2000	12.7
D	Steep glacier	5250	5900	650	900	35.8
	Overall slope	4250	5250	1000	5200	10.9

5. Discussion

The detailed analysis of optical satellite images allowed reconstructing the changes of a surging and collapsing glacier in eastern Tibet in much detail. Useful scenes are available for nearly every year and much denser time series could be used to constrain the timing of the collapses down to a week when images from Landsat 5 and 7 are combined. Whereas the 30 m spatial resolution of Landsat TM is sufficient to follow major changes of the glacier and the rock outcrops (Figure 4), the 15 m panchromatic bands from Landsat 7 and 8 reveal several details much better (Figures 5, 6 and 9) and have thus been used together with 10 m resolution images from Sentinel-2 (Figure 10) to follow the changes and to map the extents of the glacier and the deposits (Figure 11). There are likely uncertainties in the delineation, but these do not impact on the general interpretation of the events. Further in-depth insights are provided by the very high-resolution satellite images from Quickbird (in Google Earth) and the image of unknown source available in bing.com that show the situation after the second and third collapses, respectively. Both images reveal fine details of the glacier remnants (Figure 7) and deposits (Figure 8), as well as of the instable rock slope with its degrading hanging glaciers and incompetent lithology (Figure 12). Several further insights can be derived from these images but they would be more speculative.

Additional quantitative information about glacier thickness changes is derived from two DEMs representing the situation after the first (SRTM DEM from 2000) and during the fourth surge (HMA DEM from 2015) of the glacier. Despite large artifacts close to the glacier, its volume loss, the removed tributaries, and a small elevation gain in the region of the deposit are visible, confirming the observations from the satellite images. In particular, the visibility of the 2015 surge front in the difference image (that is also visible in the DEM hillshade of Figure 2b) confirms the high quality of the elevation data and reliability of the results.

As a further back-up, elevation differences have also been calculated with the ALOS DEM (AW3D30) that was merged from images acquired over the 2007 to 2011 period (not shown here). These fully confirm the description of events presented above. The images for the DEM have likely been acquired shortly after the second collapse (e.g., in 2008 or 2009) and show a completely removed valley glacier, an even clearer elevation gain in the deposit area, a smaller area affected by elevation loss in the region of RO-B, and that the artifacts marked in Figure 13 are due to the SRTM DEM. In contrast, the difference between the ALOS and HMA DEM show a small surface lowering over the deposit area (from ice melting), the fully re-developed glacier tongue up to its 2015 extent, only small elevation changes in the region of RO-A and the strong surface lowering over RO-B.

Whereas the glaciologic and geomorphologic observations described above are comparably robust, any reasoning about possible processes leading to the observed events is speculative. Hence, only the more general characteristics of the Kolka and Aru glacier collapses are compared in the following. The most striking difference is that the glacier observed here already collapsed three times (2004, 2007, and 2016) in only 12 years. Repeat collapses might have also occurred for Kolkaglacier as Kotlyakov et al. (2004) [17] wrote that the 1902 surge ended "in a catastrophic outburst of ice and water" similar to the event in 2002 and that "in a few minutes, the ice spread over 9 km through the valley". However, the Kolkaglacier "eruption" in 1902 occurred "at the height of a hot summer and after heavy showers" [17] seemingly with large amounts of water, whereas only small amounts of water could have been available for the 2004 mid-winter collapse reported here. The much longer time period between the two collapses (100 years) is also different. Moreover, all collapses reported here occurred during a surge with strong frontal advance. The Aru glaciers advanced only slightly before they collapsed, but the observed elevation change pattern is fully compliant with an ongoing surge [21]. In contrast to Kolkaglacier with its well-known surge history [17], both Aru glaciers had not been known as surging before [23]. On the other hand, Kolka glacier was retreating from 1984 to 2002 and basically stagnant before its 2002 collapse.

Whereas the 2002 collapse of Kolkaglacier was likely triggered by the additional load of rock/ice avalanches on its surface [19,26], the Aru collapses seem to be triggered by extreme water pressure in combination with specific properties of the glacier bed [22]. The events reported here have likely been facilitated by the accumulation of ice and debris from the instable rock wall. However, these likely occurred in small portions and more gradually, so that the additional load might have first caused a surge rather than a sudden collapse. A similar surge initiation has been observed for Lysiiglacier in the Ak-Shirak mountain range of the Tian-Shan, where the material excavated from the Kumtor mine has been dumped on its surface [32]. Based on the chronology of events and the DEM differences, it is speculated that the 2003/04 surge was triggered by the extra load of ice and rock from RO-A, whereas the growing RO-A and RO-B triggered the 2007 surge. The third event in Oct 2016 might have been triggered by material mostly excavated from the region of RO-B. Compared to the Kolka and Aru collapses, the events reported here were much smaller in volume. Chinese authorities estimated a volume of 36 million m^3 for the 2004 event (as written on the information board), resulting in a mean thickness of about 15 m for the deposit. They speculated that the collapse was caused by a rising snow line and freeze-thaw action.

Regarding the short time between the first and the second collapse (3 years), one has to consider that the ice mass coming down the second (and third) time was likely more a loose ice/rock mixture rather than a compact glacier. The mechanical properties for the latter two failures might thus have been different and need further investigation. More detailed studies (e.g., on ice avalanche modeling, thermal conditions, climatic trends, bedrock lithology) are also required to understand why and how the glacier has collapsed the first time, particularly when considering that the glacier did not collapse during the first surge although it was much more extended then. Some fieldwork in this region would certainly be helpful to constrain the physical properties of the glaciers and the surrounding environment.

Overall, it seems that glacier collapses are not unique events but can occur repeatedly, particularly when there is extraordinary and continuous supply of material from surrounding rock walls. As there is still some ice left to fall on the glacier and the degradation of the steep rock walls continues, the on-going but infrequent input of ice and rock will play an important role in keeping the surges (and maybe also the collapses) alive. As the rock wall and the glacier are still very active, it can be recommended to observe them more closely in the future.

6. Conclusions

This study presented a description of four glacier surges and three collapses of a small valley glacier in the Amney Machen mountain range of eastern Tibet that took place from 1987 to 1995 (surge only) and in 2004, 2007, and 2016. Whereas some characteristics of the events resemble similarities of glacier collapses reported earlier (Kolka, Aru), the combination found here is unique, because they already occurred three times with only a few years in-between and—despite the magnitude of the events and their impacts on infrastructure—because they have not been reported so far in the literature. The analysis of satellite images reveals that a developing rock-slope instability might be responsible for the last three surges by infrequently adding mass on the lower glacier. However, other reasons might play a role as well and more theoretical (numerical modelling) investigations as performed for the Kolka and Aru collapses might provide additional insights into the governing processes. The description of events presented here might help in analyzing related details in future studies. Further collapses of this glacier might occur, as the supply of material from the degrading rock wall is continuing. It is thus recommended to observe this glacier more closely in the future and collect some field data to constrain modelling efforts.

Supplementary Materials: The following are available online at http://www.mdpi.com/2072-4292/11/6/708/s1, Key images of the surges and collapses.

Author Contributions: All analysis and the writing of the paper have been performed by the author (F.P.).

Funding: This research and the APC was funded by the ESA project Glaciers_cci, grant number 4000109873/14/I-NB.

Acknowledgments: This study would not have been possible without the free access to Landsat and Sentinel data, digital elevation models, and very-high resolution satellite imagery visible in Google Earth and bing.com.

Conflicts of Interest: The author declares no conflict of interest.

Abbreviations

DEM	Digital Elevation Model
ESRI	Environmental Systems Research Institute
ETM+	Enhanced Thematic Mapper +
HMA	High Mountain Asia
GIS	Geographic Information System
GLIMS	Global Land Ice Measurements from Space
OLI	Operational Land Imager
RO	Rock Outrcrop
SRTM	Shuttle Radar Topography Mission
TM	Thematic Mapper
UTM	Universal Transverse Mercator

Appendix

Figure A1. Two images from the deposit taken by Brandon G. Butler on 12 September 2016. (**a**) A view to the north showing the avalanche deposit to the left and right of the reconstructed road crossing it. Image source: [27]. (**b**) A view to the east showing the advancing glacier (the white arrow marks its front), about 3 weeks before its next collapse. Some grass is visible on the deposit of the previous collapse in the foreground. The dark rock slope in the background (to the left of the image center) is the now huge rock outcrop RO-B. Image source: [27].

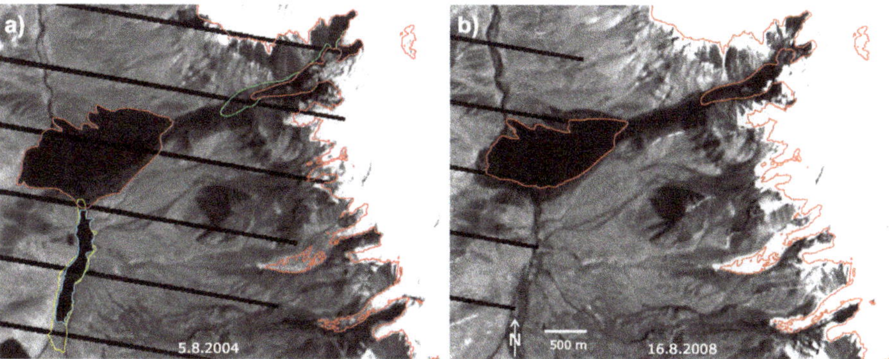

Figure A2. Landsat 7 ETM+ scenes (pan band) used to mark the extent of the deposits. (**a**) Scene from 5 August 2004 (green: maximum glacier extent before the 2004 collapse, red: extent of the deposit, light blue/yellow: lake extent from 5 August/14 September 2004. (**b**) Scene from 16 August 2008 (after the 2007 collapse) showing the new deposit in red. Image source: earthexplorer.usgs.gov.

Table A1. Overview of the satellite images used for the analysis. All Landsat scenes have path-row 133-036 and the Sentinel-2 tile used is 47SNU. Corona scenes used: #1: KH-4A scene DS1015-2164DF117, #2: KH-4B scene DS1108-2184DA110. Sensor names: L5 TM: Landsat 5 Thematic Mapper, L7 ETM+: Landsat 7 Enhanced Thematic Mapper plus, L8 OLI: Landsat 8 Operational Land Imager, S2 MSI: Sentinel-2 Multi-Spectral Instrument.

Nr.	Sensor	Date	Nr.	Sensor	Date	Nr.	Sensor	Date
1	Corona	30.12.1964	16	L7 ETM+	16.08.2002	31	L5 TM	23.09.2007
2	Corona	16.12.1969	17	L7 ETM+	03.08.2003	32	L7 ETM+	02.11.2007
3	L5 TM	15.08.1987	18	L5 TM	12.09.2003	33	L7 ETM+	16.08.2008
4	L5 TM	30.06.1988	19	L5 TM	14.10.2003	34	L5 TM	11.08.2009
5	L5 TM	21.09.1989	20	L5 TM	15.11.2003	35	L5TM	14.08.2010
6	L5 TM	06.07.1990	21	L5 TM	18.01.2004	36	L7 ETM+	24.07.2011
7	L5 TM	11.09.1991	22	L7 ETM+	26.01.2004	37	L7 ETM+	12.09.2012
8	L5 TM	31.08.1993	23	L5 TM	03.02.2004	38	L8 OLI	16.04.2013
9	L5 TM	14.05.1994	24	L7 ETM+	11.02.2004	39	L8 OLI	06.06.2014
10	L5 TM	20.07.1995	25	L7 ETM+	05.08.2004	40	L8 OLI	12.08.2015
11	L5 TM	24.09.1996	26	L5 TM	14.09.2004	41	L8 OLI	29.07.2016
12	L5 TM	10.08.1997	27	L7 ETM+	09.09.2005	42	S2 MSI	30.07.2016
13	L5 TM	31.07.1999	28	L7 ETM+	26.07.2006	43	S2 MSI	28.09.2016
14	L7 ETM+	25.07.2000	29	L5 TM	20.09.2006	44	S2 MSI	18.10.2016
15	L7 ETM+	13.08.2001	30	L7 ETM+	15.09.2007	45	S2 MSI	04.08.2017

References

1. Qiu, J. Ice on the run. *Science* **2017**, *358*, 1120–1123. [CrossRef] [PubMed]
2. Bevington, A.; Copland, L. Characteristics of the last five surges of Lowell Glacier, Yukon, Canada, since 1948. *J. Glaciol.* **2014**, *60*, 113–123. [CrossRef]
3. Quincey, D.J.; Glasser, N.F.; Cook, S.J.; Luckman, A. Heterogeneity in Karakoram glacier surges. *J. Geophys. Res. Earth Surf.* **2015**, *120*, 1288–1300. [CrossRef]
4. Yasuda, T.; Furuya, M. Dynamics of surge-type glaciers in West Kunlun Shan, Northwestern Tibet. *J. Geophys. Res. Earth Surf.* **2015**, *120*, 2393–2405. [CrossRef]
5. Häusler, H.; Ng, F.; Kopecny, A.; Leber, D. Remote-sensing-based analysis of the 1996 surge of Northern Inylchek Glacier, central Tien Shan, Kyrgyzstan. *Geomorphology* **2016**, *273*, 292–307. [CrossRef]
6. Herreid, S.; Truffer, M. Automated detection of unstable glacier flow and a spectrum of speedup behavior in the Alaska Range. *J. Geophys. Res. Earth Surf.* **2016**, *121*, 64–81. [CrossRef]
7. Pitte, P.; Berthier, E.; Masiokas, M.H.; Cabot, V.; Ruiz, L.; Ferri Hidalgo, L.; Gargantini, H.; Zalazar, L. Geometric evolution of the Horcones Inferior Glacier (Mount Aconcagua, Central Andes) during the 2002–2006 surge. *J. Geophys. Res. Earth Surf.* **2016**, *121*, 111–127. [CrossRef]
8. Bhambri, R.; Hewitt, K.; Kawishwar, P.; Pratap, B. Surge-type and surge-modified glaciers in the Karakoram. *Sci. Rep.* **2017**, *7*, 15391. [CrossRef]
9. Paul, F.; Strozzi, T.; Schellenberger, T.; Kääb, A. The 2015 surge of Hispar Glacier in the Karakoram. *Remote Sens.* **2017**, *9*, 888. [CrossRef]
10. Round, V.; Leinss, S.; Huss, M.; Haemmig, C.; Hajnsek, I. Surge dynamics and lake outbursts of Kyagar Glacier, Karakoram. *Cryosphere* **2017**, *11*, 723–739. [CrossRef]
11. Strozzi, T.; Paul, F.; Wiesmann, A.; Schellenberger, T.; Kääb, A. Circum-Arctic changes in the flow of glaciers and ice caps from satellite SAR data between the 1990s and 2017. *Remote Sens.* **2017**, *9*, 947. [CrossRef]
12. Wendt, A.; Mayer, C.; Lambrecht, A.; Floricioiu, D. A glacier surge of Bivachny glacier, Pamir mountains, observed by a time series of high-resolution digital elevation models and glacier velocities. *Remote Sens.* **2017**, *9*, 388. [CrossRef]
13. Sund, M.; Lauknes, T.R.; Eiken, T. Surge dynamics in the Nathorstbreen glacier system, Svalbard. *Cryosphere* **2014**, *8*, 623–638. [CrossRef]
14. Sevestre, H.; Benn, D.I. Climatic and geometric controls on the global distribution of surge-type glaciers: Implications for a unifying model of surging. *J. Glaciol.* **2015**, *61*, 646–659. [CrossRef]

15. Dunse, T.; Schellenberger, T.; Hagen, J.O.; Kääb, A.; Schuler, T.V.; Reijmer, C.H. Glacier-surge mechanisms promoted by a hydro-thermodynamic feedback to summer melt. *Cryosphere* **2015**, *9*, 197–215. [CrossRef]
16. Haeberli, W.; Huggel, C.; Kääb, A.; Zgraggen-Oswald, S.; Polkvoi, A.; Galushkin, I.; Zotikov, I.; Osokin, N. The Kolka–Karmadon rock/ice slide of 20 September 2002: An extraordinary event of historical dimensions in North Ossetia, Russian Caucasus. *J. Glaciol.* **2004**, *50*, 533–546. [CrossRef]
17. Kotlyakov, V.M.; Rototaeva, O.V.; Nosenko, G.A. The September 2002 Kolka Glacier catastrophe in North Ossetia, Russian Federation: Evidence and analysis. *Mt. Res. Dev.* **2004**, *24*, 78–83. [CrossRef]
18. Huggel, C.; Zgraggen-Oswald, S.; Haeberli, W.; Kääb, A.; Polkvoi, A.; Galushkin, I.; Evans, S.G. The 2002 rock/ice avalanche at Kolka/Karmadon, Russian Caucasus: Assessment of extraordinary avalanche formation and mobility, and application of QuickBird satellite imagery. *Nat. Hazards Earth Syst. Sci.* **2005**, *5*, 173–187. [CrossRef]
19. Evans, S.G.; Tutubalina, O.V.; Drobyshev, V.N.; Chernomorets, S.S.; McDougall, S.; Petrakov, D.A.; Hungr, O. Catastrophic detachment and high-velocity long-runout flow of Kolka Glacier, Caucasus Mountains, Russia in 2002. *Geomorphology* **2009**, *105*, 314–321. [CrossRef]
20. Tian, L.; Yao, T.; Gao, Y.; Thompson, L.; Mosley-Thompson, E.; Muhammad, S.; Zong, J.; Wang, C.; Jin, S.; Li, Z. Two glaciers collapse in western Tibet. *J. Glaciol.* **2017**, *63*, 194–197. [CrossRef]
21. Kääb, A.; Leinss, S.; Gilbert, A.; Bühler, Y.; Gascoin, S.; Evans, S.G.; Bartelt, P.; Berthier, E.; Brun, F.; Chao, W.; et al. Massive collapse of two glaciers in western Tibet in 2016 after surge-like instability. *Nat. Geosci.* **2018**, *11*, 114–120. [CrossRef]
22. Gilbert, A.; Leinss, S.; Kargel, J.; Kääb, A.; Yao, T.; Gascoin, S.; Leonard, G.; Berthier, E.; Karki, A. Mechanisms leading to the 2016 giant twin glacier collapses, Aru range, Tibet. *Cryosphere* **2018**, *12*, 2883–2900. [CrossRef]
23. Zhang, Z.; Liu, S.; Zhang, Y.; Wei, J.; Jiang, Z.; Wu, K. Glacier variations at Aru Co in western Tibet from 1971 to 2016 derived from remote-sensing data. *J. Glaciol.* **2018**, *64*, 397–406. [CrossRef]
24. Faillettaz, J.; Funk, M.; Vincent, C. Avalanching glacier instabilities Review on processes and early warning perspectives. *Rev. Geophys.* **2015**, *53*, 203–224. [CrossRef]
25. Drobyshev, V.N. Glacial catastrophe of 20 September 2002 in North Osetia. *Russ. J. Earth. Sci.* **2006**, *8*, ES4004. [CrossRef]
26. Ye, Q.; Zong, I.; Tian, L.; Cogley, J.G.; Song, C.; Guo, W. Glacier changes on the Tibetan Plateau derived from Landsat imagery: Mid-1970s-2000-13. *J. Glaciol.* **2017**, *63*, 273–287. [CrossRef]
27. Brando Abroad. Available online: http://www.brandoabroad.com/single-post/2017/09/22/The-life-of-Chinese-construction-workers-a-million-miles-from-home-Days-9-10 (accessed on 9 March 2019).
28. Wenying, W. Glaciers in the north-eastern part of the Ch'ing-hai-hsi-tsang (Qinghai–Xizang) Plateau (Tibet) and their variations. *J. Glaciol.* **1983**, *29*, 383–391. [CrossRef]
29. Gruber, S. Derivation and analysis of a high-resolution estimate of global permafrost zonation. *Cryosphere* **2012**, *6*, 221–233. [CrossRef]
30. Shean, D. *High Mountain Asia 8-Meter DEM Mosaics Derived from Optical Imagery*; Version 1, Tile 391; NASA National Snow and Ice Data Center Distributed Active Archive Center: Boulder, CO, USA, 2017. [CrossRef]
31. Paul, F.; Winsvold, S.H.; Kääb, A.; Nagler, T.; Schwaizer, G. Glacier remote sensing using Sentinel-2. Part II: Mapping glacier extents and surface facies, and comparison to Landsat 8. *Remote Sens.* **2016**, *8*, 575. [CrossRef]
32. Jamieson, S.S.R.; Ewertowski, M.W.; Evans, D.J.A. Rapid advance of two mountain glaciers in response to mine-related debris loading. *J. Geophys. Res. Earth Surf.* **2015**, *120*, 1418–1435. [CrossRef]

 *remote sensing*

Article

A Novel Fully Automated Mapping of the Flood Extent on SAR Images Using a Supervised Classifier

Abdelhakim Benoudjit * and Raffaella Guida

Surrey Space Centre, University of Surrey, Guildford GU2 7XH, UK; r.guida@surrey.ac.uk
* Correspondence: a.benoudjit@surrey.ac.uk

Received: 28 February 2019; Accepted: 27 March 2019; Published: 1 April 2019

Abstract: When a populated area is inundated, the availability of a flood extent map becomes vital to assist the local authorities to plan rescue operations and evacuate the premises promptly. This paper proposes a novel automatic way to rapidly map the flood extent using a supervised classifier. The methodology described in this paper is fully automated since the training of the supervised classifier is made starting from water and land masks derived from the Normalized Difference Water Index (NDWI), and without any intervention from the human operator. Both a pre-event Synthetic Aperture Radar (SAR) image and an optical Sentinel-2 image are needed to train the supervised classifier to identify the inundation on the flooded SAR image. The entire flood mapping process, which consists of preprocessing the images, the extraction of the training dataset, and finally the classification, was assessed on flood events which occurred in Tewkesbury (England) in 2007 and in Myanmar in 2015, and were captured by TerraSAR-X and Sentinel-1, respectively. This algorithm was found to offer overall a good compromise between computation time and precision of the classification, making it suitable for emergency situations. In fact, the inundation maps produced for the previous two flood events were in agreement with the ground truths for over 90% of the pixels in the SAR images. Besides, the latter process took less than 5 min to finish the flood mapping from a SAR image of more than 41 million pixels for the dataset capturing the flood in Tewkesbury, and around 2 min and 40 s for an image of 19 million pixels of the flood in Myanmar.

Keywords: flood extent mapping; supervised classification; NDWI; synthetic aperture radar (SAR); web application

---

1. Introduction

According to a recent report from an important insurance company, flooding was at the same time the costliest and the deadliest natural disaster in 2016, with considerable human fatalities [1]. In fact, more than half of the natural hazards in 2016 were hydrological, which were the most devastating type of disaster financially with 59 billion USD worth of damages. This latter category of disasters is dominated by floods with 164 floods occurrences against 13 landslides. During the same year, floods caused the greatest loss of life among all the disasters, with 4731 deaths [2]. Therefore, there is a growing need for a quantification of the impact of the flood to help response authorities to mitigate the damages and prioritize at the time of the emergency, as well as supporting insurance companies in working out an assessment of the losses sustained by each property. A thorough understanding of the potential flood risk can also assist development agencies to build resilient communities.

In the event of flooding, a clear cloud-free image acquired instantaneously is necessary to have a synoptic view of the affected area. In this context, remotely sensed images are suitable to map inundations, particularly when harsh climatic conditions are encountered and the access to the affected site is impractical [3]. Moreover, satellite-borne Synthetic Aperture Radar (SAR) sensors have been extensively used in the last decade to monitor many flooding events by taking advantage of their ability to operate independently of the sunlight, and in cloudy conditions which are common during inundations. Thanks to the important number of SAR satellites in orbit, the user has a wide choice of datasets which come in different wavelengths and resolutions. For the time being, X-band sensors like TerraSAR-X and COSMO-SkyMed provide the highest spatial resolution among all SAR sensors [4]. With this metric-resolution configuration, floods could be detected even in complex scenarios, such as urban settlements where streets are relatively narrow [5]. Furthermore, the systematic acquisition plan of the Sentinel-1 C-band satellite increases the likelihood to find a reference image in the Sentinel Hub archive. A deeper understanding of the flood hazard is achieved by extracting the extent and depth flood features as well as assessing the velocity of the floodwater, which will help to efficiently manage the inundation risk.

Operational flood mapping aims to reduce the delay between the acquisition of the satellite image and the diffusion of the flood extent map produced from it to the civil protection authorities for instantaneous relief efforts. This objective is achieved with a fully automated flood detection service [6]. The flood mapping service developed in [6] is an improvement of [7], where the workflow was adapted to operational situations covered by TerraSAR-X. Briefly, the service is triggered when the TerraSAR-X product is downloaded to the FTP server, and at the end of the process the flood extent map is available to visualize online via a Web interface. It should be mentioned that the radar pulse from COSMO-SkyMed, which operates in the X-band like TerraSAR-X, was found to be attenuated by the precipitation due to its relatively short wavelength [8]. Besides, due to the time delay between the tasking of TerraSAR-X during an emergency flood situation and the actual acquisition of the image, the peak of the inundation might be missed since this satellite needs at least 2.5 days to access the requested site [9]. In this case, the systematic acquisition mode of the Sentinel-1 constellation would be of great help. In [10], the flood mapping service proposed in [6] was modified to process Sentinel-1 SAR images. In particular, [6] was improved by adding a post-processing step which consists in eliminating from the flood map areas higher than the nearest drainage network, using a thresholding strategy. The process in [10] is in principal completely unsupervised yet the latter threshold was determined empirically. The Sentinel-1 images were automatically downloaded and preprocessed using the Sentinel Application Platform (SNAP), and then the classification carried on in a similar way to [7]. It was found in [10] that among the polarizations offered by the Sentinel-1 sensor, VV-polarized SAR images result in more accurate flood maps than cross-polarized (VH) products. In the same context, [11] suggested that HH polarization realizes the highest accuracy in terms of flood mapping. In the same prospect of an operational mapping of the flood, [12] proposed to detect the flood in vegetated, forested and built-up areas, besides the normal low backscatter flood (open water), using a Fuzzy logic approach which has permitted to combine data stemming from different sources (a DEM, a Land Cover Map). The threshold values are retrieved from three selected backscattering models (for agricultural, forested, and urban areas) applied with varying radar parameters to a number of land covers. To keep the process simple, the threshold values are calculated by considering only a few specific flood scenarios. As a result, it will be challenging to map the flood when, for instance, the plant characterisitics change and the backscattering model's preconditions become unsatisfied. Although, this issue has been addressed by allowing the user to adjust manually the values of the default thresholds, this leads to a lack of automation in the process as a consequence. An automated flood mapping method based on the approximation of the Probability Density Function (PDF) of the backscatter of the water was proposed by [13]. The threshold value is then defined as the point where the PDF of the backscatter and the gamma distribution modelling the water, and having backscatter values lower than this threshold, start to diverge. The Thresholding is followed with a region growing and a pixel-based change detection. The authors in [14] addressed the challenging task of the urban flood mapping in an unsupervised way, by improving the process introduced in [13] with a more objective estimation of the region growing's tolerance criterion. In the context of image segmentation, the region growing cannot be generalized as its cost function is not defined a priori, but is set empirically according to the specific application instead. Furthermore, it fails in practice when the edges of the object to detect are too smooth [15]. Another way the issue of the automation of the flood mapping was addressed in the literature is with a service running regularly as in [16]. In this project, the multiyear ENVISAT ASAR dataset is tiled into splits of 1° longitude by 1° latitude, and the training step is carried out by making use of the SRTM-derived water mask (SWBD) and the features extracted from the tile, which consist mainly of the backscatter and the incidence angle. Subsequently, pixels from nonlabeled images are classified using Bayes' theorem to get a probability map of water and land, after estimating the probability distributions for each class from trained histograms. Nevertheless, the low spatial resolution of the water mask which is crucial to the training phase, could impact negatively the precision of the classification, especially for smaller rivers.

Supervised and unsupervised learning methods were already applied, in a few studies, to tackle SAR flood extent mapping problems. The work in [17] used a self-organizing map (SOM), which is essentially an unsupervised artificial neural network, to segment then classify a flooded SAR image. SOM being originally a dimensionality reduction technique, a moving window centered around SAR image pixels forms a vector of neighbouring pixels that are passed as input into the neural network to train it. At the end of the learning process, the central pixel of each sliding window is mapped onto one of the neurons of a 2D grid, with multiple image pixels possibly being assigned to the same neuron. This results in the flooded SAR image being segmented, with each neuron representing a cluster. However, the eventual classification of the neurons on the grid into water and non-water was performed with the help of ground truth pixels extracted manually. The authors in [18] presented several semi-automatic and manual methods to map inundations, by exploiting free satellite multispectral and SAR data. Similarly to the current paper, the authors took advantage of water and vegetation indices like the Normalized Difference Vegetation Index (NDVI) and the Modified Normalized Difference Water Index (MNDWI), although it was the variation in these indices that was expected to reveal the presence of flooding. In another experiment, a supervised classifier was also investigated for the same purpose by choosing samples manually from different types of land cover. However, the latter two methods were applied separately and on multispectral optical images that could suffer from the cloud cover. When the flood mapping was carried out on SAR images, the threshold was manually adjusted either on a single flooded image or on the log ratio between a pair of images captured before and after the flood. With the aim of mapping urban flooding in [19], first an active contour model (snake) is employed to detect the flood in rural areas. Then, a supervised Bayesian classification is carried out on adjacent flooded urban areas, where the training data for the flood and the non-flood classes is chosen from the previously obtained rural flood map and from urban areas situated higher than the rural water level on the LiDAR DSM (Digital Surface Model), respectively. This method is semi-automated since the initialization of the snake and the selection of the training dataset are both done manually.

This study will focus on the mapping of the flood extent characteristic by proposing a fully automated classifier trained on a dataset retrieved from a pre-flood SAR image with the help of an optical Sentinel-2 image. The availability of an optical image allows to derive a water-mask without any human intervention from the Normalized Difference Water Index (NDWI), which, when multiplied by the pre-flood SAR image, on a pixel basis, permits to build a training dataset of backscatter values for the water and non-water classes. The labelled training dataset is thereby extracted from the optical and the pre-flood SAR images in an automated fashion. The preprocessing of the dataset and the classification are invoked from an online web application to map the extent of the inundation present on a post-flood SAR image. This application is intended mainly for emergency situations. As a consequence, it is of extreme importance to extract the flood map as quickly as possible and in an unsupervised way. The current paper is structured as follows. In the next section (Section 2), two datasets acquired with X-band TerraSAR-X and C-band Sentinel-1 of the inundations in Tewkesbury in 2007 and in Myanmar in 2015, respectively, are presented. The SAR images depicting these flood events will serve later on to assess the algorithm introduced relatively to a validation dataset. Afterwards, the theory behind the supervised classifier used specifically to cluster the flooded

SAR image into two classes and the post-processing utilized for refining the classified flood map, as well as the entire automated flood mapping process, are explained in detail in Section 3. The results of the flood mapping using the proposed method are validated and discussed in the following section (Section 4). Eventually, this paper closes with a conclusion about the strengths and the constraints of this algorithm.

2. Case Studies and Datasets

2.1. Tewkesbury 2007

This case study concerns the town of Tewkesbury (South-West England), which was flooded in July 2007. The town of Tewkesbury is situated at the junction of the Severn and the Avon rivers, and consequently the damages caused by the inundation there were considerable with the water propagating to the town center. Both a pre-flood TerraSAR-X image (Figure 1a) and a Sentinel-2 optical image of the studied area shown in Figure 1c with the ground truth superimposed on it, are required to train the supervised algorithm. The trained classifier will be subsequently used to classify a post-flood TerraSAR-X image (Figure 1b) of the flooded town of Tewkesbury into inundated and non-inundated pixels. This SAR image captured the flood in Tewkesbury on the 25 July 2007, while TerraSAR-X was still in its commissioning phase. The dataset used a pre-flood SAR image taken actually a year after the flooded image on the 22 July 2008 in dry conditions. The availability of a pair of SAR images acquired in the same configuration (3 m-resolution Stripmap and HH-polarized, as reported in Table 1) over the same area and processed as Single Look Slant Range Complex (SSC) products, makes this dataset suitable to conduct change detection analysis. The cloud-free Sentinel-2 optical image was acquired in the same season and the same month as the pair of SAR images, but eight years later than the pre-flood SAR image (19 July 2016).

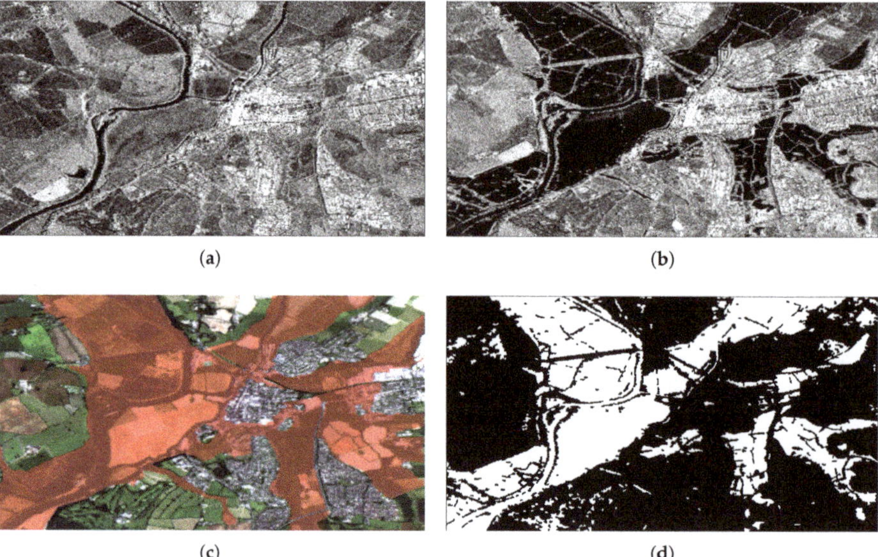

Figure 1. Subset of the town of Tewkesbury (South-West of England) on (**a**) the pre-flood TerraSAR-X image (22 July 2008), © DLR (2008) (**b**) the post-flood TerraSAR-X image (25 July 2007), © DLR (2007) (**c**) the Environment Agency's ground truth in red (20–24 July 2007) [20] superimposed on the Sentinel-2 optical image (19 July 2016), © Copernicus data (2016) (**d**) the resulting flood map obtained (white: flood, black: no-flood).

To confirm that no change in the land cover occurred between the time the pre-flood SAR image was taken and the acquisition of the Sentinel-2 optical image, an optical image captured by the Thematic Mapper (TM) sensor of Landsat-5 one month before the pre-flood SAR image and in the same season (8 June 2008), was compared visually with the latter Sentinel-2 image. Overall, the Landsat-5 and the Sentinel-2 images showed consistency in terms of water bodies, at least for the subset studied. For future flood events, thanks to the 5-days revisit time of the Sentinel-2 constellation currently in orbit, it should be easier to find a cloud-free optical image acquired in the same season and the same year as the reference SAR image to avoid any potential change in the land cover. In the current case study, the previous Landsat-5 product could not be considered for the subsequent flood mapping methodology due to its coarse 30 m spatial resolution.

Table 1. Radar parameters of the Tewkesbury 2007 Synthetic Aperture Radar (SAR) images.

Sensor	Polarization	Pixel Spacing after Terrain-Correction	Type of Product
TerraSAR-X	HH	2.2 m	SSC (complex)

2.2. Myanmar 2015

Flooding hit Myanmar during the monsoon season between July and September 2015. The area studied in this paper is situated in the South-East of the country, where the Salween River burst its banks in the month of August of the same year. Sentinel-1 took a SAR image during the flooding on the 6th of August 2015 (Figure 2b). Moreover, thanks to the systematic acquisition plan of Sentinel-1, a pre-flood SAR image taken on the 19th of March 2015 was also provided in the same acquisition parameters as the flooded SAR image (Figure 2a), to capture the same area in normal conditions. Both 20 m-resolution SAR images were acquired in VV polarization and processed as Ground Range Detected (GRD) products (Table 2). A cloud-free optical image was acquired a few years after the flooding by Sentinel-2 (Figure 2c) in the same season as the dry SAR image, to avoid inconsistencies between this pair in terms of presence or dryness of water bodies. Thanks to the open data policy of the Copernicus programme, the Sentinel-1 and Sentinel-2 images in this dataset are distributed online for free. Conversely, the commercial images from TerraSAR-X used in the previous case study can only be made freely available, in a limited number, for scientific purposes after a research proposal has been accepted. This last point introduces a constraint in the access to the SAR data following a flood disaster, and shows the advantages satellites belonging to the Copernicus programme have over commercial ones in the same context.

Table 2. Radar parameters of the Myanmar 2015 SAR images.

Sensor	Polarization	Pixel Spacing	Type of Product
Sentinel-1	VV	10.0 m	GRD (detected)

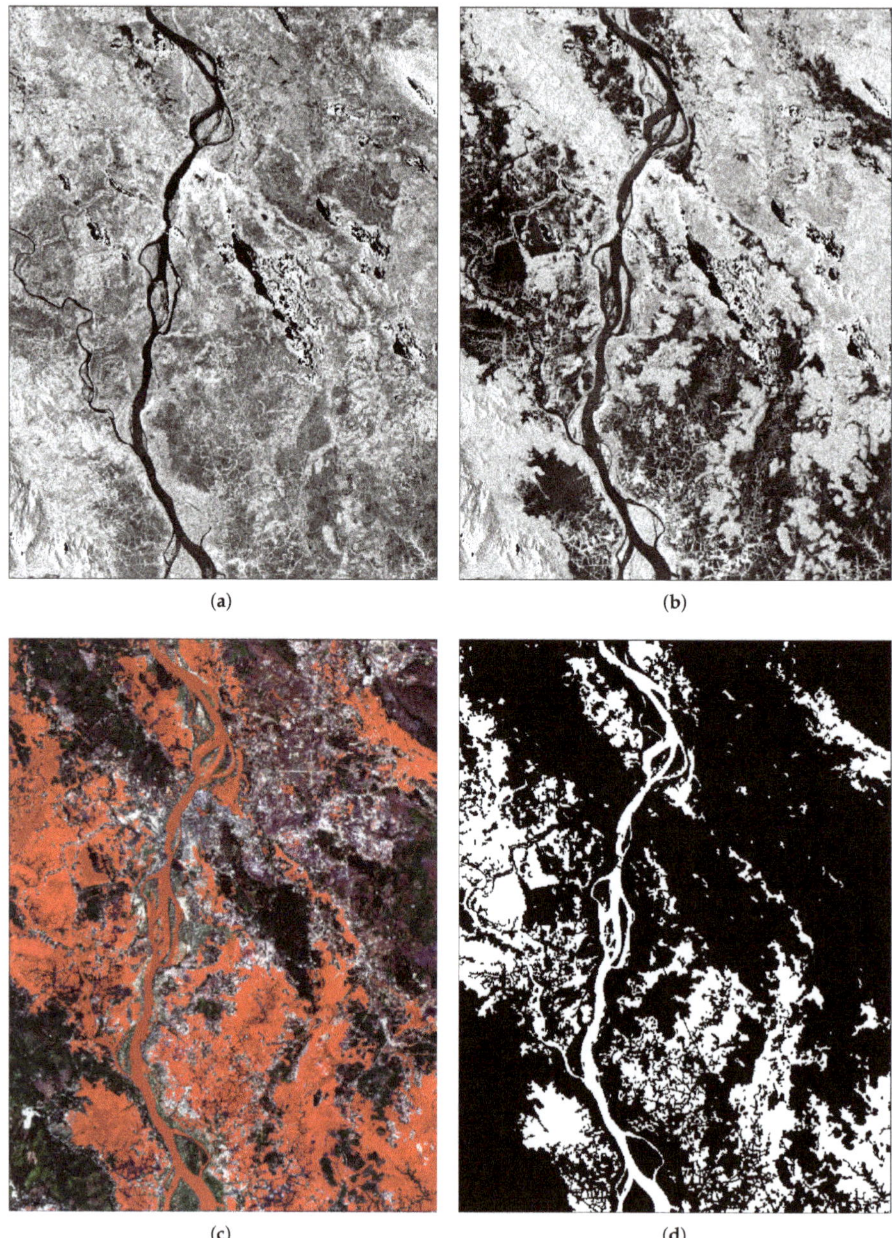

(a)

(b)

(c)

(d)

Figure 2. Subset of the city of Mawlamyine (southeastern Myanmar) on (**a**) the pre-flood Sentinel-1 Synthetic Aperture Radar (SAR) image (15 March 2015), © Copernicus data (2015) (**b**) the post-flood Sentinel-1 SAR (06 August 2015), © Copernicus data (2015) (**c**) the United Nations's ground truth vector in red (06 August 2015) [21] superimposed on the Sentinel-2 optical image (04 March 2018), © Copernicus data (2018) (**d**) the resulting flood map obtained (white: flood, black: no-flood).

3. Methodology

3.1. Stochastic Gradient Descent

The Gradient Descent (GD) is an iterative optimization algorithm which aims to find the minimum of a loss (cost) function [22]. It is based on the rationale that the cost function is minimized by moving in the opposite direction to its gradient. The learning rate η in Equation (1) serves in this particular case to regulate the steps taken down the slope (i.e., the negative of the gradient):

$$\omega_{i+1} = \omega_i - \eta \nabla_{\omega_i} L(\omega_i) \tag{1}$$

where:

ω_{i+1}: The model parameters to estimate,
ω_i: The model parameters estimated in the previous iteration,
η: The learning rate,
L: The loss (cost) function.

The learning phase in the standard Gradient Descent (called also the Batch Gradient Descent, BGD) requires that the derivative is calculated for all the samples in the training dataset in every iteration. The Batch Gradient Descent is consequently computationally intensive especially when the training set is too large [22]. The Stochastic Gradient Descent (SGD) used in this paper is another variant of the Gradient Descent, which is trained instead on a single randomly chosen training sample at a time. This online learning faculty of SGD makes it more scalable and quicker to train by allowing it to be unconstrained in terms of the execution time by the size of the training dataset [23]. SGD was in fact adopted in this study because in an operational context the objective is to produce the flood map as quickly as possible, and therefore, satellite images being generally quite large, the classifier chosen needs to fit rapidly to the training dataset regardless of the number of samples in it. The cost function used to train the classifier in this paper is the Hinge loss function given by:

$$L(x_j, y_j) = max(0, 1 - y_j \cdot (\omega x_j + b)) \tag{2}$$

where:

ω and b: The predicted model parameters,
x_j: The input sample.
y_j: The target class.

This function acts as a classification metric which appraises the linear model predicted with SGD in every iteration of the learning phase, and modifies its two parameters (ω, b) accordingly using Equation (1). In the case of the SGD classifier, ω corresponds to the weight assigned to the backscatter feature in the decision function, and b is its intercept. An interesting property of the Hinge loss function is that it punishes both misclassified samples and those who were correctly classified but with a low confidence, in order to maximize the margin between the classes.

A regularization term is also added to the loss function L in Equation (1) to help the predicted model to generalize to unlabeled data. The idea is to penalize complex models prone to overfitting, which are characterized by larger values for the parameters ω_i. The equations for the regularization functions commonly used are:

$$L1 = \sum_{i=1}^{m} |\omega_i| \tag{3}$$

$$L2 = \sum_{i=1}^{m} \omega_i^2 \tag{4}$$

The optimal values for the hyperparameters (model's parameters) in Table 3, which are the number of iterations of the SGD, the loss function, the regularization term and its coefficient α, can be determined by cross-validation where the classifier is trained on one part of the training dataset with different values of these hyperparameters, and then validated on the rest of this same dataset. The set of hyperparameters values producing the best accuracy scores during the cross-validation can be used for the subsequent training on the whole learning dataset. Because the training of the classifier and the test are performed in this case on two distinct datasets (the pre-flood and the post-flood SAR images respectively), no improvement in terms of accuracy was observed with the cross-validation. Furthermore, for the sake of a quicker computation time, the default hyperparameters values of the SGD were kept and are reported in Table 3, except for the number of iterations which was increased to a 1000 iterations.

Table 3. The values of the hyperparameters used during the training of the Stochastic Gradient Descent (SGD) classifier.

Number of Iterations	Loss Function	Regularization Term	Alpha
1000	Hinge loss	L2	0.0001

3.2. Graph Cuts

A graph $\mathcal{G} = \langle \mathcal{V}, \mathcal{E} \rangle$ is defined by its vertices (nodes) $\mathcal{V}$ linked by directed edges $\mathcal{E}$, where each edge connecting two nodes has a weight. Two terminal nodes, a source s and a sink t, are added to the two extremities of the previous directed and weighted graph $\mathcal{G}$ to form a flow network. In the context of a binary image segmentation, the image pixels are represented by the non-terminal graph nodes, while the terminal nodes (s and t) are the labels to assign to each pixel [24]. With this in mind, an s-t cut divides the flow network into two subsets $\mathcal{S}$ and $\mathcal{T}$, in such a way that the source s belongs to $\mathcal{S}$ and the sink t to $\mathcal{T}$. This kind of cut can be seen as a binary classification or a labeling task, where each non-terminal node (pixel) is assigned either to $\mathcal{S}$ or $\mathcal{T}$, depending on whether the pixel belongs to the foreground or the background for instance. For edges too, two types can be distinguished in flow networks. N-links join together pixel nodes located in the same neighborhood on the image, whereas t-links connect each pixel node to both terminals (s and t). The weight of a t-link allows to penalize a node that was mislabeled compared to a prior knowledge, and that of an n-link ensures a smooth and consistent labeling by encouraging pixels in the same neighborhood to be in the same class, after the graph is split. The weights of n-links and t-links can be defined mathematically by the energy terms $E_{smooth}(L)$ and $E_{data}(L)$ respectively, which appear in the energy function to minimize [24]:

$$E(L) = E_{data}(L) + E_{smooth}(L) \tag{5}$$

Graph cut methods intend to find the labeling L that minimizes the energy function E in the previous equation. When each energy term is replaced with its respective expression, the energy function becomes [24]:

$$E(L) = \sum_{p \in \mathcal{P}} D_p(L_p) + \sum_{(p,q) \in \mathcal{N}} V_{p,q}(L_p, L_q) \tag{6}$$

where:

D_p: The data penalty term for a pixel p,
$V_{p,q}$: The smoothness or regularization term between neighboring pixels p and q,
$\mathcal{P}$: The set of image pixels p,
$\mathcal{N}$: The set of pairs of adjacent pixels (p, q),
L_p: The labeling to assign to a pixel p.

The network flow in Figure 3 shows two pixel nodes p_0 and p_1, besides the two terminal nodes s and t, and illustrates how each edge connecting a pair of nodes is assigned the adequate energy term.

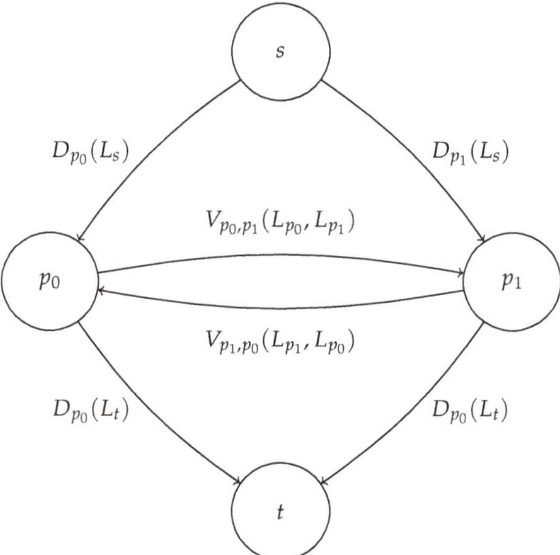

Figure 3. A flow network with a source node s, a sink node t, two non-terminal pixel nodes p_0 and p_1, and the energies used as the edges' weights.

Graphs cuts based energy minimization techniques were already used in the past to restore noisy binary images in [25], who demonstrated that simulated annealing could get stuck in a local minimum during the minimization of the energy. The Bokov-Kolmogorov min-cut/max-flow algorithm (BKA), used in the current study, has proved effective in many image processing applications like image segmentation, stereo matching, and image restoration [24]. It extends the idea in [25] by proposing a graph cut algorithm that performs quickly and is even able to generalize to N-dimensional image segmentation problems [26]. Analogously to when they were first proposed in [25], graph cuts will be integrated in this paper's methodology as a post-processing to enforce the spatial continuity between pixels assigned to the same class. In other words, the objective of graph cuts as a post-processing in this case is to take into account the spatial proximity between the pixels in the binary flood map, by reclassifying the noisy pixels resulting from the previous pixel-based classification.

3.3. Flood Extent Mapping

The process chain shown in the flowchart in Figure 4 consists in a supervised classifier trained automatically to build the model that maps the floodwater on the post-flood SAR image. Briefly, a labelled training dataset of water and land classes is gathered in an automated way from a Sentinel-2 optical image and a pre-flood SAR image, and is next fed into the classification algorithm during the learning phase. Although the classifier is normally supervised, in the sense that a labelled training dataset needs to be provided beforehand to train it, the selection and the labelling of the training samples is carried out automatically in this paper. Ultimately, the trained classifier will be ready to discriminate the pixels individually in the flooded SAR image into flood and non-flood pixels. Each phase of the process will be explained in more detail below.

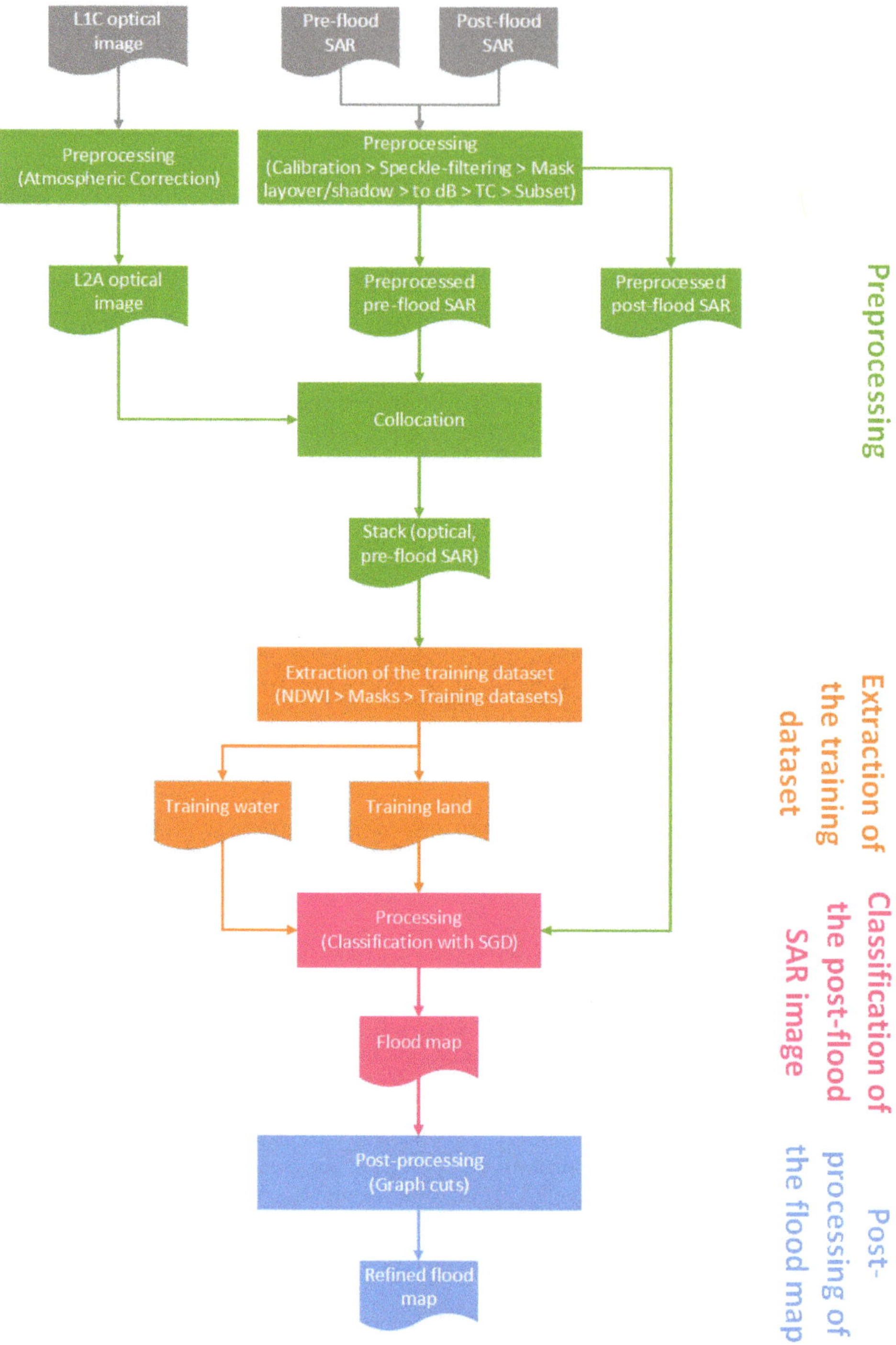

Figure 4. Flowchart of the automatic flood mapping process including the preprocessing, the extraction of the training dataset, and the classification.

3.3.1. Preprocessing

The SAR images taken prior and after the flooding were radiometrically calibrated and speckle-filtered with a 5 × 5 Gamma Map filter to avoid having a noisy flood map later. Afterwards, masks of shadow and layover were extracted automatically from the SAR images with the help of the free SRTM DEM. Pixels where these geometric distortions appear will be systematically disregarded during the training and classified as unflooded after that. The shadow and layover pixels detected on the pair of SAR images showing the studied area in Myanmar appear in a homogeneous black color on the mountains located to the right of the river (Figure 2a,b). The SAR images with masked shadow and layover were eventually converted to dB and projected to the ground-range with a terrain-correction (TC) using the same DEM. The subsetting operation was left till the very end of the preprocessing to get perfectly rectangular images after the terrain-correction.

The Sentinel-2 optical image used to calculate the NDWI (Normalized Difference Water Index) has to be atmospherically corrected by processing it to a Level-2A product on the user side. This preprocessing step aims to remove the effect of the atmosphere (i.e., clouds, aerosols, gases...) from optical satellite images, and is necessary before computing the NDWI. Recently, the Sentinel Hub started distributing online products already processed as Level-2A Bottom-Of-Atmosphere (BOA) Sentinel-2 products.

The pre-flood SAR image needed also to be collocated with the Sentinel-2 image using the Sentinel Application Platform (SNAP) Python API, prior to the extraction of the SAR training samples from it.

3.3.2. Extraction of the Training Dataset

The Normalized Difference Water Index (NDWI) was initially presented in [27] to detect water bodies from multispectral optical images. The NDWI mathematical expression is given by:

$$NDWI = \frac{Green - NIR}{Green + NIR} \tag{7}$$

where, similarly to [28]:

$Green$: Band 3 in the Sentinel-2 product,
NIR: Band 8 in the Sentinel-2 product.

It is based on the idea that water has at the same time a high reflectance in the green band and a low one in the near-infrared (NIR) one, while other types of land cover to disregard (soil and vegetation) appear brighter in the latter band [29]. The NDWI was originally calculated from multispectral images captured by Landsat's Multispectral Scanner (MSS) [27], but was estimated in further works from Landsat's ETM+ bands [29], high-resolution Quickbird ones [30], and even from Sentinel-2 images [28]. In rural areas, pixels with a positive NDWI were expected to correspond to water. However, this could lead to false alarms in urban areas where houses rooftops for instance were found to have positive NDWI values, although lower than that of water. Consequently, a higher threshold was determined in [30] (Equation (8)) by the same author who proposed the NDWI, in order to identify the water surfaces in swimming pools which constitute a common place for mosquitoes to lay their eggs. The NDWI threshold could nevertheless fail to detect water surfaces that are concealed by protruding vegetation or shadowed by trees or buildings.

$$Class = \begin{cases} Water, & \text{if } NDWI \geq 0.3 \\ Land, & \text{otherwise} \end{cases} \tag{8}$$

Other water indices were also proposed in the literature like the Modified Normalized Difference Water Index (MNDWI) [29], which was calculated by replacing the near-infrared (NIR) band in Equation (7) with the shortwave infrared (SWIR) one. The MNDWI was proposed to prevent the obtained water mask from including false positives from urban areas, based on the observation that the spectral response from built-up land was higher in the SWIR band compared to the green band. Therefore, built-up areas would be expected to result in negative MNDWI values, on the contrary of the NDWI ones. However, the SWIR band used to calculate the MNDWI has a 20 m spatial resolution in Sentinel-2 products, as opposed to the 10 m spatial resolution of NIR and visible bands (Table 4). As a result, the calculation of the MNDWI from Sentinel-2 bands should be preceded either by a downsampling of the green band to 20 m-resolution, or by a sharpening of the SWIR band to 10 m-resolution [28]. Besides, according to a few experiments carried out on the Sentinel-2 datasets used in this paper, the MNDWI still required a threshold greater than zero to identify the water, although it should be lower than the NDWI one according to [29]. Lastly, most high-resolution optical satellite sensors currently in orbit have a NIR band but lack a SWIR band.

Table 4. The Sentinel-2 bands used to calculate the Normalized Difference Water Index (NDWI) and their resolutions.

Band Name	Band Number	Resolution [m]
Green	Band 3	10
NIR	Band 8	10

In the current study, the NDWI index is calculated using Equation (7) from the green and near-infrared (NIR) bands of the Sentinel-2 Bottom-Of-Atmosphere (BOA) level-2A optical image, which was collocated with the preprocessed pre-flood SAR one in the previous step. Then, by applying the threshold in Equation (8) to the NDWI, a water mask is produced. The land mask is simply the logical negation of the water mask. Both the NDWI-derived water and land masks produced in the previous step were separately multiplied with the pre-flood SAR image present in the same stacked product, to extract from the latter image the pixels belonging to water and land classes, respectively. The previous step depends on the accurate collocation between the optical and the pre-flood SAR image, so that the location of one class (water or land) in the former product matches its location in the latter one. Afterwards, an equal number of samples (1000 samples/class) was taken randomly from the extracted water and land pixels, and was used as a training dataset. Having training classes with the same number of samples helps to avoid favoring the majority class in the subsequent step. One pre-requisite is that the training samples are randomly shuffled prior to the learning phase [23], to avoid that the classifier recognizes in the post-flood SAR image the last class it was trained on better than the first one.

3.3.3. Classification of the Post-Flood SAR Image

The learning dataset serves to train the supervised classifier to recognize water and land in the post-flood SAR image using algorithms such as the Stochastic Gradient Descent (SGD) included in the Scikit-learn library. Thanks to the online learning ability of the SGD, recalled in Section 3.1, the classification is performed very quickly even when the training dataset is quite large. It is not necessary in this case that the training and test datasets are normally distributed with zero mean and unit variance, since there is only one feature (the backscatter in dB). Eventually, the trained classifier was employed to segment the post-flood SAR image into water and non-water (land) classes. In summary, the classifier is trained on water and land pixels from the pre-flood SAR image, and used to categorize the post-flood SAR image pixels into the same two classes. The classification is thereby based on the assumption that the flood has the same low backscatter signature on the SAR image as permanent water bodies and rivers. This supposition can be confirmed visually, as their similar low radar return makes them easily distinguishable from the rest of the land cover. Therefore, the trained classifier does not differentiate floodwater from permanent water bodies on the flooded SAR image, since both of them are classified as water. The water map produced is specific only to the post-flood SAR image in the sense that potential changes in rivers morphologies, relatively to the date of acquisition of the pre-flood SAR image, are captured by the latter thematic map.

3.3.4. Post-Processing of the Flood Map

The Bokov–Kolmogorov min-cut/max-flow algorithm (BKA) [24] used during the post-processing of the flood map requires to set the values of both the penalty and the regularization energy terms. Knowing that the input flood map image has binary values (p_i), the penalty (data) term is null when the pixel is assigned the same label it got after the previous classification, otherwise it is equal to one:

$$D = \begin{cases} p_i & , \text{if } x_i = 0 \\ 1 - p_i & , \text{if } x_i = 1 \end{cases} \tag{9}$$

where:

p_i: The value of the pixel i in the input image,
x_i: The label assigned to the pixel i.

For the smoothness term, an 8-connectivity neighborhood was considered so that each pixel is connected with an n-link to the 8 pixels around it. The weights of n-links were set empirically to a constant ($V = 1$) that matches the weights of the t-links (Equation (9)) in terms of order of magnitude ($D \in \{0, 1\}$):

$$V = 1 \tag{10}$$

The idea behind this regularization term is that a strong edge with a high weight connecting a pair of adjacent nodes will not be broken during the cut, and these adjacent pixels will therefore be encouraged to take the same label [31]. In a way, the smoothness term balances out the penalty one during the minimization of the total energy. This post-processing of the flood map allows to take into account the spatial contiguity between the image pixels in order to remove the noise created by the previous pixel-based classification. In fact, the higher the value of the regularization term, the smoother the segmentation is.

3.3.5. Implementation

The flood extent mapping algorithm can be called from a web application which was built with the Django web framework. The preprocessing of the SAR images calls the requested SNAP operators using its Python API. As for the SGD classification algorithm, it is implemented in Scikit-learn, which is a machine learning library written in Python. The implementation of the max-flow algorithm (BKA) utilized for the post-processing of the flood map is the one provided by a python wrapper (PyMaxflow) available in [32], which is itself based on [24]. The flood mapping is followed by the visualization of the produced flood map raster served by GeoServer on an Open Street Map rendered using Mapbox. The developed application is cost-effective since all its dependencies are open-source libraries (SNAP, Python, Scikit-learn, Django, PyMaxflow). It is also cross-platform and can be deployed on a server and queried remotely, thanks to its ability to be called from the internet browser.

4. Results and Discussion

4.1. Accuracy Metrics

The overall accuracy of the classification was calculated for the resulting flood maps relatively to the ground truth associated with them, according to the equation given in [33]. Besides the overall accuracy, two other accuracy metrics are commonly used in remote sensing to assess the classification of each class separately. These are the producer's accuracy and the user's accuracy which are the complements of the omission error and the commission error, respectively [34]. In addition to these accuracy metrics, the area flooded was also easily calculated by multiplying the number of pixels classified as flooded by the pixel spacing of the SAR image:

$$area = n \cdot s \qquad (11)$$

where:

> *area*: The flooded area in km^2,
> *n*: The number of pixels flooded,
> *s*: The pixel area in m^2, which is equal to the squared pixel spacing.

4.2. Tewkesbury 2007

The extent of the flooding mapped with the SGD in the town of Teweskbury is shown in Figure 1d. The flood map obtained was validated against a ground truth vector (the red mask in Figure 1c), which was collected by the Environment Agency using in situ surveys and aerial photography [20] in the same town of Tewkesbury at least a day before the acquisition of the post-flood SAR image, between the 20th and the 24th of July 2007. The ground truth vector had to be rasterized and confined to the area of interest first, then the flood map produced in this paper was compared with it on a pixel level. After an assessment of the results obtained with the SGD classifier (Table 5), the accuracy of the classification was found to be around 77%. Moreover, from the producer's accuracy of the flood class (61.18%) in the same table, it can be inferred that there is an underestimation of the inundation and consequently an overestimation of the non-flooded areas. The under-estimation of the floodwater is also clear from the flooded area presented in km^2 in Table 6. One type of land cover possibly responsible for the missed classifications are flooded urban areas (clearly flooded in aerial images in [19]) characterized by an increase in the backscatter, while the classifier was not trained to recognize their signatures. However, it should be noted that the flooded SAR image was acquired on a different day (25 July 2007) than the ground truth (20–24 July 2007), and therefore a flood recession should not be ruled out between the two dates.

Table 5. The producer's and user's accuracies for the flood and non-flood classes and the overall accuracy of the classification for the Tewkesbury 2007 dataset.

Class	Producer's Accuracy	User's Accuracy	Overall Accuracy
Non-flood	94.15%	69.19%	77.03%
Flood	61.18%	91.87%	

Table 6. Area suffering from the flooding in km^2 on the obtained flood map and on the ground truth for the Tewkesbury 2007 dataset.

Product	Flooded Area
Flood map	7.58 Km^2
Ground truth	11.39 Km^2

4.3. Myanmar 2015

For the Myanmar 2015 dataset, the validation was performed against flood maps obtained from the same Sentinel-1 dataset by the United Nations Institute for Training and Research (UNITAR) [21]. The preprocessing step, consisting of masking out geometric distortions SAR images suffer from, allowed to get rid of parts of the topographic shadows which otherwise could be misclassified as water due to the similar dark appearance on the SAR image. The results in Table 7 were superior to those realized on the previous dataset, reaching an accuracy over 90%. Due to the larger pixel spacing of Sentinel-1 SAR images (10 m in Table 2), the missed classification translates into a larger difference between the area inundated in km^2 on the flood map and on the ground truth in Table 8.

Table 7. The producer's and user's accuracies for the flood and non-flood classes and the overall accuracy of the classification for the Myanmar 2015 dataset.

Class	Producer's Accuracy	User's Accuracy	Overall Accuracy
Non-flood	98.96%	91.97%	93.47%
Flood	82.06%	97.43%	

Table 8. Area suffering from the flooding in km^2 on the obtained flood map and on the ground truth for the Myanmar 2015 dataset.

Product	Flooded Area
Flood map	407.43 Km^2
Ground truth	483.74 Km^2

4.4. Classification of Urban and Non-Urban Areas in Tewkesbury 2007

The validation of the results for Tewkesbury 2007 went one step further and the flood maps were assessed separately in the urban and non-urban areas. With this in mind, the urban mask was obtained by thresholding the 25 m-resolution land cover map for the UK which was downloaded from [35], while the non-urban mask was simply taken as the logical negation of the latter mask. The same process described in the previous sections was applied on the SAR image as before, but the validation was carried out on the two types of land use separately.

Tables 9 and 10 give an assessment of the results obtained by the SGD classifier in the urban and the non-urban regions of the SAR image in Figure 1b, respectively. As expected, the classification is slightly more accurate in non-urban areas compared to urban areas (almost 77.68% against 74.7%, respectively). Nevertheless, in terms of False Negatives the classification in urban settlements did a lot worse than in non-urban ones (close to 5% of producer's accuracy against 68%, respectively). As stated in the previous section, this might be explained by the fact that the classifier missed flooded urban settlements characterized by an increase in the backscatter, while the training dataset identifies only dark open-water on the pre-flood SAR image. Alternatively, the same reason also referred to in the previous section regarding the different acquisition dates of the flooded SAR image and the ground truth could possibly be responsible for the high missed detection in urban areas.

Table 9. The producer's and user's accuracies for the flood and non-flood classes and the overall accuracy of the classification for the urban areas of Tewkesbury 2007.

Class	Producer's Accuracy	User's Accuracy	Overall Accuracy
Non-flood	99.1%	74.87%	74.7%
Flood	4.88%	65.6%	

Table 10. The producer's and user's accuracies for the flood and non-flood classes and the overall accuracy of the classification for the non-urban areas of Tewkesbury 2007.

Class	Producer's Accuracy	User's Accuracy	Overall Accuracy
Non-flood	91.67%	66.46%	77.68%
Flood	68.0%	92.19%	

4.5. Maps of False Negatives and False Positives

The maps of False Negatives (FNs) and False Positives (FPs) for Tewkesbury and Myanmar datasets (Figures 5 and 6, respectively) were produced in QGIS from the flood and the ground truth binary maps using this equation:

$$Error = \begin{cases} FN, & \text{if } (predict - valid) \cdot mask = -1 \\ FP, & \text{if } (predict - valid) \cdot mask = 1 \end{cases} \tag{12}$$

where:

Error: The type of misclassification in the masked area,
FN: A false negative,
FP: A false positive,
predict: The predictions,
valid: The ground truth,
mask: The urban or non-urban mask, if applicable.

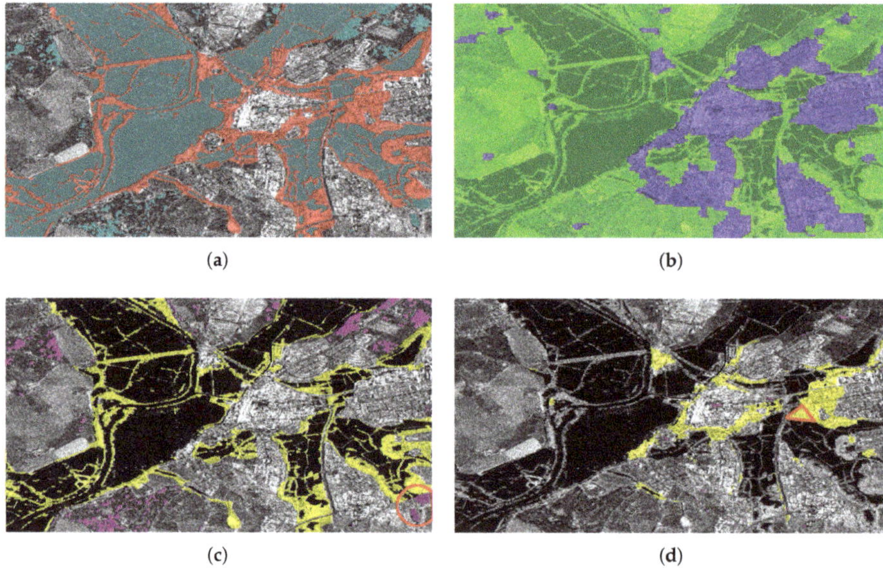

Figure 5. Superimposed on the post-flood SAR image are: (**a**) the flood map produced from Tewkesbury 2007 TerraSAR-X image in cyan superimposed on the ground truth in red (**b**) The urban mask extracted from the UK Land Cover map [35] in blue and the non-urban mask (negation of the urban mask) in green (**c**) The false negatives (FN) in yellow and the false positives (FP) in magenta, in non-urban areas (**d**) The false negatives (FN) in yellow and the false positives (FP) in magenta, in urban areas.

For the Tewkesbury dataset, the FNs and FPs were extracted separately for urban and non-urban areas thanks to the availability of the UK land cover map [35]. The yellow field inside the red triangle in Figure 5d was clearly not flooded on the post-flood SAR image of Tewkesbury (Figure 1b). This means that its inclusion with the missed classifications was probably due to a flood recession in this area. There were also some False Negatives in the urban regions of the same image (Figure 5d) appearing in yellow, but the resolution of the SAR image was not high enough to detect the flood in narrow roads. As for the False Positives appearing in magenta inside the red circle in Figure 5c, this rural area is clearly flooded since it appears in dark due to the specular reflection from the water. One possible explanation for it being considered a False Positive is that because the ground truth and the SAR image were not acquired on the same date, different areas were flooded on the two dates of acquisition. Other false alarms appearing in magenta in the same image (Figure 5c) might come from low grow grass or bare fields which exhibit the same low backscatter that characterizes water on radar imagery, whereas in the lower-left corner of the image the shadow from the trees could have resulted in some false positives.

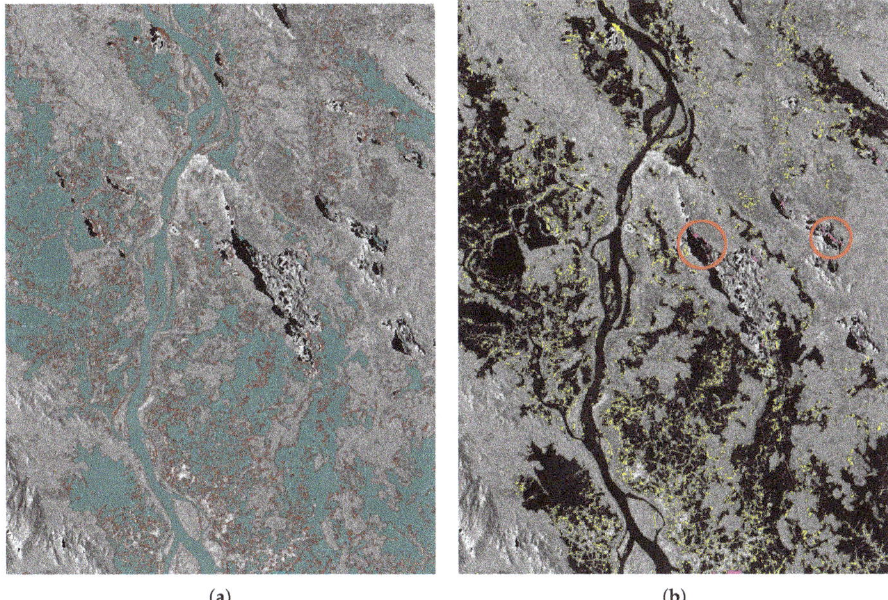

(a)   (b)

Figure 6. Superimposed on the post-flood SAR image are: (**a**) the flood map produced from Myanmar 2015 Sentinel-1 image in cyan superimposed on the ground truth in red (**b**) The false negatives (FN) in yellow and the false positives (FP) in magenta.

Regarding the Myanmar 2015 flood map, the misclassification caused by False Negatives (yellow pixels in Figure 6) is generally located on the boundaries of the flooded zones. It is suspected that the speckle-filter smoothed out the boundaries of the flooding on the post-event SAR image, which led the pixels constituting them to be wrongly classified as dry. The persistent dark topographic shadow which exists even after masking out the geometric distortions, creates false positives visible in magenta inside the two red circles in Figure 6. This might be explained by the relatively coarse resolution (around 30 m) of the DEM employed during the extraction of the shadow and layover from the SAR images.

4.6. Computation Time

Each operation performed during the flood mapping process was profiled individually, by measuring the time it takes to run. The time to upload the optical and SAR images to process to the Web server as well as the time to load the web page are included in the execution time. The upload time is not considerable when working locally, however it very much depends on the internet speed and the size of the images if the flood mapping Web application is deployed remotely. The computation times for the preprocessing of the SAR images, the generation of the training dataset, and the classification plus the post-processing are shown in Tables 11 and 12 when these operations were executed on the Myanmar 2015 and the Tewkesbury 2007 datasets, respectively. Overall, the processing time of the whole flood mapping chain for the Myanmar 2015 dataset is around 2 min and 40 s, considering that the input SAR subsets have over 19 millions pixels each prior to the preprocessing. As for the Tewkesbury 2007 dataset, the total processing time is nearly 1 min for SAR subsets of 5 million pixels (Table 12).

Table 11. The execution time for each operation of the flood mapping process for the Myanmar 2015 dataset.

Operation	Execution Time [seconds]
Uploading SAR pair & optical image	0.87
Preprocessing SAR pair	96.82
Building training dataset	27.61
Training, classification, and post-processing	35.59
Total	160.89

Table 12. The execution time for each operation of the flood mapping process for the the Tewkesbury 2007 dataset.

Operation	Execution Time [seconds]
Uploading SAR pair & optical image	1.58
Preprocessing SAR pair	36.91
Building training dataset	7.97
Training, classification, and post-processing	12.1
Total	58.56

4.7. Comparison with the Literature

The results in [7,10] and [36] were chosen to benchmark the flood mapping chain presented in this paper, since these methods were also tested on the Tewkesbury 2007 dataset. However, the performances of this paper's method in terms of computation time cannot be objectively compared relatively to these methodologies, since they were run on machines with different characteristics and on image subsets of different dimensions. In absolute terms, this paper's method finished the whole processing in under 5 min (Table 13) on a large subset of Tewkesbury 2007 SAR dataset of more than 41 million pixels (Figure 7), which is more than 8 times larger than the subset used in the previous section (Section 4.6). Most of the execution time (more than 3 min) was in fact dedicated to the preprocessing of the pair of SAR images, while the post-processing with graph cuts also required a considerable amount of time to improve the flood map (over 1 min) because the image was quite large. The current approach is thus suitable to operate in emergency situations, when it is necessary to have a quick overview of the flood situation. In fact, the time taken to produce the flood map is in the same order of magnitude as the targeted response time for critical calls in the UK. According to the legislation, the Ambulance and the Fire & Rescue Services are expected to arrive at the location of the reported incident in 8 and 10 min, respectively, 75% of the time [37]. By way of comparison, a method with a similar purpose of mapping the inundation in near real-time in [7] took around 2.67 min on a smaller subset of 5.4 million pixels. Furthermore, for [10] who adapted the process in [7] to Sentinel-1 SAR images, the whole chain took 45 min to map the flood on the entire Ground Range Detected (GRD) image, but included the downloading of the SAR images which is normally dependent on the internet speed. Accordingly, the process described in the current study extends the literature on operational flood mapping from SAR images with a novel method suitable mainly for emergency situations.

The precision of the classification was assessed by running this algorithm on a workstation equipped with a 6-core 2.67GHz Intel Xeon X5650 CPU and 24.0 GB of RAM. The overall accuracy (90.66% in Table 13) was found to be comparable to other studies in the literature of operational flood mapping (95.44% in [7]). The accuracy results were also compared by land use (urban and rural) to [36], who performed the flood mapping on a subset of the Teweskbury 2007 dataset of roughly the same size. The accuracy obtained in rural areas with the algorithm described in this paper (90.81% in Table 13) is quite close to what [36] achieved on a similar subset (89%). As for the urban areas, the accuracy in [36] dropped from 75% to 57%, depending on whether the shadow and the layover where the flood cannot be detected were masked or not. The accuracies for rural and urban areas achieved in this paper are not too far from one another as can be seen in Table 13 (90.81% for non-urban areas and 88.97% for urban ones), considering that the DEM is too coarse to detect the geometric distortions in Tewkesbury's town center and that the surface is quite flat.

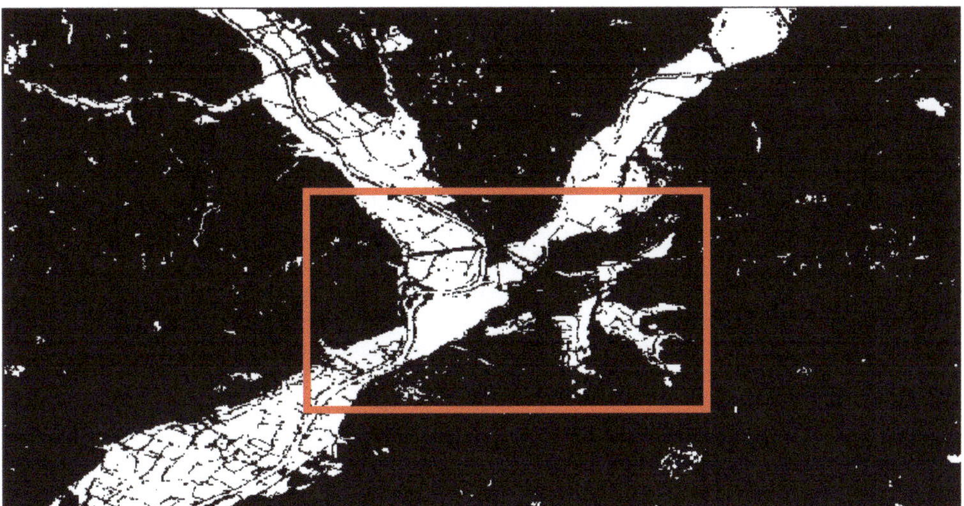

Figure 7. The flood map produced from a large subset of the the Tewkesbury 2007 dataset. The subset assessed in Section 4.2 is inside the red rectangle.

Table 13. Comparison between the results obtained with this paper's method on the large Tewkesbury 2007 subset in Figure 7, and those reported in [36] on the same dataset with a similar subset size.

Paper	Overall Accuracy	Urban/non-Urban Accuracy	Execution Time [minutes]
This paper	90.66%	Non-urban areas: 90.81% Urban areas: 88.97%	4.74
[36]	-	Non-urban areas: 89% Urban areas: 75% if the shadow and layover are ignored, 57% otherwise	19.1

4.8. Training Dataset

Besides the water and non-water classes, the automated process proposed in this paper gives the possibility to train the classifier on other classes, if additional training subsets can be extracted in an unsupervised way. However, it may prove difficult to collect training datasets, for example, for flooded urban or vegetated areas at the time of the disaster. In a separate experiment, the flood mapping classifier was effectively trained with a dry vegetation class retrieved similarly to the water class, by thresholding the Normalized Difference Vegetation Index (NDVI) according to [38]. The vegetation threshold ($NDVI \geq 0.2$) normally corresponds to the NDVI values exhibited by shrubs and grassland. The previous heterogeneous non-water subset was thus split again into vegetation and non-vegetation. But the accuracy results for the large subset of Tewkesbury in particular decreased by almost 10%, after considering the vegetation as a separate class. For this reason, only water and non-water classes were maintained in the previous tests.

Finally, it was found that when the water is not well represented in the pre-flood SAR image and the optical image, the classification using the method in this paper might later fail to efficiently recognize the flooding on the post-event SAR image. Another detail to point out concerns the muddy water present at the mouth of the River Severn near the city of Bristol, which was not completely detected by the NDWI threshold. The river mouth is located on the same Sentinel-2 optical image but to the southwest of the subset taken around Tewkesbury.

5. Conclusions

The novel method proposed in this paper proved capable of a quick and automatic mapping of the extent of the inundation, which can assist response authorities to prioritize during rescue operations. The rapidity of execution is a very important factor, especially when the main purpose of the flood maps is to support relief efforts at the time of the disaster.

This approach was evaluated on two flood events captured by SAR sensors in different wavelengths. The first SAR pair was taken in X-band by TerraSAR-X, while the other one was captured by the Sentinel-1 C-band sensor. Thanks to the availability of a ground truth in both cases, it was possible to assess the algorithm proposed and discuss the results obtained. The flood classifier showed satisfying results, however it might fail if the water bodies and rivers are too small to appear clearly in the pre-flood SAR image and the Sentinel-2 product. In this particular case, the classifier cannot be trained efficiently to recognize the water class. Moreover, when the optical and the pre-event SAR images are acquired a few years from each other, inconsistencies in terms of the presence and absence of water bodies might exist between these two images, even if they were both taken during the same season. The dates of acquisition of this pair of images is therefore a crucial parameter worth considering, to avoid having pixels belonging to one class being labeled in the other one in the training dataset. In this case, a change in the land cover between the pre-flood SAR image and the Sentinel-2 optical image is responsible for the mislabeling, when for instance the date of acquisition of the latter image is almost a decade after the pre-flood SAR image (e.g., the Tewkesbury 2007 dataset). Cloud coverage could also lead to a similar wrong labeling of the learning dataset, particularly when parts of the water bodies are hidden by clouds. Nevertheless, the NDWI threshold utilized for extracting the water mask from the optical bands appeared to generally provide a good compromise between over- and under-estimations, although a thorough validation of this threshold on other datasets, preferably from different climates, could become required in the future.

This process chain assumes the availability of a reference non-flooded SAR image and a dry optical image, besides the flooded SAR image. However, this does not constitute a serious constraint since areas at risk of a flooding can ensure that these reference images are at their disposal beforehand and regularly updated. Furthermore, the systematic acquisition mode of the Satellites launched for the Copernicus Programme (Sentinel-1 and Sentinel-2) increases the chances of finding suitable reference images (SAR and optical products) in the archive. Like the DEM utilized to mask out the geometric distortions (shadow and layover), Sentinel-1 and Sentinel-2 images are distributed for free, in contrast to the images acquired by commercial satellite missions such as TerraSAR-X and COSMO-SkyMed, and cover most of the Earth's surface.

In the future, the proposed method could benefit from recent commercial optical sensors, capable of delivering metric or even sub-metric resolution images, to address the abovementioned issue concerning narrow river channels that cannot be distinguished with the 10 m-resolution optical bands of Sentinel-2. Moreover, it is expected that SAR images of a higher spatial resolution could improve the accuracy of the classification in flooded urban areas. With this in mind, both TerraSAR-X with its staring spotlight mode and its future successor TerraSAR-X NG [39] can take SAR images with up to 25 cm resolution. Finally, centimetric-resolution airborne LiDAR DEMs with finer horizontal and vertical accuracies compared to the SRTM DEM used, should result in a more precise terrain-correction and a more efficient masking out of the shadowed pixels from the SAR images.

Author Contributions: The first author, A.B., was responsible of developing the methodology, implementing the software, validating the results, and writing the original draft. The research conducted was supervised by the second author, R.G., who provided comments all along the project about the process and the draft.

Funding: This research was funded by Surrey Satellite Technology Ltd (SSTL). The APC was funded by Surrey Space Centre, University of Surrey.

Acknowledgments: The authors would like to thank the UK Environment Agency and UNITAR (United Nations Institute for Training and Research) for producing the validation data for Tewkesbury and Myanmar flood events respectively, which are both available online free of any charge. The authors would also like to thank the German Aerospace Centre for providing the TerraSAR-X datasets within the LAN3308 proposal.

Conflicts of Interest: The authors declare no conflict of interest.

References

1. AON. Global Climate and Catastrophe. In *2016 Annual Report*; AON: London, UK, 2017.
2. Guha-Sapir, D.; Hoyois, P.; Wallemacq, P.; Below, R. Annual Disaster Statistical Review 2016: The Numbers and Trends. *Brussels: CRED*, 31 October 2017.
3. Refice, A.; Capolongo, D.; Lepera, A.; Pasquariello, G.; Pietranera, L.; Volpe, F.; D'Addabbo, A.; Bovenga, F. SAR And Insar For Flood Monitoring: Examples With Cosmo/Skymed Data. *IEEE Int. Geosci. Remote Sens. Symp.* **2013**, *7*, 703–706.
4. Voormansik, K.; Praks, J.; Antropov, O.; Jagomagi, J.; Zalite, K. Flood mapping with terraSAR-X in forested regions in estonia. *IEEE J. Sel. Top. Appl. Earth Observ. Remote Sens.* **2014**, *7*, 562–577, doi:10.1109/JSTARS.2013.2283340. [CrossRef]
5. Mason, D.; Giustarini, L.; Garcia-Pintado, J.; Cloke, H. Detection of flooded urban areas in high resolution Synthetic Aperture Radar images using double scattering. *Int. J. Appl. Earth Observ. Geoinf.* **2014**, *28*, 150–159, doi:10.1016/j.jag.2013.12.002. [CrossRef]
6. Martinis, S.; Kersten, J.; Twele, A. A fully automated TerraSAR-X based flood service. *ISPRS J. Photogramm. Remote.Sens.* **2015**, *104*, 203–212, doi:10.1016/j.isprsjprs.2014.07.014. [CrossRef]
7. Martinis, S.; Twele, A.; Voigt, S. Towards operational near real-time flood detection using a split-based automatic thresholding procedure on high resolution TerraSAR-X data. *Nat. Hazards Earth Syst. Sci.* **2009**, *9*, 303–314, doi:10.5194/nhess-9-303-2009. [CrossRef]
8. Pulvirenti, L.; Marzano, F.S.; Pierdicca, N.; Mori, S.; Chini, M. Discrimination of water surfaces, heavy rainfall, and wet snow using COSMO-SkyMed observations of severe weather events. *IEEE Trans. Geosci. Remote Sens.* **2014**, *52*, 858–869. [CrossRef]
9. Herrera-Cruz, V.; Koudogbo, F. TerraSAR-X rapid mapping for flood events. In Proceedings of the International Society for Photogrammetry and Remote Sensing (Earth Imaging for Geospatial Information), Hannover, Germany, 2–5 June 2009; pp. 170–175.
10. Twele, A.; Cao, W.; Plank, S.; Martinis, S. Sentinel-1-based flood mapping: A fully automated processing chain. *Int. J. Remote Sens.* **2016**, *37*, 2990–3004. [CrossRef]
11. Henry, J.; Chastanet, P.; Fellah, K.; Desnos, Y. Envisat multi-polarized ASAR data for flood mapping. *Int. J. Remote Sens.* **2006**, 1921–1929, doi:10.1080/01431160500486724. [CrossRef]
12. Pulvirenti, L.; Pierdicca, N.; Chini, M.; Guerriero, L. An algorithm for operational flood mapping from Synthetic Aperture Radar (SAR) data using fuzzy logic. *Nat. Hazards Earth Syst. Sci.* **2011**, *11*, 529–540, doi:10.5194/nhess-11-529-2011. [CrossRef]
13. Matgen, P.; Hostache, R.; Schumann, G.; Pfister, L.; Hoffmann, L.; Savenije, H. Towards an automated SAR-based flood monitoring system: Lessons learned from two case studies. *Phys. Chem. Earth Parts A/B/C* **2011**, *36*, 241–252, doi:10.1016/j.pce.2010.12.009. [CrossRef]
14. Giustarini, L.; Hostache, R.; Matgen, P.; Schumann, G.J.; Bates, P.D.; Mason, D.C. A Change Detection Approach to Flood Mapping in Urban Areas Using TerraSAR-X. *IEEE Trans. Geosci. Remote Sens.* **2013**, *51*, 2417–2430, doi:10.1109/TGRS.2012.2210901. [CrossRef]
15. Boykov, Y.; Funka-Lea, G. Graph cuts and efficient ND image segmentation. *Int. J. Comput. Vis.* **2006**, *70*, 109–131. [CrossRef]
16. Westerhoff, R.S.; Kleuskens, M.P.H.; Winsemius, H.C.; Huizinga, H.J.; Brakenridge, G.R.; Bishop, C. Automated global water mapping based on wide-swath orbital synthetic-aperture radar. *Hydrol. Earth Syst. Sci.* **2013**, *17*, 651–663, doi:10.5194/hess-17-651-2013. [CrossRef]
17. Skakun, S. A neural network approach to flood mapping using satellite imagery. *Comput. Inform.* **2012**, *29*, 1013–1024.
18. Notti, D.; Giordan, D.; Caló, F.; Pepe, A.; Zucca, F.; Galve, J. Potential and Limitations of Open Satellite Data for Flood Mapping. *Remote Sens.* **2018**, *10*, 1673, [CrossRef]

19. Mason, D.; Speck, R.; Devereux, B.; Schumann, G.P.; Neal, J.; Bates, P. Flood Detection in Urban Areas Using TerraSAR-X. *IEEE Trans. Geosci. Remote Sens.* **2010**, *48*, 882–894, doi:10.1109/TGRS.2009.2029236. [CrossRef]

20. UK Environment Agency. Recorded Flood Outlines for Past Inundation Events. 2017. Available online: http://environment.data.gov.uk/ds/catalogue/#/catalogue (accessed on 1 November 2017).

21. UNITAR. Flood Waters Over Kayin and Mon State, Myanmar. 2015. Available online: www.unitar.org/unosat/node/44/2268 (accessed on 23 April 2018).

22. Géron, A. *Hands-On Machine Learning with Scikit-Learn and TensorFlow*, 1st ed.; O'Reilly Media: Sebastopol, CA, USA, 2017.

23. Bottou, L. *Stochastic Gradient Descent Tricks*; Springer: Brelin, Germany, 2012; Volume 7700, p. 430–445.

24. Boykov, Y.; Kolmogorov, V. An experimental comparison of min-cut/max-flow algorithms for energy minimization in vision. *IEEE Trans. Pattern Anal. Mach. Intell.* **2004**, *26*, 1124–1137. [CrossRef] [PubMed]

25. Greig, D.M.; Porteous, B.T.; Seheult, A.H. Exact maximum a posteriori estimation for binary images. *J. R. Stat. Soci. Ser. B (Methodological)* **1989**, *51*, 271–279. [CrossRef]

26. Boykov, Y.Y.; Jolly, M.P. Interactive graph cuts for optimal boundary & region segmentation of objects in ND images. In Proceedings of the ICCV 2001, Eighth IEEE International Conference on Computer Vision, Vancouver, BC, Canada, 7–14 July 2001.

27. McFeeters, S.K. The use of the Normalized Difference Water Index (NDWI) in the delineation of open water features. *Int. J. Remote Sens.* **1996**, *17*, 1425–1432. [CrossRef]

28. Du, Y.; Zhang, Y.; Ling, F.; Wang, Q.; Li, W.; Li, X. Water bodies' mapping from Sentinel-2 imagery with modified normalized difference water index at 10-m spatial resolution produced by sharpening the SWIR band. *Remote Sens.* **2016**, *8*, 354. [CrossRef]

29. Xu, H. Modification of normalised difference water index (NDWI) to enhance open water features in remotely sensed imagery. *Int. J. Remote Sens.* **2006**, *27*, 3025–3033. [CrossRef]

30. McFeeters, S.K. Using the Normalized Difference Water Index (NDWI) within a Geographic Information System to Detect Swimming Pools for Mosquito Abatement: A Practical Approach. *Remote Sens.* **2013**, *5*, 3544–3561, doi:10.3390/rs5073544. [CrossRef]

31. Boykov, Y.; Veksler, O. *Graph Cuts in Vision and Graphics: Theories and Applications*; Springer: Boston, MA, USA, 2006; pp. 79–96.

32. PyMaxflow. Python Library for Graph Construction and Maxflow Computation. 2019. Available online: https://github.com/pmneila/PyMaxflow (accessed on 6 February 2019).

33. Jensen, J.R. *Introductory Digital Image Processing: A Remote Sensing Perspective*, 3rd ed.; Pearson: London, UK, 2005; p. 507.

34. Banko, G. *A Review of Assessing the Accuracy of Classifications of Remotely Sensed Data and of Methods Including Remote Sensing Data in Forest Inventory*; International Institute for Applied Systems Analysis: Laxenburg, Austria, 1998.

35. Rowland, C.; Morton, R.; Carrasco, L.; McShane, G.; O'Neil, A.; Wood, C. Land Cover Map 2015 (25 m raster, GB). 2017. Available online: https://doi.org/10.5285/bb15e200-9349-403c-bda9-b430093807c7 (accessed on 1 April 2019).

36. Mason, D.C.; Davenport, I.J.; Neal, J.C.; Schumann, G.J.P.; Bates, P.D. Near Real-Time Flood Detection in Urban and Rural Areas Using High-Resolution Synthetic Aperture Radar Images. *IEEE Trans. Geosci. Remote Sens.* **2012**, *50*, 3041–3052, doi:10.1109/TGRS.2011.2178030. [CrossRef]

37. Coles, D.; Yu, D.; Wilby, R.L.; Green, D.; Herring, Z. Beyond 'flood hotspots': Modelling emergency service accessibility during flooding in York, UK. *J. Hydrol.* **2017**, *546*, 419–436. [CrossRef]

38. Bhandari, A.; Kumar, A.; Singh, G. Feature extraction using Normalized Difference Vegetation Index (NDVI): A case study of Jabalpur city. *Procedia Technol.* **2012**, *6*, 612–621. [CrossRef]

39. Janoth, J.; Gantert, S.; Schrage, T.; Kaptein, A. From TerraSAR-X towards TerraSAR next generation. In Proceedings of the 10th European Conference on Synthetic Aperture Radar, Berlin, Germany, 3–5 June 2014; pp. 1–4.

Article

Glacier Monitoring Using Frequency Domain Offset Tracking Applied to Sentinel-1 Images: A Product Performance Comparison

Donato Amitrano [1,*]**, Raffaella Guida** [1]**, Gerardo Di Martino** [2] **and Antonio Iodice** [2]

[1]   Surrey Space Centre, University of Surrey, Guildford GU2 7XH, UK; r.guida@surrey.ac.uk
[2]   University of Napoli Federico II, Department of Electrical Engineering and Information Technology,
     Via Claudio 21, 80125 Napoli, Italy; gerardo.dimartino@unina.it (G.D.M.); iodice@unina.it (A.I.)
*   Correspondence: d.amitrano@surrey.ac.uk

Received: 4 April 2019; Accepted: 29 May 2019; Published: 1 June 2019

Abstract: The Sentinel-1 mission has now reached its maturity, and is acquiring high-quality images with a high revisit time, allowing for effective continuous monitoring of our rapidly changing planet. The purpose of this work is to assess the performance of the different synthetic aperture radar products made available by the European Space Agency through the Sentinels Data Hub against glacier displacement monitoring with offset tracking methodology. In particular, four classes of products have been tested: the medium resolution ground range detected, the high-resolution ground range detected, acquired in both interferometric wide and extra-wide swath, and the single look complex. The first are detected pre-processed images with about 40, 25, and 10-m pixel spacing, respectively. The last category, the most commonly adopted for the application at issue, represents the standard coherent synthetic aperture radar product, delivered in unprocessed focused complex format with pixel spacing ranging from 14 to 20 m in azimuth and from approximately 2 to 6 m in range, depending on the acquisition area and mode. Tests have been performed on data acquired over four glaciers, i.e., the Petermann Glacier, the Nioghalvfjerdsfjorden, the Jackobshavn Isbræ and the Thwaites Glacier. They revealed that the displacements estimated using interferometric wide swath single look complex and high-resolution ground range detected products are fully comparable, even at computational level. As a result, considering the differences in memory consumption and pre-processing requirements presented by these two kinds of product, detected formats should be preferred for facing the application.

Keywords: synthetic aperture radar; offset tracking; displacements; Sentinel-1; glacier monitoring

---

1. Introduction

Under the aegis of the Copernicus Programme, the Sentinel-1 (S1) mission of the European Space Agency (ESA) has been providing high quality synthetic aperture radar (SAR) images since 2014. The short revisit time and the free and open data distribution policy is substantially affecting the remote sensing downstream sector, with more and more users involved in an increasing number of applications [1].

The S1 mission acquires data in different modes resulting in products at different spatial resolutions. The most commonly used are the interferometric wide (IW) swath and the interferometric extra-wide (EW) swath, in which data are acquired in swaths using the Terrain Observation with Progressive Scanning SAR (TOPSAR) imaging technique [2]. Raw products are then processed and uploaded on the Sentinels Data Hub (SDH) in two different formats: the single look complex (SLC) data format, and the ground range detected (GRD) data format [3].

SLC images represent the standard SAR product [4]. The products are in zero-doppler geometry. Each row of pixels represents points along a line perpendicular to the sub-satellite track. The products include a single look in each dimension, using the full available signal bandwidth and complex samples preserving the phase information. For IW and EW acquisition modes, each sub-swath consists of a series of bursts. Each burst is processed as a separate SLC image, and all processed bursts are finally assembled into the final product [3]. In the first case, the pixel spacing is approximately 14 × 2 m in azimuth/slant range directions. In the latter, it is approximately 6 × 20 m.

GRD products are detected, multi-looked and projected onto ground range (using an Earth ellipsoid model) images. Ground range coordinates are the slant range coordinates projected onto the Earth's ellipsoid. The phase information originally contained in SLC products is not preserved. The resulting product has approximately square resolution cell and square pixel spacing with reduced speckle at a cost of reduced geometric resolution. These products are provided in three different formats: the full resolution ground range detected (GRDF), the high-resolution ground range detected (GRDH), and the medium resolution ground range detected (GRDM).

GRDF products present a pixel spacing of 3.5 m in azimuth/slant range directions and derive only from stripmap (SM) acquisitions, which are scarcely tasked at the time of this research. GRDH products have a pixel spacing of approximately 10 m in azimuth/slant range directions when derived from SM or IW acquisitions, and up to about 25 m when images come from EW data. Finally, GRDM products present 40-m pixel spacing roughly when acquired both in IW and EW mode. For a complete reference concerning S1 products and acquisition modality, the reader should refer to [3,5].

Given the variety of S1 products, it is important to identify the best among them in feeding information extraction algorithms and to assess each performance in relation to a specific application [6] or image quality parameter [7]. This work has two main objectives: the first one is to assess the performance of the three most available products in the SDH, namely the SLCs, the GRDHs, and the GRDMs, with regard to a classic SAR remote sensing application, which is glacier displacement monitoring with offset tracking (OT) [8–10]. The second one is to provide users and scientists interested in this application with some guidelines concerning the exploitation of different products and the best parameters setting based on quantitative analysis of the estimated displacements. The test cases here concern four of the largest worldwide glaciers i.e., the Petermann Glacier (PG), the Nioghalvfjerdsfjorden (NI), the Jackobshavn Isbræ (JAK) and the Thwaites Glacier (THW). The first three are in the Greenland and the last in Antarctica. Displacements have been measured by processing 24 images per product class in pairs, approximately a couple per month, during the year 2017.

Remote sensing technologies have been widely exploited for monitoring glaciers at continental scale since the 1980s [11,12]. At that time, displacements were estimated through manual feature tracking of natural color images acquired with long temporal baseline. From the 1990s, the literature started to propose methodologies able to automatically track displacements exploiting correlation of multispectral images [13–15].

Due to the large availability of free images, multispectral sensors were the privileged source of data. Landsat 4 and 5 acquisitions were exploited to map displacements of the Ross [16] and Larsen [17] ice shelves in Antarctica. Landsat 8 images were preferred for large scale mapping of ice flows of Greenland, Antarctica and Alaska [18,19]. For estimation of ice velocity in Antarctica MODIS-based mosaics were employed [20]. Historical trends of the Pamir-Karakoram-Himalaya system were retrieved from archive Landsat 5 and 7 data [21]. The suitability of data acquired by the recently launched Sentinel-2 satellite with glacier monitoring (not limited to displacements) was investigated in [22]. A data performance comparison between Sentinel-2 and Landsat was provided in [23].

Due to the dependence upon light and weather conditions of passive sensors (which is particularly severe at very high/very low latitudes [24]) SAR sensors have been widely preferred for glaciers monitoring since the 1990s with the launch of the ERS-1 satellite. As explained in Joughin et al. [25], initially SAR interferometry was exploited to map displacements [26–28]. However, these techniques have limitations when applied to large displacements as discussed in the following Section. Therefore, intensity tracking methods, originally developed for multispectral data, started being developed to deal with large displacements. OT methods have been successfully applied to map movements of terrains due to natural phenomena such as landslides, earthquakes [29–31], human activities (e.g., mining) [32,33] and glaciers [8–10,34–37]. In this paper, the technique is applied to pairs of S1 images acquired in different modalities to assess the performance of the different product classes and determine the best in terms of accuracy of the estimated displacements and computational demand.

The work is organized as follows. Materials and methods are introduced in Section 2. Experimental results are presented and commented in Section 3 according to inter-product comparison and comparison with literature. It is worth stressing that the analysis has been performed from a data product perspective only. In other words, the paper does not contain any physical interpretation of the phenomenon considered (i.e., why glaciers move in one or another direction) but can advise an expert in the field of the best S1 products to infer similar conclusions.

2. Materials and Methods

2.1. Offset Tracking

OT methods allow for large displacements estimation without the need to exploit the SAR phase information. In this sense, they can be seen as complementary to classic differential synthetic aperture radar interferometry (DInSAR) [38]. As is well-known, DInSAR-based estimations are limited in the maximum observable displacement gradient (depending on the signal wavelength) and on the preservation of the interferometric coherence. Hence, it is typically applied only on areas exhibiting high temporal stability with respect to the signal phase (e.g., slow subsidence or buildings) [29]. Moreover, classic DInSAR can measure just displacements in the slant range direction, so no information concerning horizontal movements can be provided unless special techniques (such as the one proposed in [39]) are applied.

OT methods, instead, are applied to the SAR amplitude channel, thus being less sensitive to atmospheric effects and insensitive to phase instability of targets. They allow for measuring South-North and East-West displacements without any limitation on the observable gradient and even in areas typically characterized by low interferometric coherence, such as those highly vegetated [40]. This means that, by using just a couple of SAR images, movements of several meters can be detected with a good degree of approximation [31].

The block diagram of the implemented frequency-domain OT technique is depicted in Figure 1. It starts with a couple of SAR images as input to a coregistration block. The image acquired first is the reference for the displacements estimation and it is called master. The image in which displacements are evaluated is referred to as the slave image. They are co-registered with standard techniques [4]. When S1 data are selected, the application of precise orbit [41], if available, is advised before co-registration.

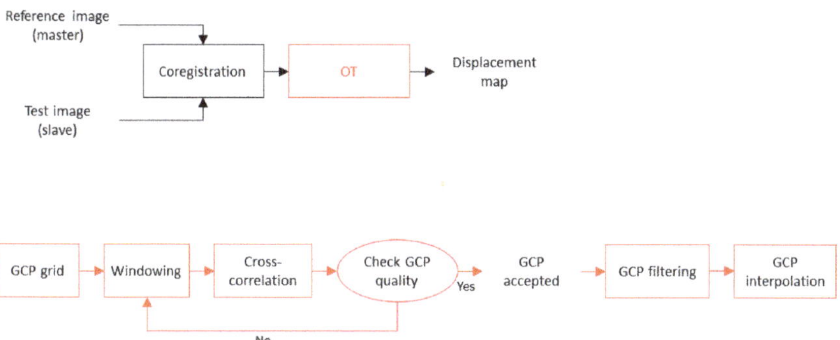

Figure 1. Block diagram of the implemented frequency-domain offset tracking (OT) technique. The lower diagram is an exploded view of the OT processing block.

Co-registered data feed the OT algorithm depicted in the lower diagram of Figure 1 [29]. It exploits cross-correlation (CC) calculated on several windows extracted from the image pair to estimate the shift between the master patch and the slave patch. Windows are extracted around grid-points (or ground control points, GCPs) usually regularly distributed across the images.

The CC matrix C between two null-mean patches M and S from, respectively, the master and slave image, is computed as follows:

$$C = \frac{\text{IFFT}\{\text{FFT}\{M\} \times \text{FFT}\{S\}^*\}}{\sqrt{\langle M^2 \rangle \times \langle S^2 \rangle}}, \; C \in [0, 1], \tag{1}$$

in which FFT and IFFT stand for the fast Fourier and the inverse fast Fourier transforms, respectively, the apex * represents the complex conjugation operation, and the symbol <*> the mean operator. In Equation (1), M and S are oversampled by a factor f (which must be a power of two in order to optimize the FFT calculation) to take into account the sub-pixel movements, being the minimum detectable displacement (in pixel units), i.e., the technique sensitivity, equal to $1/f$.

The peak value of the matrix C, c_{max}, identifies the amount of the shift to be applied to the slave patch to be superimposed to the master one. The higher the peak, the more reliable the estimated shift is. Note that C is a circular matrix, therefore the maximum detectable shift is equal to $\pm d/2$, where d is the dimension of the patches.

In order to identify reliable shifts, two quality parameters for the selection of reliable GCPs, i.e., the peak value c_{max} previously introduced and the ratio $q = c_{max} / \langle C \rangle$ between the matrix peak and the background, are considered [9]. For both parameters, a pre-determined user-defined threshold is adopted to exclude invalid GCPs. Spatial smoothing filter is usually applied to minimize noisy displacement patterns and reduce high-frequency noise [31]. Accepted GCPs are finally interpolated to produce the displacement map. Due to interpolation it is possible that they show displacement values under the sensitivity of the method dictated by the oversampling factor.

2.2. Algorithm Set-Up

In Table 1, the OT algorithm parameters setting for the implemented experiments are reported. They vary as a function of the product class. The objective was to obtain output maps with comparable sensitivity, i.e., with a minimum detectable displacement of the same order. Accordingly, due to the different pixel spacing of the input products, a different parameters set-up was chosen for the grid spacing, the oversampling factor and the cross-correlation window size. In the table, when the values relevant to the sensitivity and the CC window size are in square brackets they refer to a range of possible values used to set-up different experiments, which details are reported in Section 4.

As for the grid spacing, it was slightly increased as the product pixel spacing decreases in order to reduce the computational load. In fact, the number of GCPs to be evaluated is given by the size of the analyzed subset divided by the grid spacing. Intuitively, as the image resolution improves, the subset size enclosing the glacier increases as well. Therefore, a less dense grid was preferred when running the OT on the products with higher resolution.

The oversampling factor f (which determines the sensitivity with respect to the minimum detectable displacement) is related to the phenomenon under observation and the product resolution. As far as glaciers are concerned, a resolution in the order of a few meters is fit-for-purpose. Therefore, f is set in such way to have a sensitivity ranging from 0.9 to 2.5 m in both azimuth and range directions. Most of the experiments based on IW SLC images have been performed using an oversampling factor of 8 in azimuth direction and 4 in range direction. As for IW GRDH, f was set to 8 in both azimuth and range directions. Finally, the value for f in the case of EW GRDH and EW GRDM images was 16.

As a general guideline, it is a good practice to check the specific pixel spacing for each product as reported in the respective metadata. The use of nominal values, those declared by ESA for a certain product class in the product specification documents (see [3]), in place of the specific ones would bring an error in the estimated displacements.

Table 1. OT algorithm parameters for the different product classes and test sites.

Parameter	Unit	SLC	GRDH$_{10}$	GRDH$_{25}$	GRDM	Glacier
Pixel spacing azimuth/range	*m*	13.9/3.79	10.1/10.0	25.4/25.3	40.6/40.4	PG
		13.8/3.49	10.1/10.0	na	40.5/40.0	NI
		13.8/7.04	10.1/10.0	25.4/25.3	40.6/40.6	JAK
		14.03/4.27	10.1/10.0	na	40.3/40.1	THW
Sensitivity azimuth/range	*m*	[0.9, 1.7]/0.9	[1.2, 2.5]/[1.2, 2.5]	1.6/1.6	2.5/2.5	PG
		[0.9, 1.7]/0.9	[1.2, 2.5]/[1.2, 2.5]	na	2.5/2.5	NI
		1.7/1.7	[1.2, 2.5]/[1.2, 2.5]	1.6/1.6	2.5/2.5	JAK
		[0.9, 1.7]/0.9	[1.2, 2.5]/[1.2, 2.5]	na	2.5/2.5	THW
CC window size azimuth/range	*m*	[64, 128]/[64, 256]	[64, 256]/[64, 256]	[64, 128]/[64, 128]	[64, 128]/[64, 128]	PG
		[64, 128]/[64, 256]	[64, 256]/[64, 256]	na	[64, 128]/[64, 128]	NI
		[64, 128]/[64, 128]	[64, 128]/[64, 128]	[64, 128]/[64, 128]	[64, 128]/[64, 128]	JAK
		[64, 128]/[64, 256]	[64, 256]/[64, 256]	na	[64, 128]/[64, 128]	THW
Grid size azimuth/range	*Pixel*	30/100 30/50 (JAK)	40/40	30/30	20/20	
CC threshold				0.1		
q threshold				4		
Max flow speed	*m/day*			8		PG
				8		NI
				40		JAK
				20		THW
Subset size	$km^2 \times 10^3$	37.8	25.0	53.8	47.4	PG
		21.3	15.0	na	43.8	NI
		9.65	4.43	13.42	7.22	JAK
		23.5	13.6	na	22.8	THW

2.3. Test Areas and Data in Use

2.3.1. Petermann Glacier

The PG flows within the Hall Basin in the Nares Strait in Northwestern Greenland, constituting one of the largest glaciers of the region. It drains ice from the center of the Greenland ice sheet to the ocean through the Petermann Fjord, which is approximately 90 km long. The thickness of the ice shelf is higher than 100 m, while its width is of approximately 15 km. This glacier is characterized by low resistive stresses along flow because of the limited cohesion with the fjord walls [42]. It flows like a river towards the sea with a velocity of approximately 1.1 km/year at its grounding line since the 1990s [43,44], delivering annually 12 tons of ice into the ocean [42].

2.3.2. Nioghalvfjerdsfjorden

NI is located in the Northeastern part of the Greenland, of which it represents the largest ice shelf. Its length and width at mid-distance measure more than 70 km and approximately 20 km, respectively. Together with the adjacent Zachariæ Isstrøm and Storstrømmen, it drains more than the 10% of the Greenland ice sheet [45]. Potentially, this region alone could rise the global sea level of more than one meter in the unlikely event of complete loss of the ice sheet [46]. The maximum velocities for NI, in the order of 5 m/day, are found near the grounding line and have been quite stable in recent years [47].

2.3.3. Jackobshavn Isbræ

JAK is located in the Western part of Greenland and is the fastest glacier draining the ice sheet [48]. During the late 1990s, the ice tongue was involved in several break-up events causing the glacier to increase its velocity. This phenomenon continued until recent years [49] and is influenced by the warming of the water in the adjacent ocean [50]. The literature pointed out that this glacier is subject to high velocity variability over the time [51]. Nowadays its peak velocity is, on average, more than 20 m/day [47].

2.3.4. Thwaites Glacier

THW is an extremely large and fast-moving Antarctic glacier flowing into Pine Island Bay. As highlighted in [52], it is contributing to the global sea level rise for about one millimeter per year.

The THW has a wide ice front (about 120 km long), and its ice flow speed increased, in its lower part, from 50 to 100 m/year since 2009 [52]. Authors of recent studies believe that the THW has the greatest potential for further near-term increases in ice flux, thus causing a rapid sea-level rise [53].

2.3.5. Data

The datasets used in this study have been acquired by the S1 constellation in 2017. For each test site and for each product class (when available), 24 images organized in 12 pairs (approximately a couple per month) have been analyzed for displacement estimation using OT. Overall, 168 couples have been considered. In the case of the PG and JAK, all the product classes (IW SLC, IW GRDH, EW GRDH, and EW GRDM) were available in the SDH. In the case of NI and THW, the EW GRDH product class was not accessible, therefore the processing for EW acquisitions was limited to GRDM products.

Reference data for glaciers displacements along selected transects were provided in [47]. They were retrieved using the OT algorithm implemented in the GAMMA-SAR software suite [54] and applied to couples of S1 SLC images [54]. As declared by the authors, the spatial resolution of the output velocity maps is of 388 m in ground range and 320 m in azimuth. After post-processing and filtering, the final product was resampled on a 100 m × 100 m cartographic grid [47] representing the source of information used in this study in the assessment phase. The number of sampling points for the considered transects are the same as in [47]. Their values are: PG—900 along flow, 1107 across flow; NI—800 along flow, 736 across flow; JAK—900 along flow, 699 across flow; THW—1800 along flow, 2135 across flow.

In Table 2, the monthly average flow velocity across the selected transects for the four considered glaciers as extracted from the reference literature [47] is reported. The PG shows a quite uniform behavior all over the year, with exceptions in the months of July (for both along and across flows), and August (across flow only).

Table 2. Monthly average flow velocities for the four considered glaciers extracted from the reference literature [47]. Values are expressed in m/day.

Time	PG		NI		JAK		THW	
	Along	**Across**	**Along**	**Across**	**Along**	**Across**	**Along**	**Across**
January	2.49	2.95	1.94	2.86	10.6	3.45	9.64	4.75
February	2.58	2.86	1.92	2.87	9.76	3.26	9.88	5.62
March	2.57	2.87	1.93	2.90	9.71	3.30	9.86	5.26
April	2.45	2.91	1.88	2.88	9.26	3.29	9.75	4.87
May	2.47	2.97	1.92	2.87	10.3	2.47	9.68	5.07
June	2.60	2.99	1.89	3.04	12.3	3.17	9.64	5.20
July	3.56	3.22	3.16	3.12	8.80	3.48	9.67	5.05
August	2.87	3.21	1.97	3.08	13.0	3.54	9.08	5.09
September	2.69	2.96	1.90	2.99	11.0	3.25	9.08	4.68
October	2.73	2.90	1.94	2.90	10.4	3.25	9.39	4.68
November	2.73	2.93	1.88	2.92	10.0	3.25	9.71	4.73
December	2.72	2.93	1.93	2.91	9.65	3.24	10.3	4.75

As for the NI, an abrupt change in the flow velocity (more than one meter per day above the average over the rest of the year) is registered in the month of July for the along flow. As for the across flow, it is more stable, with a slight increase during the summer.

The average monthly flow speed for JAK is more variable. As for the other Greenland glaciers, a speed-up (especially in the along flow direction) is registered during summer, likely due to ice melting effects, with exception made for the month of July, when the flow speed drops at its annual minimum.

As regards the THW, as reported in the last column of Table 2, the flow speed is almost uniform all over the year, i.e., there are no remarkable seasonal effects in the glacier behavior.

3. Results

The CC window dimension is determined by a trade-off between different needs. First, it must be large enough to estimate the maximum expected displacement, since the maximum detectable one (in pixel units) is equal to half the size of the correlation window. Large windows can affect the preservation of the edges of the features of interest [55], present a higher likelihood of including changing areas (thus causing a drop in the correlation peak) and tend to increase the computational load. In the literature, some techniques are proposed for a feature-based selection of the CC window size (see [56] as an example) but, in most of the cases, the simplest and safest way to operate is with a trial-and-error application-oriented approach. Conversely, it is known that, if the correlation window size is too small, the signal-to-noise ratio is low and this leads to a noisy estimate of the shifts between the master and slave patches [57]. The tests performed revealed that the minimum window dimension allowing for reliable results is 64 x 64 square pixel. As explained in the following, different windows have been tested in order to find the best match between the displacements estimated through the implemented OT and those available from past literature.

For a GCP to qualify as a good candidate, the quality parameters c_{max} and q need to be greater than empirically pre-determined thresholds. In this study the threshold value for c_{max} has been set to 0.1 which is in line with similar choices in literature (as an example, in Reference [40] a threshold of 0.2 was suggested). However, in order to enforce the requirements for valid GCPs, both the conditions on the maximum correlation and on the ratio with respect to the background, here estimated to be at least 4, have been imposed.

3.1. Comparison with Literature Data

First, the results of the implemented OT algorithm were compared with available literature data provided in [47] and relevant to selected transects in along flow and across flow directions. A graphic overview of the four test sites and of the corresponding analyzed transects (AA—along flow, BB—across flow) is provided in Figure 2a–d for the PG, NI, JAK and THW, respectively. Greenland data have been geocoded using a polar stereographic projection with origin latitude of 70 decimal degrees and origin longitude of −45 decimal degrees. As for Antarctica data, the origin latitude was of −71 decimal degrees and the origin longitude of 0 decimal degrees.

Data reported in the subsequent tables represent the root mean square error (RMSE) of the estimated flow velocities against reference data calculated as follows:

$$\text{RMSE} = \sqrt{\sum_{i=1}^{N} \frac{(\hat{v}_i - v_i)^2}{N}}, \tag{2}$$

where $\hat{v}_i$ is the estimated flow speed (expressed in m/day) and v_i is reference flow speed for a given point i along one of the considered transects.

3.1.1. Petermann Glacier

In Tables 3–5, the results obtained on the PG for the IW SLC, IW GRDH, and EW GRDH and EW GRDM product classes, are shown respectively. Each table entry represents the root mean square error (RMSE), expressed in meters per day, against reference data along the along flow and across flow transects introduced above. Boxes with grey shading indicate the best performance with respect to the considered literature data, while those with orange shading indicate the same result achieved with a higher computational time. Each column corresponds to a selected combination of the CC window size and of the oversampling factor, both declared on the top of the column (w stands for the dimension of the CC window and f for the oversampling factor). When only one value for these two parameters is present, it refers to both azimuth and slant range directions. Two numbers separated by the slash character means that an asymmetric CC window and/or oversampling factor has been used for running that particular experiment. This notation is kept for all the subsequent tables reporting the results of the experiments concerning the other analyzed glaciers.

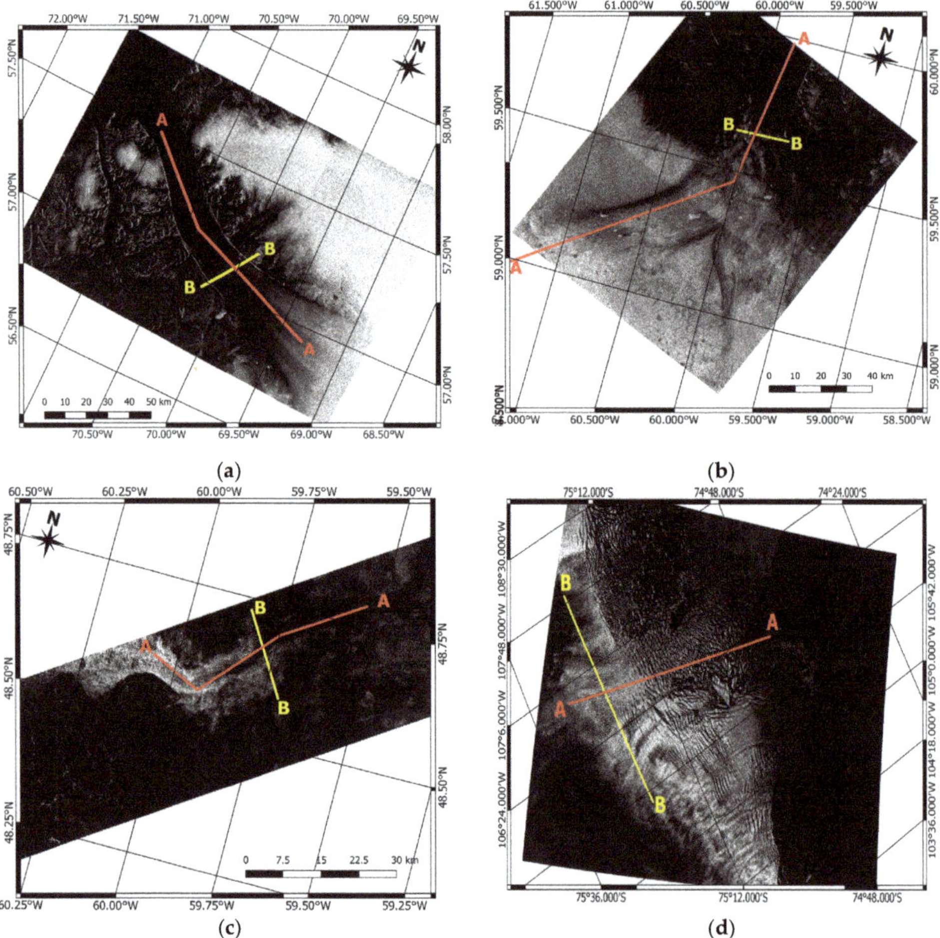

Figure 2. SAR images (all acquired in right ascending orbit) of the test sites with the corresponding analyzed transects (AA—along flow, BB—across flow). (**a**) Petermann glacier, (**b**) Nioghalvfjerdsfjorden, (**c**) Jakobshavn Isbræ and (**d**) Thwaites glacier.

From the results we can see that, for IW products, the change in the window size does not influence significantly the performance with respect to reference data which shows, in most cases, a discrepancy of less than 20 cm per day, with a peak of 0.86 m/day for the pair 11–17 June using GRDH data. The most significant effect is that of the CC window size on the computational time. Especially for GRDH images, an increase in the dimension of the CC window has a negative impact on it, with negligible variations of the estimated flow speed.

In Figure 3a–d, an example of the results obtained running the OT algorithm, respectively on PG IW SLC, IW GRDH, EW GRDH and EW GRDM images, is shown. Examples of the quality parameter maps, i.e., maximum CC and q, are reported in Figure 4a,b, respectively. Only the IW SLC case is shown for brevity.

Table 3. PG, IW SLC product class experiments, RMSE (expressed in m/day) with respect to available literature data measured along selected along flow and across flow transects.

		Petermann–IW SLC							
Observation	Track	$w = 64, f = 16/4$		$w = 64/128, f = 8/4$		$w = 128, f = 8/4$		$w = 64/256, f = 8/4$	
		Along	Across	Along	Across	Along	Across	Along	Across
6–12 Jan	26 AA	0.14	0.10	0.14	0.14	0.14	0.17	0.10	0.14
11–17 Feb	26 DD	0.30	0.26	0.28	0.28	0.28	0.28	0.28	0.26
1–7 Mar	26 DD	0.14	0.14	0.14	0.14	0.10	0.14	0.10	0.14
6–12 Apr	26 DD	0.20	0.14	0.22	0.14	0.22	0.14	0.20	0.14
18–24 May	26 DD	0.17	0.10	0.14	0.17	0.17	0.14	0.14	0.14
11–17 Jun	26 DD	0.14	0.10	0.28	0.24	0.17	0.35	0.20	0.28
11–17 Aug	26 DD	0.20	0.26	0.20	0.35	0.17	0.28	0.17	0.33
10–16 Aug	26 DD	0.26	0.45	0.22	0.45	0.22	0.44	0.24	0.35
9–15 Sep	26 DD	0.10	0.10	0.10	0.10	0.00	0.10	0.10	0.10
9–15 Oct	26 DD	0.14	0.17	0.14	0.17	0.14	0.17	0.14	0.17
8–14 Nov	26 DD	0.10	0.14	0.10	0.17	0.14	0.20	0.10	0.17
8–14 Dec	26 DD	0.24	0.32	0.26	0.35	0.24	0.35	0.26	0.36
Time		≈1.5 h		≈1.7 h		≈3.5 h		≈4 h	

Table 4. PG, IW GRDH product class experiments, RMSE (expressed in m/day) with respect to available literature data measured along selected along flow and across flow transects.

		Petermann–IW GRDH					
Observation	Track	$w = 64, f = 8$		$w = 128, f = 8$		$w = 256, f = 4$	
		Along	Across	Along	Across	Along	Across
6–12 Jan	26 AA	0.17	0.24	0.14	0.17	0.14	0.32
11–17 Feb	26 DD	0.28	0.26	0.24	0.24	0.22	0.33
1–7 Mar	26 DD	0.26	0.24	0.24	0.22	0.22	0.30
6–12 Apr	26 DD	0.28	0.28	0.24	0.28	0.22	0.48
18–24 May	26 DD	0.28	0.32	0.26	0.30	0.17	0.45
11–17 Jun	26 DD	0.87	0.48	0.28	0.47	0.32	0.46
11–17 Aug	26 DD	0.36	0.32	0.32	0.28	0.28	0.37
10–16 Aug	26 DD	0.24	0.24	0.22	0.20	0.20	0.30
9–15 Sep	26 DD	0.17	0.17	0.17	0.28	0.42	0.30
9–15 Oct	26 DD	0.17	0.28	0.10	0.17	0.10	0.41
8–14 Nov	26 DD	0.24	0.20	0.24	0.20	0.14	0.33
8–14 Dec	26 DD	0.33	0.22	0.33	0.22	0.26	0.37
Time		≈1 h		≈7.5 h		≈8.8 h	

Table 5. PG, EW GRDH and EW GRDM product classes experiments, RMSE (expressed in m/day) with respect to available literature data measured along selected along flow and across flow transects.

		Petermann–EW GRDH				Petermann–EW GRDM			
Observation	Track	$w = 64, f = 16$		$w = 128, f = 16$		$w = 64, f = 16$		$w = 128, f = 16$	
		Along	Across	Along	Across	Along	Across	Along	Across
1–7 Jan	41 AA	na	na	na	na	na	na	na	na
12–18 Feb	41 DD	na	na	na	na	na	na	na	na
8–1 Apr	41 AA	na	na	na	na	na	na	na	na
1–13 Apr	41 AA	0.47	0.46	0.42	0.75	0.69	0.66	0.67	1.02
7–19 May	41 AA	0.45	0.62	0.44	0.62	0.67	0.91	0.73	1.17
10–22 Jun	41 AA	na	na	0.35	0.68	0.61	0.98	0.57	1.21
6–18 Jul	41 AA	0.88	0.82	0.62	0.68	0.49	0.68	0.75	1.07
11–23 Aug	41 AA	0.50	0.59	0.37	0.63	0.48	0.59	0.69	1.10
4–16 Sep	41 AA	0.36	0.40	0.32	0.41	0.62	0.94	0.60	1.14
10–22 Oct	41 AA	0.48	0.52	0.36	0.69	0.70	0.81	0.48	1.10
3–15 Nov	41 AA	0.33	0.35	0.35	0.45	0.71	0.71	0.42	1.07
9–21 Dec	41 AA	0.33	0.41	0.33	0.66	0.33	0.63	0.22	1.04
Time		≈0.8 h		≈22.4 h		≈1.7 h		≈15.6 h	

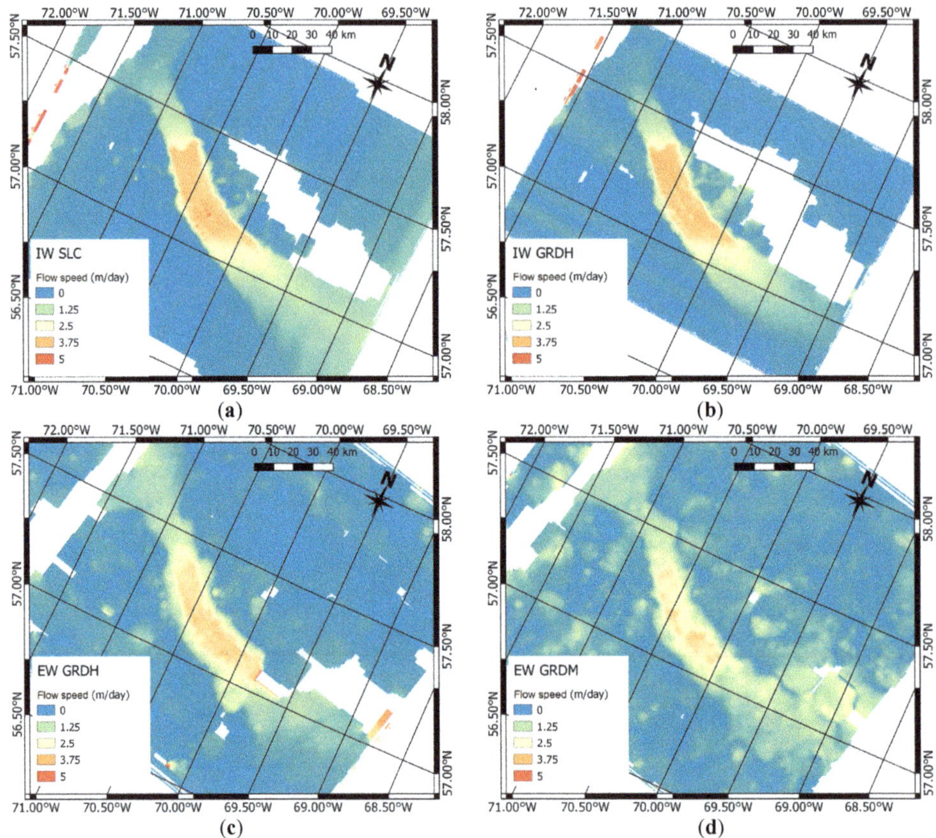

Figure 3. PG, flow speed maps obtained using (**a**) IW SLC, (**b**) IW GRDH, (**c**) EW GRDH and (**d**) EW GRDM images. Observation periods are 6–12 January 2017 for IW images and 1–13 April for EW images.

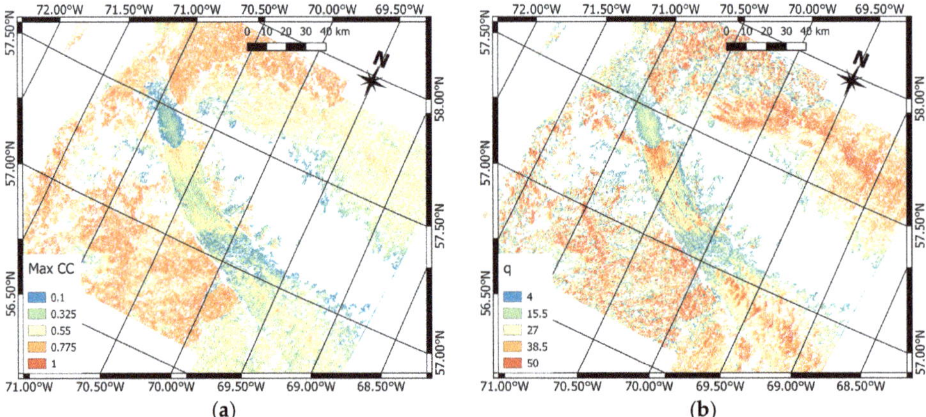

Figure 4. PG, quality parameters maps obtained by processing the IW SLC pair 6–12 January 2017. (**a**) Maximum correlation. (**b**) Ratio between the peak and the background of the correlation matrix (q-parameter).

Qualitatively, the flow speed maps obtained by processing IW products are quite similar. In both cases, the velocity fields are very homogeneous outside the glacier area, where they are expected to be almost null. Within it, there are no significant differences in the delineation of edges and features due to the resolution change of the input product.

Considering the maps obtained by processing EW images, it is evident that, in both cases, the lower resolution of the input product causes a blurring and a fragmentation of the estimated velocity field. Outside the glacier area, many regions exhibiting non-null velocity appear. Inside the glacier, the estimated velocity field tends to be flatter and the edges less defined. This effect is more pronounced when using GRDM images.

The correlation map depicted in Figure 4a exhibits acceptable values all over the glacier, with no significant presence of rejected GCPs which are mainly concentrated at the North edge of the glacier. As a general comment, the correlation resulted higher south of the glacier, in areas where the ice melted and, consequently, the bare land got exposed. Similar considerations hold for the q parameter and, overall, for the quality parameters of the GRDH experiments (not shown for brevity) and this supports the similarity of the outputs.

In Figure 5, the difference map (expressed in m/day) between the estimated velocity field and reference data for the observation period 6–12 January 2017 is shown. In particular, Figure 5a is relevant to SLC image input, while Figure 5b concerns to the IW GRDH experiment. Reference data in form of map are available only for the central part of the glacier, therefore these pictures refer to a subset of the whole study area. Data concerning EW GRDH and EW GRDM are not displayed for brevity. As explained before, they are affected by the lower resolution of the input products, which shows significant differences with respect to reference data, especially at the edges of the glacier.

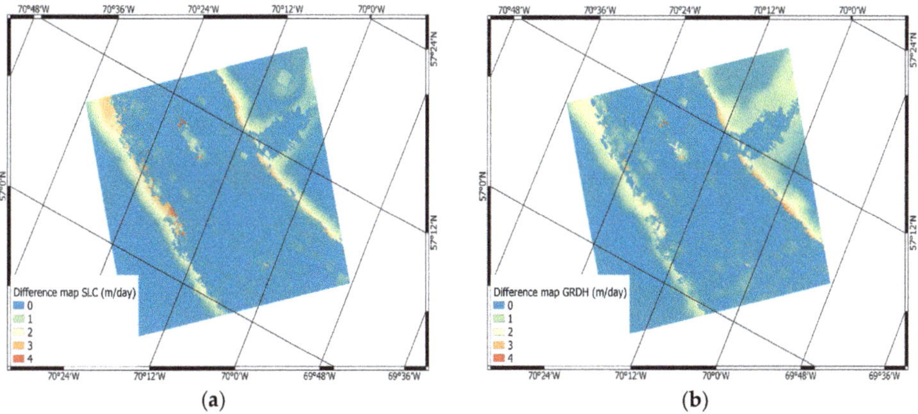

(a)   (b)

Figure 5. PG, velocity difference map relevant to the observation period 6–12 January 2017 between the implemented frequency domain OT and reference data for (**a**) SLC and (**b**) GRDH image input.

In both cases, very small differences can be appreciated in the two velocity fields, in particular within the glacier area, where the maps show almost null differences. The higher values characterize the edges of the glacier. This can be partially due to both the diverse windowing used by the two different algorithms for the calculation of the CC and the resampling necessary to overlap the different maps. Another area exhibiting moderate discrepancies is at the right-hand side of the glacier, where it forms a sort of branch. Here, the differences are slightly more pronounced when using IW GRDH images.

3.1.2. Nioghalvfjerdsfjorden

Results concerning the NI are shown in Tables 6 and 7 respectively for the IW SLC and GRDH product classes. Those concerning EW GRDH images are omitted for brevity but will be discussed afterwards together with aggregated data for all of the performed experiments.

Table 6. NI, IW SLC product class experiments, RMSE (expressed in m/day) with respect to available literature data measured along selected along/across flow transects.

		Nioghalvifierdsfjorden–IW SLC							
Observation	Track	$w = 64, f = 16/4$		$w = 64/128, f = 8/4$		$w = 128, f = 8/4$		$w = 256, f = 8/4$	
		Along	Across	Along	Across	Along	Across	Along	Across
3–9 Jan	74 AA	0.36	0.47	0.33	0.32	0.33	0.39	0.33	0.44
2–8 Feb	74 AA	0.24	0.20	0.30	0.26	0.30	0.24	0.30	0.30
4–10 Mar	74 AA	0.28	0.26	0.47	0.45	0.45	0.47	0.37	0.41
15–21 Apr	74 AA	0.14	0.22	0.10	0.22	0.10	0.24	0.10	0.30
3–9 May	74 AA	0.28	0.41	0.42	0.40	0.37	0.45	0.39	0.52
2–8 Jun	74 AA	0.52	0.59	0.70	0.66	0.49	0.48	0.49	0.53
2–8 Jul	74 AA	0.28	0.53	1.05	1.02	0.22	0.37	0.22	0.42
1–7 Aug	74 AA	0.42	0.40	1.35	1.04	0.24	0.37	0.26	0.40
6–12 Sep	74 AA	0.33	0.32	0.32	0.33	0.28	0.20	0.56	0.28
6–12 Oct	74 AA	0.14	0.20	0.14	0.30	0.10	0.14	0.14	0.20
5–11 Nov	74 AA	0.14	0.17	0.20	0.50	0.10	0.20	0.14	0.26
5–11 Dec	74 AA	0.14	0.22	0.22	0.17	0.22	0.20	0.22	0.17
Time		$\approx$1 h		$\approx$1.4 h		$\approx$2.8 h		$\approx$15.8 h	

Table 7. NI, IW GRDH product class experiments, RMSE (expressed in m/day) with respect to available literature data measured along selected along/across flow transects.

		Nioghalvifierdsfjorden–IW GRDH					
Observation	Track	$w = 64, f = 8$		$w = 128, f = 8$		$w = 256, f = 4$	
		Along	Across	Along	Across	Along	Across
3–9 Jan	74 AA	0.24	0.33	0.22	0.32	0.17	0.26
2–8 Feb	74 AA	0.35	0.54	0.32	0.57	0.28	0.47
4–10 Mar	74 AA	0.30	0.56	0.24	0.52	0.22	0.45
15–21 Apr	74 AA	0.28	0.48	0.24	0.48	0.20	0.42
3–9 May	74 AA	0.24	0.33	0.24	0.33	0.24	0.39
2–8 Jun	74 AA	0.30	0.40	0.30	0.41	0.33	0.47
2–8 Jul	74 AA	0.17	0.24	0.20	0.26	0.14	0.28
1–7 Aug	74 AA	0.20	0.44	0.14	0.30	0.20	0.37
6–12 Sep	74 AA	0.61	0.50	0.32	0.24	0.63	0.22
6–12 Oct	74 AA	0.14	0.22	0.10	0.20	0.26	0.35
5–11 Nov	74 AA	0.10	0.10	0.10	0.36	0.22	0.36
5–11 Dec	74 AA	0.10	0.24	0.10	0.24	0.20	0.33
Time		$\approx$0.7 h		$\approx$4.5 h		$\approx$5.3 h	

As for the PG, the flow velocities estimated using the EW GRDH product class are highly affected by the lower resolution. As better explained below, the RMSE with respect to reference data is significant especially in the across flow direction, for which it is close to one meter per day.

Overall, the results are similar to those discussed above for the PG. Even in this case, the performance of IW SLC and IW GRDH products are fully comparable against the reference literature. Using SLC images, the higher values of the RMSE tend to occur during the summer when, as reported in Table 2, the flow velocity is higher. The increase in the CC window size results in a more stable RMSE, at the expense of computational times. Conversely, IW GRDH results are not significantly affected by seasonal effects.

3.1.3. Jackobshavn Isbræ

The results concerning JAK are reported in Tables 8 and 9 for the processed IW SLC and IW GRDH images, respectively. Based on literature data, during 2017 this glacier reached peak velocities of more than 30 m/day [47]. Therefore, over an observation period of 6 days, displacements in the order of 200 m are expected. Using SLC images with range pixel spacing in the order of 3.5 m, the range window size of 64 or 128 pixels is not enough to detect this order of displacements. Moreover, the use of windows of 256 pixels in range direction (or bigger) leads to wide decorrelation areas preventing a reliable estimate of the displacements. Therefore, for this glacier, a multilook factor of two in range direction was applied to SLC products. Being an incoherent mean [4], multilooking increased the range pixel spacing of input products to approximately 7 m (see Table 1) allowing, consequently, the adoption of smaller CC windows.

Table 8. JAK, IW SLC product class experiments, RMSE (expressed in m/day) with respect to available literature data measured along selected along/across flow transects.

<table>
<tr><td colspan="10" align="center">Jackobshavn–IW SLC</td></tr>
<tr><td rowspan="2">Observation</td><td rowspan="2">Track</td><td colspan="2">$w = 64, f = 8/4$</td><td colspan="2">$w = 64/128, f = 8/4$</td><td colspan="2">$w = 128, f = 8/4$</td><td colspan="2">$w = 128/256, f = 8/4$</td></tr>
<tr><td>Along</td><td>Across</td><td>Along</td><td>Across</td><td>Along</td><td>Along</td><td>Along</td><td>Across</td></tr>
<tr><td>4–10 Jan</td><td>90 AA</td><td>1.67</td><td>0.33</td><td>1.29</td><td>0.38</td><td>1.65</td><td>0.41</td><td>2.52</td><td>0.34</td></tr>
<tr><td>15–21 Feb</td><td>90 AA</td><td>1.57</td><td>0.31</td><td>1.35</td><td>0.22</td><td>1.44</td><td>0.26</td><td>2.15</td><td>0.22</td></tr>
<tr><td>5–11 Mar</td><td>90 AA</td><td>1.55</td><td>0.26</td><td>1.56</td><td>0.24</td><td>1.57</td><td>0.26</td><td>1.97</td><td>0.28</td></tr>
<tr><td>4–10 Apr</td><td>90 AA</td><td>1.58</td><td>0.38</td><td>1.10</td><td>0.41</td><td>1.20</td><td>0.42</td><td>1.90</td><td>0.37</td></tr>
<tr><td>4–10 May</td><td>90 AA</td><td>1.45</td><td>0.86</td><td>1.08</td><td>0.30</td><td>1.30</td><td>0.28</td><td>2.43</td><td>0.30</td></tr>
<tr><td>3–9 Jun</td><td>90 AA</td><td>1.59</td><td>0.47</td><td>1.22</td><td>0.26</td><td>1.83</td><td>0.26</td><td>2.97</td><td>0.24</td></tr>
<tr><td>2–2 Aug</td><td>90 AA</td><td>1.12</td><td>0.22</td><td>1.35</td><td>0.20</td><td>1.22</td><td>0.22</td><td>2.32</td><td>0.28</td></tr>
<tr><td>20–26 Aug</td><td>90 AA</td><td>2.25</td><td>0.65</td><td>2.39</td><td>0.20</td><td>1.40</td><td>0.26</td><td>3.40</td><td>0.24</td></tr>
<tr><td>13–19 Sep</td><td>90 AA</td><td>1.32</td><td>0.66</td><td>1.44</td><td>0.30</td><td>1.80</td><td>0.34</td><td>4.01</td><td>0.36</td></tr>
<tr><td>7–13 Oct</td><td>90 AA</td><td>1.75</td><td>0.48</td><td>1.27</td><td>0.31</td><td>1.55</td><td>0.36</td><td>2.20</td><td>0.33</td></tr>
<tr><td>6–12 Nov</td><td>90 AA</td><td>1.62</td><td>0.64</td><td>1.17</td><td>0.28</td><td>1.40</td><td>0.17</td><td>2.70</td><td>0.20</td></tr>
<tr><td>6–12 Dec</td><td>90 AA</td><td>1.88</td><td>0.55</td><td>1.51</td><td>0.24</td><td>1.63</td><td>0.26</td><td>2.35</td><td>0.28</td></tr>
<tr><td align="center">Time</td><td></td><td colspan="2" align="center">≈0.2 h</td><td colspan="2" align="center">≈0.3 h</td><td colspan="2" align="center">≈0.6 h</td><td colspan="2" align="center">≈1.6 h</td></tr>
</table>

Table 9. JAK, IW GRDH product class experiments, RMSE (expressed in m/day) with respect to available literature data measured along selected along/across flow transects.

<table>
<tr><td colspan="12" align="center">Jackobshavn–IW GRDH</td></tr>
<tr><td rowspan="2">Observation</td><td rowspan="2">Track</td><td colspan="2">$w = 64, f = 8$</td><td colspan="2">$w = 64/128, f = 8$</td><td colspan="2">$w = 128/64, f = 8$</td><td colspan="2">$w = 128, f = 8$</td><td colspan="2">$w = 256, f = 4$</td></tr>
<tr><td>Along</td><td>Across</td><td>Along</td><td>Across</td><td>Along</td><td>Across</td><td>Along</td><td>Across</td><td>Along</td><td>Across</td></tr>
<tr><td>4–10 Jan</td><td>90 AA</td><td>1.99</td><td>0.79</td><td>2.03</td><td>0.82</td><td>1.89</td><td>0.85</td><td>1.45</td><td>0.42</td><td>3.07</td><td>0.45</td></tr>
<tr><td>15–21 Feb</td><td>90 AA</td><td>1.44</td><td>0.24</td><td>1.82</td><td>0.26</td><td>1.27</td><td>0.24</td><td>1.42</td><td>0.24</td><td>2.46</td><td>0.49</td></tr>
<tr><td>5–11 Mar</td><td>90 AA</td><td>1.57</td><td>0.53</td><td>1.76</td><td>0.32</td><td>1.35</td><td>0.35</td><td>1.43</td><td>0.32</td><td>2.24</td><td>0.55</td></tr>
<tr><td>4–10 Apr</td><td>90 AA</td><td>1.81</td><td>0.77</td><td>1.37</td><td>0.37</td><td>1.17</td><td>0.40</td><td>1.10</td><td>0.35</td><td>1.98</td><td>0.49</td></tr>
<tr><td>4–10 May</td><td>90 AA</td><td>1.61</td><td>0.41</td><td>1.79</td><td>0.41</td><td>1.28</td><td>0.42</td><td>1.76</td><td>0.44</td><td>3.08</td><td>0.42</td></tr>
<tr><td>3–9 Jun</td><td>90 AA</td><td>1.52</td><td>0.79</td><td>1.69</td><td>0.39</td><td>1.06</td><td>0.45</td><td>1.24</td><td>0.36</td><td>3.42</td><td>0.41</td></tr>
<tr><td>27–2 Aug</td><td>90 AA</td><td>1.66</td><td>1.10</td><td>1.09</td><td>0.57</td><td>0.85</td><td>0.37</td><td>0.89</td><td>0.33</td><td>2.47</td><td>0.45</td></tr>
<tr><td>20–26 Aug</td><td>90 AA</td><td>1.86</td><td>0.58</td><td>1.88</td><td>0.57</td><td>1.20</td><td>0.37</td><td>1.17</td><td>0.28</td><td>2.91</td><td>0.32</td></tr>
<tr><td>13–19 Sep</td><td>90 AA</td><td>2.14</td><td>0.92</td><td>1.65</td><td>0.48</td><td>1.30</td><td>0.45</td><td>1.57</td><td>0.45</td><td>3.95</td><td>0.47</td></tr>
<tr><td>7–13 Oct</td><td>90 AA</td><td>1.82</td><td>0.91</td><td>1.57</td><td>0.20</td><td>1.29</td><td>0.20</td><td>1.31</td><td>0.17</td><td>2.63</td><td>0.42</td></tr>
<tr><td>6–12 Nov</td><td>90 AA</td><td>1.97</td><td>0.71</td><td>2.20</td><td>0.70</td><td>1.72</td><td>0.63</td><td>1.35</td><td>0.44</td><td>3.24</td><td>0.33</td></tr>
<tr><td>6–12 Dec</td><td>90 AA</td><td>1.94</td><td>0.72</td><td>1.90</td><td>0.36</td><td>1.51</td><td>0.35</td><td>1.55</td><td>0.35</td><td>2.80</td><td>0.37</td></tr>
<tr><td align="center">Time</td><td></td><td colspan="2" align="center">≈0.2 h</td><td colspan="2" align="center">≈0.4 h</td><td colspan="2" align="center">≈0.4 h</td><td colspan="2" align="center">≈1.3 h</td><td colspan="2" align="center">≈1.2 h</td></tr>
</table>

The RMSE values reported in the tables indicate that, even for JAK, there is no significant difference in the performance of IW SLC and IW GRDH products. In the first case, the best fit with reference data was obtained using a window size of 64 pixels in azimuth and 128 pixels in range. In the second case, the best performance against the literature was given by an asymmetric window of 128 pixels in azimuth and 64 pixels in range.

As for EW GRDH and EW GRDM products, the estimated flow velocities were significantly affected by the image lower resolution. Detailed results are not shown for brevity. Aggregated results will be provided at the end of this section.

3.1.4. Thwaites Glacier

The last analyzed glacier was the THW. The RMSE for the estimated flow velocities calculated against the reference literature for the 12 processed pairs is shown in Tables 10 and 11 for IW SLC and IW GRDH experiments. As for the other glaciers, the results are quite similar. Some pairs exhibited serious decorrelation issues, especially in the SLC case, where data on the pairs acquired in February, March and August were not retrievable. These couples (in addition to that acquired in November) were affected by the same problems using GRDH images but, in this case, the increase of the CC window had positive influence on the correlation, thus allowing for a reliable retrieval of the velocity field. The lack of any clear link to seasonal effects, such as ice melting during the Antarctic summer, suggests that correlation patterns can be significantly modified by varying the product type and that the pre-processing applied to data (standard TOPSAR, eventually followed by detection and resampling in the case of GRD images) can play a significant role in the calculation of the velocity field.

Table 10. THW, IW SLC product class experiments, RMSE (expressed in m/day) with respect to available literature data measured along selected along/across flow transects.

		Thwaites–IW SLC							
Observation	**Track**	$w = 64, f = 16/4$		$w = 64/128, f = 16/4$		$w = 128, f = 8/4$		$w = 128/256, f = 8/4$	
		Along	**Across**	**Along**	**Across**	**Along**	**Across**	**Along**	**Across**
9–15 Jan	65 AA	0.62	0.62	0.67	0.70	0.73	0.74	0.71	0.73
20–26 Feb	65 AA	na	na	na	na	na	na	na	na
10–16 Mar	65 AA	na	na	na	na	na	na	na	na
9–15 Apr	65 AA	0.51	0.52	0.48	0.50	0.50	0.49	0.47	0.49
15–21 May	65 AA	1.79	1.56	1.78	1.61	1.68	1.52	1.65	1.52
2–8 Jun	65 AA	1.17	1.28	1.13	1.37	1.11	1.34	1.08	1.35
2–8 Jul	65 DD	1.10	1.15	0.98	1.09	0.89	1.05	0.89	1.05
1–7 Aug	65 DD	na	Na	na	na	na	na	na	na
6–12 Sep	65 DD	1.64	1.10	1.75	1.27	1.77	1.25	1.65	1.09
6–12 Oct	65 DD	0.71	0.59	0.61	0.53	0.66	0.56	0.66	0.55
5–11 Nov	65 DD	0.99	0.68	0.94	0.63	0.96	0.69	0.78	0.59
5–11 Dec	65 DD	1.68	0.83	1.37	0.95	1.42	1.00	1.52	0.98
Time		≈0.75 h		≈1.5 h		≈2 h		≈4.9 h	

Table 11. THW, IW GRDH product class experiments, RMSE (expressed in m/day) with respect to available literature data measured along selected along/across flow transects.

		Thwaites–IW GRDH					
Observation	**Track**	$w = 64, f = 8$		$w = 128, f = 8$		$w = 256, f = 4$	
		Along	**Across**	**Along**	**Across**	**Along**	**Across**
9–15 Jan	65 AA	0.95	0.90	0.84	0.87	1.07	1.00
20–26 Feb	65 AA	na	na	1.64	1.51	1.71	1.51
10–16 Mar	65 AA	na	na	na	na	1.29	1.27
9–15 Apr	65 AA	1.55	1.20	1.41	1.21	1.47	1.26
15–21 May	65 AA	1.55	1.28	1.70	1.45	1.49	1.30
2–8 Jun	65 AA	1.52	1.51	1.55	1.47	1.45	1.39
2–8 Jul	65 DD	1.33	0.99	1.43	1.31	0.84	0.73
1–7 Aug	65 DD	na	na	na	na	1.38	1.26
6–12 Sep	65 DD	1.97	1.70	2.01	1.67	2.12	1.72
6–12 Oct	65 DD	1.28	1.28	1.20	1.20	1.26	1.23
5–11 Nov	65 DD	na	na	1.16	1.36	1.42	1.49
5–11 Dec	65 DD	1.49	1.43	1.57	1.56	1.73	1.36
Time		≈0.5 h		≈2.5 h		≈3.5 h	

3.1.5. Aggregated Results

Aggregated results for all of the performed experiments and test sites are provided in Table 12 and in the four successive pictures. Data reported in the table represent the annual RMSE of the estimated velocities against reference data calculated as described in Equation (2) using now the annual average of the estimated and reference flow speeds. The average annual flow speed along the considered transects for all the products involved in the comparison is depicted in Figures 6–9 for PG, NI, JAK and THW, respectively.

Table 12. RMSE (expressed in m/day) against reference literature for the selected along/across flow transects for the four analyzed glaciers and product classes calculated by averaging one year of observations.

Glacier	IW SLC		IW GRDH		EW GRDH		EW GRDM	
	Along	Across	Along	Across	Along	Across	Along	Across
Petermann	0.10	0.20	0.17	0.24	0.32	0.50	0.53	0.66
Nioghalvifierdsfjorden	0.17	0.20	0.14	0.20	na	na	0.57	0.91
Jackobshavn	1.69	0.17	1.55	0.22	7.37	1.24	8.39	1.37
Thwaites	1.10	0.81	1.41	1.17	na	na	1.63	1.01

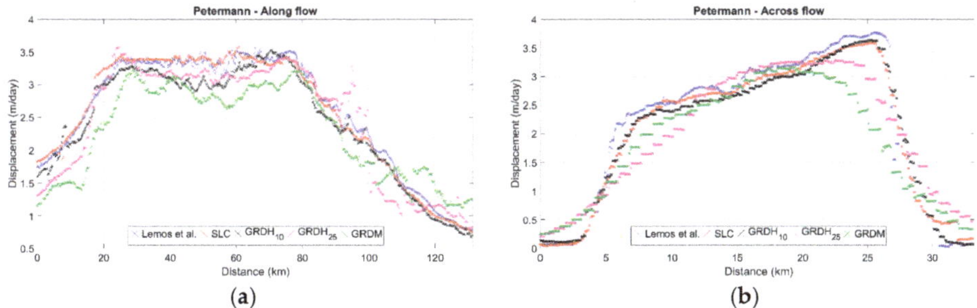

Figure 6. PG, average annual flow speed for reference data (blue curve) and all the analyzed products (IW SLC red curve, IW GRDH black curve, EW GRDH magenta curve, EW GRDM green curve). (**a**) Along flow speed. (**b**) Across flow speed.

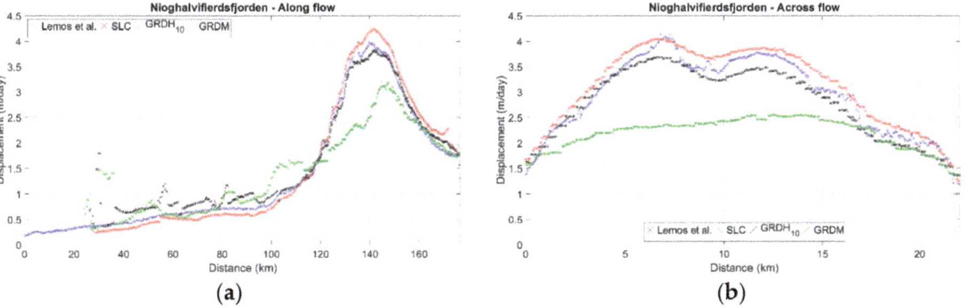

Figure 7. NI, average annual flow speed for reference data (blue curve) and all the analyzed products (IW SLC red curve, IW GRDH black curve, EW GRDM green curve). (**a**) Along flow speed. (**b**) Across flow speed.

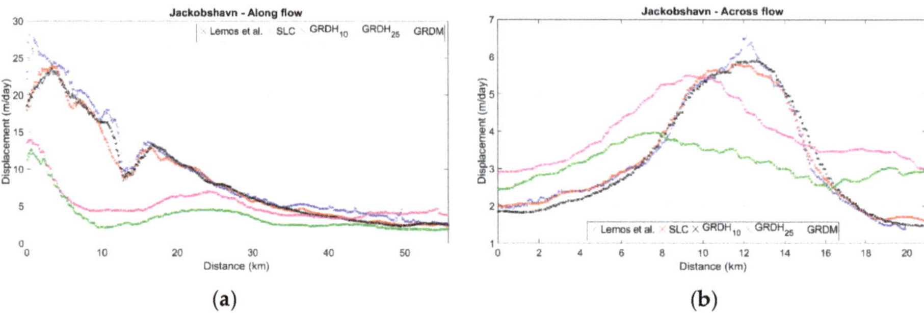

(a)   **(b)**

Figure 8. JAK, average annual flow speed for reference data (blue curve) and all the analyzed products (IW SLC red curve, IW GRDH black curve, EW GRDH magenta curve, EW GRDM green curve). (**a**) Along flow speed. (**b**) Across flow speed.

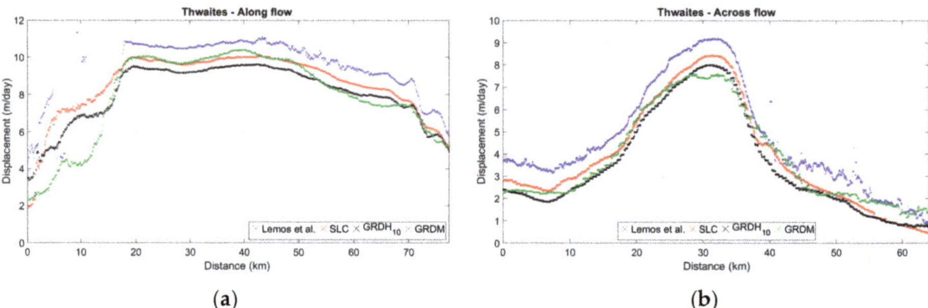

(a)   **(b)**

Figure 9. THW, average annual flow speed for reference data (blue curve) and all the analyzed products (IW SLC red curve, IW GRDH black curve, EW GRDM green curve). (**a**) Along flow speed. (**b**) Across flow speed.

From Table 12, it arises that for PG the performance of IW SLC and IW GRDH is fully comparable with respect to reference data, with registered RMSE of few tens of centimeters per day on the annual average in both along and across flow directions. The performance of EW products is still acceptable, although the retrieved velocity field is noisier at the edge of the glacier (see Figure 6a,b). Overall, a slight underestimation of the displacement velocity is registered whatever the product class is used if compared with reference data.

As for NI, the displacement velocity estimated using EW GRDH products is not fitting the reference distribution, as shown in Figure 7a,b. This is also confirmed by the RMSE values reported in Table 12, which are higher with respect to those calculated for IW SLC and IW GRDH images. As a general comment, the reference curves for the flow velocity for the along and across flow transects are almost in the middle of those retrieved using these products, with those relevant to IW SLC images placed slightly upper.

In the case of JAK, the differences between reference data and the results obtained by processing IW SLC and IW GRDH images are concentrated in the along flow direction, with particular reference to the first kilometers of the considered transect (see Figure 8a). Concerning the across flow direction, the velocity field here estimated does almost perfectly match the values provided by the reference literature (Figure 8b) as shown by the average RMSE in that direction, which is for both products of few centimeters per day.

As for EW products, both the GRDH and GRDM product class failed in the reconstruction of the velocity field of the JAK. As witnessed by the RMSE values reported in Table 12 and by the displacement velocity distributions depicted in Figure 8a,b, these product classes are not suitable for the monitoring of this glacier.

In the case of THW, the behavior of the analyzed product classes is quite consistent. For all of them, the trend is an underestimation of the displacement velocity with respect to reference data, with the underestimation increasing as the product resolution decreases. As reported in Table 12, the best registered performance against reference data is that of IW SLC products. Differences with respect to the results obtained using IW GRDH images arise especially in the along flow direction (see Figure 9a). In the across flow direction, the curves depicting the displacement velocity along the transect almost overlap, especially where the glacier exhibits its peak velocity (see Figure 9a).

The same considerations can be made for the EW GRDM product class. The discrepancy with respect to reference data is higher in the along flow direction, especially in the first and last kilometers of the transect. If the across flow direction is considered, the performance of this product class is in line with those of higher resolution data.

3.2. Same Algorithm Comparison

In the previous section, the velocity fields estimated by an in-house implemented OT algorithm were compared with the outputs from a commercial software to test the reliability of the solution. Here, the results are discussed fixing the algorithm and the parameter of the output flow velocity maps. In other words, the comparison will be implemented between algorithm runs leading to maps having comparable sensitivity and obtained with similar computational times.

In Table 13, data concerning PG (left panel) and NI (right panel) are reported. The reference velocity field is that obtained using IW SLC images as input for the in-house OT. In the case of PG, the reference experiments have been performed by setting the CC window dimension to 64×64 square pixel and the oversampling factor to 16/4 in the azimuth/slant range directions, respectively. The same settings were applied to reference NI experiments.

Table 13. RMSE (expressed in m/day) for the displacement velocities calculated with the in-house OT algorithm with respect to reference IW SLC results for the PG (left panel) and NI (right panel).

Petermann				Nioghalvifierdsfjorden			
		IW GRDH				**IW GRDH**	
Observation	**Track**	**Along**	**Across**	**Observation**	**Track**	**Along**	**Across**
6–12 Jan	26 AA	0.17	0.36	3–9 Jan	74 AA	0.47	0.39
11–17 Feb	26 DD	0.26	0.20	2–8 Feb	74 AA	0.47	0.59
1–7 Mar	26 DD	0.32	0.30	4–10 mar	74 AA	0.42	0.49
6–12 Apr	26 DD	0.26	0.35	15–21 Apr	74 AA	0.30	0.54
18–24 May	26 DD	0.36	0.30	3–9 May	74 AA	0.30	0.67
11–17 Jun	26 DD	0.44	0.92	2–8 Jun	74 AA	0.46	0.66
11–17 Jul	26 DD	0.35	0.73	2–8 Jul	74 AA	0.28	0.30
10–16 Aug	26 DD	0.26	0.69	1–7 Aug	74 AA	0.54	0.30
9–15 Sep	26 DD	0.20	0.20	6–12 Sep	74 AA	0.64	0.32
9–15 Oct	26 DD	0.14	0.22	6–12 Oct	74 AA	0.17	0.24
8–14 Nov	26 DD	0.22	0.14	5–11 Nov	74 AA	0.20	0.36
8–14 Dec	26 DD	0.33	0.17	5–11 Dec	74 AA	0.26	0.20
Mean		0.20	0.10	Mean		0.24	0.33

Comparison data are shown only for IW GRDH images for brevity, also because this product class, together with the IW SLC, is showing the best agreement with the reference literature data. For both the PG and the NI, the OT setting for these experiments was CC window of 64×64 pixels and oversampling factor of 8 in both azimuth and slant range directions.

The table confirms that, as previously discussed, there are no significant differences in the displacement velocities estimated for the two product classes. In particular, for the PG, a peak

RMSE of 0.44 m/day was recorded in the along flow direction, while for the across flow the highest MSE was of 0.92 m/day. Average values for the along flow and across flow RMSE were of 0.20 and 0.10 m/day respectively.

Similarly, for the NI, the peak values for the RMSE were of 0.64 and 0.67 m/day for the along flow and across flow directions, respectively. On average, the registered RMSE was of 0.24 and 0.33 m/day for the two reference directions.

Data concerning JAK and THW are reported in the left and right panel of Table 14, respectively. As for the previous glaciers, only the comparison between reference IW SLC results and IW GRDH outputs is considered for brevity.

Table 14. RMSE (expressed in m/day) for the displacement velocities calculated with the in-house OT algorithm with respect to reference IW SLC results for JAK (left panel) and THW (right panel).

Jackobshavn				Thwaites			
		GRDH$_{10}$				GRDH$_{10}$	
Observation	Track	Along	Across	Observation	Track	Along	Across
4–10 Jan	90 AA	1.36	0.75	9–15 Jan	65 AA	0.62	0.33
15–21 Feb	90 AA	1.19	0.20	20–26 Feb	65 AA	na	na
5–11 Mar	90 AA	1.23	0.32	10–16 Mar	65 AA	na	na
4–10 Apr	90 AA	1.79	0.30	9–15 Apr	65 AA	1.12	1.18
4–10 May	90 AA	0.95	0.42	15–21 May	65 AA	0.62	0.30
3–9 Jun	90 AA	0.95	0.56	2–8 Jun	65 AA	0.64	0.24
27–2 Aug	90 AA	1.34	0.40	2–8 Jul	65 DD	0.79	0.47
20–26 Aug	90 AA	2.36	0.42	1–7 Aug	65 DD	na	na
13–19 Sep	90 AA	1.38	0.36	6–12 Sep	65 DD	0.57	0.69
7–13 Oct	90 AA	1.24	0.35	6–12 Oct	65 DD	0.74	0.85
6–12 Nov	90 AA	1.18	0.73	5–11 Nov	65 DD	0.92	1.04
6–12 Dec	90 AA	0.90	0.26	5–11 Dec	65 DD	0.72	0.45
Mean		0.67	0.17	Mean		0.75	0.45

As for JAK, data relevant to IW SLC experiments have been collected by setting the CC window to 64/128 pixels and the oversampling factor to 8/4 in the azimuth/slant range directions, respectively. In the IW GRDH experiments, the CC window was 128/64 pixels in azimuth/slant range direction, respectively, and the oversampling factor of 8 in both directions.

The highest performance difference between the two product classes is in the long track direction, with a peak RMSE of 2.36 m/day for the observation period 20–26 August 2017. For almost all the other pairs, the discrepancies are minor, leading to an average RMSE for the entire year of 0.67 m/day. The agreement between the results obtained using the two different product types is better in the across track direction, where the RMSE peak and yearly average recorded are 0.75 and 0.17 m/day, respectively.

The results on the THW were obtained, in the case of IW SLC images by setting the CC window to 128/256 pixels and the oversampling factor to 8/4 in azimuth/slant range directions, respectively. As for IW GRDH products, the CC window was set to 256 × 256 pixels and the oversampling factor to 8 for both directions.

The RMSE values calculated for both the along track and the across track directions are within the expected values. In the first case, the highest registered RMSE is 1.12 m/day for the pair 9–15 April 2017. All the other values are much lower than one meter per day, leading to a yearly average of 0.75 m/day.

In the across track direction, the peak RMSE, in the order of 1.18 m/day was registered for the same pair. In this case, the yearly average was of 0.45 m/day.

3.3. Storage and Computational Times

As discussed in the previous Sections, the computational time necessary to produce displacement maps with a sensitivity in the order of one meter in both the azimuth and slant range directions is roughly the same whatever the product class considered. This is because it is possible to balance the higher resolution of the input image with a lower oversampling factor and, on the other side, the lower resolution, requiring higher oversampling, by processing tiny subsets of the whole scan.

As an example, in the case of the PG, displacement maps with theoretical sensitivity of 0.9 m in both azimuth and slant range directions have been produced using IW SLC products as input in 1.5 h running a serial code on an 8-cores machine with 128 GB of RAM memory. Using IW GRDH products, a slightly lower theoretical sensitivity, in the order of 1.2 m in the two directions, has been obtained in approximately one-hour processing time. Computational times of the same order have been registered when producing maps with sensitivity of 1.6 and 2.5 m in the two directions using EW GRDH and EW GRDM products, respectively. For these products, however, the increase of the CC window dimension from 64×64 to 128×128 pixels had a destructive impact on computational times. Similar considerations hold for the other glaciers analyzed.

The scenario changes if storage needs reported in Table 15 are considered. It arises that, to process the four time-series using IW SLC products, approximately slightly less than one terabyte of data was archived to output one image per month. This means that, the exploitation of the full temporal resolution of the Sentinel-1 constellation, delivering up to one image every six days, would imply to multiply by four the amount of data processed in this work, with the total reaching approximately 4 terabytes for the considered glaciers.

Table 15. Storage needs (in GB) per analyzed time-series, processing step, and product class.

Glacier	Level-1	Coregistration	Subset	OT	Map	Product
PG	93.1	78.8	62.5	62.5	0.35	IW SLC
	20.2	37.8	22.0	22.0	0.24	IW GRDH
	15.2	18.2	5.66	5.66	0.77	EW GRDH
	5.74	8.92	1.95	1.95	0.67	EW GRDM
NI	106	81.6	39.4	39.4	0.20	IW SLC
	22.9	37.8	13.2	13.2	0.15	IW GRDH
	6.07	8.66	2.21	2.21	0.27	EW GRDM
JAK	80.5	80.2	17.7	17.7	0.07	IW SLC
	17.7	38.3	3.91	3.91	0.04	IW GRDH
	17.6	19.2	1.70	1.70	0.17	EW GRDH
	6.54	9.30	0.45	0.45	0.11	EW GRDM
THW	51.6	77.3	36.7	36.7	0.28	IW SLC
	12.0	37.9	11.3	11.3	0.12	IW GRDH
	6.40	9.28	1.28	1.28	0.20	EW GRDM

Using IW GRDH images, the total amount of data to be archived is less than 300 GB, about one third less than that required for the full resolution complex product class. This value further decreases using EW images up to less 100 GB exploiting the lowest resolution GRDM product class. These values do not consider any data compression technique.

4. Discussion

The SAR literature is rich in works addressing glacier monitoring using OT techniques and, currently, it is benefiting from the open data policy of ESA regarding acquisitions made under the aegis of Copernicus Programme. The improved access to images is boosting the research in this field, with particular focus on continuous monitoring. However, works addressing the problem at product level, offering insights into the different product types performance, are still missing.

OT methods, using a pair of SAR images, can detect movements of several meters with a good degree of approximation [31]. The accuracy of the estimated displacements depends on many factors including the resolution of input images, the magnitude of real displacements, the acquisition geometry and the correlation coefficient between the considered image patches [58]. These parameters are interconnected. Higher resolution images tend to maximize correlation at finer scale and should be preferred when dealing, as an example, with landslides [30,56]. Spatial decorrelation can be also due to orbital issues, i.e., differences in the acquisition geometry and/or large spatial baseline that, changing speckle patterns or the response of rugged terrains [35], can affect the amplitude and the sharpness of the correlation peak [59].

The general trend in the literature is to feed OT algorithms with complex images [25]. However, as demonstrated in the previous Section, the use of pre-processed detected images presents advantages at storage and computational level. Moreover, their performance is fully comparable with that of complex products, in particular in the case of the PG and of the NI, which are characterized by the smallest movements among the analyzed glaciers. A significant error of more than 2 m/day against reference data was registered for the JAK case study of August 2017 using SLC images with correlation windows of 64 × 64 square pixel and of 64 pixels in azimuth and 128 pixels in range. This can be due to the combination of effects given by the small correlation window and the flow velocity, which reach its peak in that month (see Table 2). A similar situation is observed in the IW GRDH case.

In some cases, it was not possible to retrieve the velocity field due to decorrelation issues. This happened for the PG using interferometric extra-wide swath products and for the THW Glacier using interferometric wide swath images, both complex and detected. In the first case, the decorrelation was likely due to orbital issues, since footprints appear quite squinted. The same did not occur for the THW Glacier, for which the acquisition geometries seem to be consistent. Moreover, an improvement of correlation was registered with the increase of the calculation window using detected images, and this allowed for a reliable estimation of the velocity field. Conversely, this was not possible exploiting complex products. This suggests that correlation patterns can be significantly modified varying the product type, and that they do not necessarily benefit from the enhancement of the resolution of the input product. The investigation of the effects of pre-processing and/or image parameters on the cross-correlation and, as a consequence, on the estimated velocity field, is an open point to be addressed with further research.

Moving to aggregated results (see Section 3.1.5), the most remarkable differences with respect to reference data arose for the JAK in the along flow direction, with particular reference to the first kilometers of the considered transect (see Figure 8a). Due to the lack of any ground data and the limited visibility of the algorithm implemented in the GAMMA-SAR software suite, it is difficult to understand which result better represents the real movement of this part of the glacier. However, the assessment of the best performing method is not the purpose of this paper which has focused on the study of the performance of different classes of SAR products. In this perspective, it is possible to argue that the performance of IW SLC and IW GRDH product classes is mostly equivalent. This is confirmed by the values of the RMSE reported in Table 12.

5. Conclusions

Since 2014, the Sentinel-1 mission of the European Space Agency has been providing high-quality synthetic aperture radar data allowing for more effective monitoring of our rapidly changing planet. Data are delivered in different formats and at different stages of the SAR pre-processing chain. However, only a few studies in literature are concerned with the assessment of the best product for a specific application. In this work, we compared the performance of all the Sentinel-1 product classes (interferometric wide swath single look complex, interferometric wide swath high-resolution ground range detected, extra-wide swath high-resolution ground range detected, and extra-wide swath medium-resolution ground range detected) in a classic radar remote sensing application, i.e., glacier monitoring using offset tracking.

To this end, four different glaciers (namely the Petermann Glacier, the Nioghalvifierdsfjorden, the Jakobshavn Isbræ and the Thwaites glacier) have been investigated using a state-of-the-art algorithm applied to couple of images acquired with short temporal baseline (6 to 12 days) during the year 2017. The obtained flow velocities were compared initially with reference literature data obtained using an offset tracking tool implemented in the commercial software suite GAMMA-SAR to test the reliability of the implemented solution. Then, the comparison moved at product level (i.e., fixing the offset tracking algorithm) to better understand each performance.

The results showed that the flow velocities retrieved along selected along/across flow transects by using interferometric wide swath products (both complex and detected) were in good agreement with literature data, although the implemented solution tend to slightly underestimate the flow velocities. However, the lack of ground data as well as of information on the algorithm implemented in the aforementioned commercial software suite makes it difficult to understand which solution better depicts the real phenomenon. As a general comment, the flow velocities obtained using complex and detected images were fully comparable. The most remarkable difference was observed in the case of the Thwaites Glacier, in both along and across flow directions, where complex product performed slightly better.

The passage to extra-wide swath images, instead, was disadvantageous, with results tending to drift more from the reference as the resolution of the input product decreases. Another point against these product classes is that the time span between two successive acquisitions on the same area is of 12 days in most of the cases (while for interferometric wide swath images is 6 days). This has a negative impact on the correlation between the image pairs and, consequently, on the estimated displacements.

The product-level comparison aimed at the analysis of the velocity field output by the same algorithm once fixed the parameters of the output map, i.e., theoretical sensitivity and the resolution, and it mainly confirmed what arose from the comparison with the literature. The performance of interferometric wide swath high-resolution ground range detected images was fully comparable with that of complex ones, with yearly averages of the root mean square errors calculated along the transects ranging from few centimeters to few tens of centimeters per day.

The major differences between the two product classes concern the pre-processing, applied to prepare the images for the ingestion in the core information process, and storage needs. Ground range detected images are ready-to-use data, which do not need any pre-processing before being ingested in the offset tracking algorithm. On the contrary, complex images must be compensated for the TOPSAR acquisition mode before being exploited for information extraction.

Storage needs greatly vary depending on the product class. Each step forward in the resolution (from the extra wide swath medium resolution to the full resolution complex) roughly implied quadrupling the storage capacity.

Summarizing, this work demonstrated that the performance of Sentinel-1 interferometric wide swath images, both complex and detected, are fully comparable, making them a perfectly interchangeable input for offset tracking procedures. However, detected images could be advantageous, for example in case of studies requiring short revisit times, due to their faster processing (as no pre-processing is required) and the lower storage demand. As for extra-wide swath images, the estimated flow velocities are highly affected by the lower resolution of the input products. Therefore, these product classes are not recommended for this kind of applications.

Author Contributions: Conceptualization, D.A.; methodology, D.A.; software, D.A. and G.D.M.; validation, D.A and G.D.M.; investigation, D.A.; resources, A.I. and R.G.; writing—original draft preparation, D.A.; writing—review and editing, R.G., G.D.M. and A.I.; supervision, A.I. and R.G.; funding acquisition, R.G.

Funding: This research received no external funding.

Conflicts of Interest: The authors declare no conflict of interest.

Abbreviations

List of the acronyms (in order of appearance):

S1	Sentinel-1
ESA	European Space Agency
SAR	Synthetic Aperture Radar
IW	Interferometric Wide swath

EW	Interferometric Extra Wide swath
TOPSAR	Terrain Observation with Progressive Scanning SAR
SDH	Sentinels Data Hub
SLC	Single Look Complex
GRD	Ground Range Detected
GRDF	Full resolution Ground Range Detected
GRDH	High-resolution Ground Range Detected
GRDM	Medium resolution Ground Range Detected
SM	Stripmap
OT	Offset Tracking
PG	Petermann Glacier
NI	Nioghalvfjerdsfjorden
JAK	Jackobshavn Isbræ
THW	Thwaites Glacier
DInSAR	Differential Synthetic Aperture Radar Interferometry
CC	Cross-correlation
RMSE	Root Mean Square Error

References

1. European Commission. *Copernicus User Uptake—Engaging with Public Authorities, the Private Sector and Civil Society*; Publications Office of the European Union: Luxembourg, 2016.
2. de Zan, F.; Guarnieri, A.M. TOPSAR: Terrain Observation by Progressive Scans. *IEEE Trans. Geosci. Remote Sens.* **2006**, *44*, 2352–2360. [CrossRef]
3. European Space Agency. *Sentinel-1: ESA's Radar Observatory Mission for GMES Operational Services*; ESA Communications: Noordwijk, The Netherlands, 2012.
4. Franceschetti, G.; Lanari, R. *Synthetic Aperture Radar Processing*; CRC Press: Boca Raton, FL, USA, 1999.
5. European Space Agency. Sentinel-1 Product Definition. Available online: http://sentinel.esa.int/documents/247904/1877131/Sentinel-1-Product-Definition (accessed on 30 May 2019).
6. Amitrano, D.; di Martino, G.; Iodice, A.; Riccio, D.; Ruello, G. Unsupervised Rapid Flood Mapping Using Sentinel-1 GRD SAR Images. *IEEE Trans. Geosci. Remote Sens.* **2018**, *56*, 3290–3299. [CrossRef]
7. Schubert, D.; Small, N.; Miranda, D.G.; Meier, E. Sentinel-1A Product Geolocation Accuracy: Commissioning Phase Results. *Remote Sens.* **2015**, *7*, 9431–9449. [CrossRef]
8. Lüttig, C.; Neckel, N.; Humbert, A. A combined approach for filtering ice surface velocity fields derived from remote sensing methods. *Remote Sens.* **2017**, *9*, 1062. [CrossRef]
9. Strozzi, T.; Luckman, A.; Murray, T.; Wegmüller, U.; Werner, C.L. Glacier motion estimation using SAR offset-tracking procedures. *IEEE Trans. Geosci. Remote Sens.* **2002**, *40*, 2384–2391. [CrossRef]
10. Riveros, N.; Euillades, L.; Euillades, P.; Moreiras, S.; Balbarani, S. Offset tracking procedure applied to high resolution SAR data on Viedma Glacier, Patagonian Andes, Argentina. *Adv. Geosci.* **2013**, *35*, 7–13. [CrossRef]
11. Lucchitta, B.K.; Ferguson, H.M. Antarctica: Measuring Glacier Velocity from Satellite Images. *Science* **1986**, *234*, 1105–1108. [CrossRef]
12. Lindstrom, D.; Tyler, D. Preliminary results of Pine island and Thwaites glaciers study. *Antarct. J. U. S.* **1984**, *19*, 53–55.
13. Bindschadler, R.A.; Scambos, T.A. Satellite-image-derived velocity field of an Antarctic ice stream. *Science* **1991**, *252*, 242–246. [CrossRef]
14. Emery, W.J.; Fowler, C.W.; Hawkins, J.; Preller, R.H. Fram strait satellite image-derived ice motions. *J. Geophys. Res. Ocean* **1991**, *96*, 4751–4768. [CrossRef]
15. Scambos, T.A.; Dutkiewicz, M.J.; Wilson, J.C.; Bindschadler, R.A. Application of image cross-correlation to the measurement of glacier velocity using satellite image data. *Remote Sen. Environ.* **1992**, *42*, 177–186. [CrossRef]
16. Bindschadler, R.A.; Vornberger, P.; Blanskenship, D.; Scambos, T.A.; Jacobel, R. Surface velocity and mass balance of Ice Streams D and E, West Antarctica. *J. Glaciol.* **1996**, *42*, 461–475. [CrossRef]
17. Bindschadler, R.A.; Fahnestock, M.A.; Skvarca, P.; Scambos, T.A. Surface-velocity field of the northern Larsen ice shelf. *Ann. Glaciol.* **1994**, *20*, 319–326. [CrossRef]

18. Fahnestock, M.; Scambos, T.; Moon, T.; Gardner, A.; Haran, T.; Klinger, M. Rapid large-area mapping of ice flow using Landsat 8. *Remote Sens. Environ.* **2016**, *185*, 84–94. [CrossRef]

19. Armstrong, W.H.; Anderson, R.S.; Fahnestock, M.A. Spatial Patterns of Summer Speedup on South Central Alaska Glaciers. *Geophys. Res. Lett.* **2017**, *44*, 9379–9388. [CrossRef]

20. Li, T.; Liu, Y.; Li, T.; Hui, F.; Chen, Z.; Cheng, X. Antarctic surface ice velocity retrieval from MODIS-based mosaic of Antarctica (MOA). *Remote Sens.* **2018**, *10*, 1045. [CrossRef]

21. Dehecq, A.; Gourmelen, N.; Trouve, E. Deriving large-scale glacier velocities from a complete satellite archive: Application to the Pamir-Karakoram-Himalaya. *Remote Sen. Environ.* **2015**, *162*, 55–66. [CrossRef]

22. Kääb, A.; Winsvold, S.H.; Altena, B.; Nuth, C.; Nagler, T.; Wuite, J. Glacier remote sensing using Sentinel-2. part I: Radiometric and geometric performance, and application to ice velocity. *Remote Sens.* **2016**, *8*, 598. [CrossRef]

23. Paul, F.; Winsvold, S.H.; Kääb, A.; Nagler, T.; Schwaizer, G. Glacier remote sensing using Sentinel-2. part II: Mapping glacier extents and surface facies, and comparison to Landsat 8. *Remote Sens.* **2016**, *8*, 575. [CrossRef]

24. Wylie, D.P.; Jackson, D.L.; Menzel, W.P.; Bates, J.J. Global cloud cover trends inferred from two decades of HIRS observations. *J. Clim.* **2005**, *18*, 3021–3031. [CrossRef]

25. Joughin, R.; Smith, B.E.; Abdalati, W. Glaciological advances made with interferometric synthetic aperture radar. *J. Glaciol.* **2010**, *56*, 1026–1042. [CrossRef]

26. Goldstein, R.M.; Engelhardt, H.; Kamb, B.; Frolich, R.M. Satellite Radar Interferometry for Monitoring Ice Sheet Motion: Application to an Antarctic Ice Stream. *Science* **1993**, *262*, 1525–1530. [CrossRef]

27. Joughin, R.; Kwok, R.; Fahnestock, M.A. Interferometric Estimation of the Three-Dimensional Ice-Flow Velocity Vector Using Ascending and Descending Passes. *IEEE Trans. Geosci. Remote Sens.* **1988**, *36*, 25–37. [CrossRef]

28. Mattar, E.; Vachon, P.W.; Geudtner, D.; Gray, A.L.; Gumming, L.G.; Brugman, M. Validation of alpine glacier velocity measurements using ERS tandem-mission SAR data. *IEEE Trans. Geosci. Remote Sens.* **1998**, *36*, 974–984. [CrossRef]

29. Singleton, A.; Li, Z.; Hoey, T.; Muller, J.P. Evaluating sub-pixel offset techniques as an alternative to D-InSAR for monitoring episodic landslide movements in vegetated terrain. *Remote Sens. Environ.* **2014**, *147*, 133–144. [CrossRef]

30. Amitrano, D.; Guida, R.; Dell'Aglio, D.; di Martino, G.; di Martire, D.; Iodice, A.; Costantini, M.; Malvarosa, F.; Minati, F. Long-Term Satellite Monitoring of the Slumgullion Landslide Using Space-Borne Synthetic Aperture Radar Sub-Pixel Offset Tracking. *Remote Sens.* **2019**, *11*, 369. [CrossRef]

31. Manconi, A.; Casu, F.; Ardizzone, F.; Bonano, M.; Cardinali, M.; de Luca, C.; Gueguen, E.; Marchesini, I.; Parise, M.; Vennari, C.; et al. Brief communication: Rapid mapping of landslide events: The 3 December 2013 Montescaglioso landslide, Italy. *Nat. Hazards Earth Syst. Sci.* **2014**, *14*, 1835–1841. [CrossRef]

32. Huang, J.; Deng, K.; Fan, H.; Yan, S. An improved pixel-tracking method for monitoring mining subsidence. *Remote Sens. Lett.* **2016**, *7*, 731–740. [CrossRef]

33. Fan, H.; Gao, X.; Yang, J.; Deng, K.; Yu, Y. Monitoring mining subsidence using a combination of phase-stacking and offset-tracking methods. *Remote Sens.* **2015**, *7*, 9166–9183. [CrossRef]

34. Joughin, I.; Smith, B.E.; Howat, I.M.; Scambos, T.; Moon, T. Greenland flow variability from ice-sheet wide velocity mapping. *J. Glaciol.* **2010**, *56*, 415–430. [CrossRef]

35. Yan, S.; Liu, G.; Wang, Y.; Ruan, Z. Accurate determination of glacier surface velocity fields with a DEM-assisted pixel-tracking technique from SAR imagery. *Remote Sens.* **2015**, *7*, 10898–10916. [CrossRef]

36. Sanchez-Gamez, P.; Navarro, F.J. Glacier Surface Velocity Retrieval Using D-InSAR and Offset Tracking Techniques Applied to Ascending and Descending Passes of Sentinel-1 Data for Southern Ellesmere Ice Caps, Canadian Arctic. *Remote Sens.* **2017**, *9*, 442. [CrossRef]

37. Wuite, J.; Rott, H.; Hetzenecker, M.; Floricioiu, D.; de Rydt, J.; Gudmundsson, G.H.; Nagler, T.; Kern, M. Evolution of surface velocities and ice discharge of Larsen B outlet glaciers from 1995 to 2013. *Cryosphere* **2015**, *9*, 957–969. [CrossRef]

38. Lanari, R.; Mora, O.; Manunta, M.; Mallorquí, J.J.; Berardino, P.; Sansosti, E. A small-baseline approach for investigating deformations on full-resolution differential SAR interferograms. *IEEE Trans. Geosci. Remote Sens.* **2004**, *42*, 1377–1386. [CrossRef]

39. Bechor, N.B.D.; Zebker, H.A. Measuring two-dimensional movements using a single InSAR pair. *Geophys. Res. Lett.* **2006**, *33*, L16311. [CrossRef]

40. Sun, L.; Muller, J.P. Evaluation of the use of sub-pixel offset tracking techniques to monitor landslides in densely vegetated steeply sloped areas. *Remote Sens.* **2016**, *8*, 659. [CrossRef]

41. Peter, H.; Jäggi, A.; Fernández, J.; Escobar, D.; Ayuga, F.; Arnold, D.; Wermuth, M.; Hackel, S.; Otten, M.; Simons, W.; et al. Sentinel-1A-First precise orbit determination results. *Adv. Space Res.* **2017**, *60*, 879–892. [CrossRef]

42. Nick, F.M.; Luckman, A.; Vieli, A.; van der Veen, C.J.; van As, D.; van de Wal, R.S.W.; Pattyn, F.; Hubbard, A.L.; Floricioiu, D. The response of Petermann Glacier, Greenland, to large calving events, and its future stability in the context of atmospheric and oceanic warming. *J. Glaciol.* **2012**, *58*, 229–239. [CrossRef]

43. Rignot, E. Tidal motion, ice velocity and melt rate of Petermann Gletscher, Greenland, measured from radar interferometry. *J. Glaciol.* **1996**, *42*, 476–485. [CrossRef]

44. Rignot, E.; Steffen, K. Channelized bottom melting and stability of floating ice shelves. *Geophys. Res. Lett.* **2008**, *35*. [CrossRef]

45. Rignot, E.; Mouginot, J. Ice flow in Greenland for the International Polar Year 2008–2009. *Geophys. Res. Lett.* **2012**, *39*. [CrossRef]

46. Mayer, C.; Schaffer, J.; Hattermann, T.; Floricioiu, D.; Krieger, L.; Dodd, P.A.; Kanzow, T.; Licciulli, C.; Schannwell, C. Large ice loss variability at Nioghalvfjerdsfjorden Glacier, Northeast-Greenland. *Nat. Commun.* **2018**, *9*, 2768. [CrossRef] [PubMed]

47. Lemos, A.; Shepherd, A.; McMillan, M.; Hogg, A.E.; Hatton, E.; Joughin, I. Ice velocity of Jakobshavn Isbræ, Petermann Glacier, Nioghalvfjerdsfjorden, and Zachariæ Isstrøm, 2015–2017, from Sentinel 1-a/b SAR imagery. *Cryosphere* **2018**, *12*, 2087–2097. [CrossRef]

48. Enderlin, E.M.; Howat, I.M.; Jeong, S.; Noh, M.-J.; van Angelen, J.H.; van den Broeke, M.R. An improved mass budget for the Greenland ice sheet. *Geophys. Res. Lett.* **2014**, *41*, 866–872. [CrossRef]

49. Joughin, I.; Smith, B.E.; Shean, D.E.; Floricioiu, D. Brief Communication: Further summer speedup of Jakobshavn Isbræ. *Cryosphere* **2014**, *8*, 209–214. [CrossRef]

50. Holland, M.; Thomas, R.H.; de Young, B.; Ribergaard, M.H.; Lyberth, B. Acceleration of Jakobshavn Isbræ trig- gered by warm subsurface ocean waters. *Nat. Geosci.* **2008**, *1*, 659–664. [CrossRef]

51. Joughin, I.; Abdalati, W.; Fahnestock, M.A. Large fluctuations in speed on Greenland's Jakobshavn Isbræ glacier. *Nature* **2004**, *432*, 608–610. [CrossRef]

52. Scambos, T.A.; Bell, R.E.; Alley, R.B.; Anandakrishnan, S.; Bromwich, D.H.; Brunt, K.; Christianson, K.; Creyts, T.; Das, S.B.; DeConto, R.; et al. How much, how fast?: A science review and outlook for research on the instability of Antarctica's Thwaites Glacier in the 21st century. *Glob. Planet. Chang.* **2017**, *153*, 16–34. [CrossRef]

53. DeConto, R.M.; Pollard, D. Contribution of Antarctica to past and future sea-level rise. *Nature* **2016**, *531*, 591–597. [CrossRef]

54. Wegnüller, U.; Werner, C.; Strozzi, T.; Wiesmann, A.; Frey, O.; Santoro, M. Sentinel-1 Support in the GAMMA Software. *Procedia Comput. Sci.* **2016**, *100*, 1305–1312. [CrossRef]

55. Heo, Y.S.; Lee, K.M.; Lee, S.U. Robust Stereo matching using adaptive normalized cross-correlation. *IEEE Trans. Pattern Anal. Mach. Intell.* **2011**, *33*, 807–822. [PubMed]

56. Cai, J.; Wang, C.; Mao, X.; Wang, Q. An adaptive offset tracking method with SAR images for landslide displacement monitoring. *Remote Sens.* **2017**, *9*, 830. [CrossRef]

57. Kanade, T.; Okutomi, M. A stereo matching algorithm with an adaptive window: Theory and Experiment. *IEEE Trans. Pattern Anal. Mach. Intell.* **1994**, *16*, 920–932. [CrossRef]

58. Sun, L.; Muller, J.P.; Chen, J. Time series analysis of very slow landslides in the three Gorges region through small baseline SAR offset tracking. *Remote Sens.* **2017**, *9*, 1314. [CrossRef]

59. Yonezawa, C.; Takeuchi, S. Decorrelation of SAR data by urban damages caused by the 1995 Hyogoken-nanbu earthquake. *Int. J. Remote Sens.* **2010**, *22*, 37–41. [CrossRef]

Article

Operational Flood Mapping Using Multi-Temporal Sentinel-1 SAR Images: A Case Study from Bangladesh

Kabir Uddin [1,*], **Mir A. Matin** [1] **and Franz J. Meyer** [2]

[1] International Centre for Integrated Mountain Development, Kathmandu 44700, Nepal
[2] Geophysical Institute, University of Alaska Fairbanks, Fairbanks, AK 757500, USA
* Correspondence: Kabir.Uddin@icimod.org; Tel.: +977-1-527-5222

Received: 1 May 2019; Accepted: 23 June 2019; Published: 3 July 2019

Abstract: Bangladesh is one of the most flood-affected countries in the world. In the last few decades, flood frequency, intensity, duration, and devastation have increased in Bangladesh. Identifying flood-damaged areas is highly essential for an effective flood response. This study aimed at developing an operational methodology for rapid flood inundation and potential flood damaged area mapping to support a quick and effective event response. Sentinel-1 images from March, April, June, and August 2017 were used to generate inundation extents of the corresponding months. The 2017 pre-flood land cover maps were prepared using Landsat-8 images to identify major land cover on the ground before flooding. The overall accuracy of flood inundation mapping was 96.44% and the accuracy of the land cover map was 87.51%. The total flood inundated area corresponded to 2.01%, 4.53%, and 7.01% for the months April, June, and August 2017, respectively. Based on the Landsat-8 derived land cover information, the study determined that cropland damaged by floods was 1.51% in April, 3.46% in June, 5.30% in August, located mostly in the Sylhet and Rangpur divisions. Finally, flood inundation maps were distributed to the broader user community to aid in hazard response. The data and methodology of the study can be replicated for every year to map flooding in Bangladesh.

Keywords: flood mapping; damage assessment; SAR image; Sentinel-1; Landsat-8; Google Earth Engine; GEE; Bangladesh

---

1. Introduction

The Ganges, Brahmaputra, Meghna (GBM) basins are one of the most flood-prone basins in the world. Due to being part of such big basins and most of the area being less than 7 m above mean sea level, Bangladesh faces the cumulative effects of floods due to water flashing from nearby hills, the accumulation of the inflow of water from upstream catchments, and locally heavy rainfall enhanced by drainage congestion [1–4]. The country has a long history of destructive flooding that has had very adverse impacts on lives and property [5–7]. Approximately 20,000 deaths have been reported due to flooding from 1954–2007 [8]. An analysis by the Bangladesh Bureau of Statistics showed that from 2009–2014, 56.62% of households have been affected by disasters at least once [9]. Among them, 24.44% were affected by flood events. More than 80% of the country is flood prone [10]. In an average year, approximately 20–25% of the area of the country is inundated by floods, while in extreme years, the inundated area makes up more than 60% of the country [11]. During the 2017 flood, more than 30% of the country's areas were inundated, causing at least 134 deaths and affecting more than 5.7 million people [12]. Widespread frequent flooding is not only a profound problem for Bangladesh. Between 1995–2015, there was at least a $166 billion economic loss caused by different floods events around the world [13].

For effective response during flooding events, the rapid monitoring of flood situations, including mapping the extent of the inundation and damage, is highly critical [14,15]. Before any flood event, flood forecasting and simulation of the inundation extent is critical for risk mitigation [16]. At present, flood early warning and monitoring information is produced by the Flood Forecasting and Warning Centre (FFWC) as water levels change in major river systems in Bangladesh [5]. The extent of inundation is also mapped by comparing the water level with the national digital elevation model (DEM). Generating inundation maps from hydrological models requires an up-to-date and accurate DEM as well as computing infrastructure to model the effects of obstacles on the flow of flood water in the floodplains. Unfortunately, a sufficiently high accurate DEM and infrastructure data are often not available [17]. Flood management based on water level forecasting is not effective in providing a spatially distributed flood area for the timely monitoring of flood events [13,18]. Satellite-based monitoring of flood extent overcomes the limitations of the hydrological model-based approach [19]. Various attempts have been made in the past to map the flood extent in Bangladesh from satellite images [20]. Rasid and Pramanik [7] attempted comprehensive flood extent mapping of Bangladesh in 1980 using National Oceanographic Atmospheric Administrative's (NOAA) advanced very high resolution radiometer (AVHRR) data. Islam, et al. [21] used MODIS surface reflectance images for flood mapping for the years 2004 and 2007. Ahmed, et al. [22] used Landsat-8 and MODIS to determine the impact of a 2017 flash flooding event on rice production in the Haor area. MODIS images also have considerable potential for daily agriculture flood mapping of Bangladesh [23].

Although optical images have great potential for mapping during good weather conditions with image analysis capacity and accuracy [16,24–26], their use is limited to flood mapping in Bangladesh due to high (approximately 80%) cloud coverage during the monsoon and flood period. The average cloud coverage for Bangladesh was found to be 88.5% in June, 90.8% in July, 78.3% in August, 78.3% in September, and 17% in December (all at latitude: 24.921°N and longitude 91.869°E) [27]. In reality, the acquisition of cloud-free MODIS and Landsat optical images for flood mapping during flooding events is almost impossible. In April 2017, there were no cloud-free Landsat-8 images available for the study area for flood mapping. Similarly, during the period of 22 March–6 April 2017, MODIS images were not usable for mapping [22].

In Bangladesh, floods often occur when the sky is covered by clouds, thus making the utilisation of optical satellite images is infeasible in providing inundation mapping during the disaster. Therefore, spaceborne synthetic aperture radar (SAR) systems are the most preferred option for monitoring the flood condition. The introduction of SAR sensors has shown great potential for flood mapping due to their independence from solar illumination and very low dependency on atmospheric conditions [28]. Some studies have demonstrated that SAR images are useful for determining flood extent during a disaster [29–31]. An analysis of the benefits of using SAR images for flood mapping in Bangladesh was conducted using RADARSAT images from 1998–2004 and indicated strong applicability to flood response [32,33]. Using six scenes of RADARSAT images from 1988 (July–September) and available GIS database, Dewan, et al. [34] conducted a study on flood hazard mapping in Greater Dhaka. In support of ground data, Hoque, Nakayama, Matsuyama and Matsumoto [32] analyzed RADARSAT images to produce inundation maps from 2000–2004 and proposed unique flood hazard maps of the north-eastern region of Bangladesh. Using the same RADARSAT images of 1998 and 2004, another flood mapping study was conducted for the Kurigram district of Bangladesh [35]. Considering the published literature, none of the studies produced flood maps (2005 onward) with national coverage for damage assessment using Sentinel-1 and ALOS PALSAR images. However, the availability of free SAR data through the European Space Agency's (ESA) Sentinel-1 C-band SAR mission created a major opportunity for flood extent monitoring in developing countries such as Bangladesh.

For an estimation of flood damaged areas, pre-flood national level land cover is also essential. In the region, knowledge-based and object-based image analysis methods were used for the national land cover change assessment of Bangladesh [36], Bhutan [37], and Nepal [38,39]. These studies were mostly conducted at decadal intervals for addressing national environmental issues but are not useful for analyzing flood damage assessment. Considering the amount of image downloading and processing required, the desktop based systems are not suitable for providing rapid processing services critical for a flood response. The cloud-based image processing platform from Google Earth Engine (GEE) is enabling the rapid processing of such big datasets covering a large area [40,41]. The GEE has publicly made available large amounts of remote sensing satellite data collection and provides image analysis functionality at large spatial scales [42].

In 2017, unpredicted early heavy rain caused flooding in several parts of Bangladesh and damaged pre-harvested crops in April [22]. The flood started in April and continued until the last week of August, causing substantial damage to housing, property, and infrastructure. As part of the rapid response, the disaster management agencies were in urgent need of information about the inundated areas to prioritize their relief and rescue activities [43,44]. This study aims to develop an operational methodology to support the response agencies by providing timely information on such inundated areas so as to prioritize emergency response activities. This study also aims to develop a methodology for potential damage assessment by analyzing pre-flood land cover maps automatically on the GEE platform. The inundation and flood damage data have been made publicly available for decision making processes related to flood management.

2. Materials and Methods

2.1. Study Area

Figure 1 shows the study area for rapid flood mapping and potential damage assessment in Bangladesh. Bangladesh encompasses the world's largest delta system located in the southern part of the foothills of the Himalayan mountain region and in the northern part of the Bay of Bengal, with a boundary between 20°N to 26°N and 88°N to 92°E with an area of 147,570 km^2 [45]. The southern part of the delta is occupied by the world's largest mangrove forest (the Sundarbans). The south-eastern part includes Bangladesh's main hilly region, while low hills characterize the north-east of Bangladesh and the remaining area is mainly plain land.

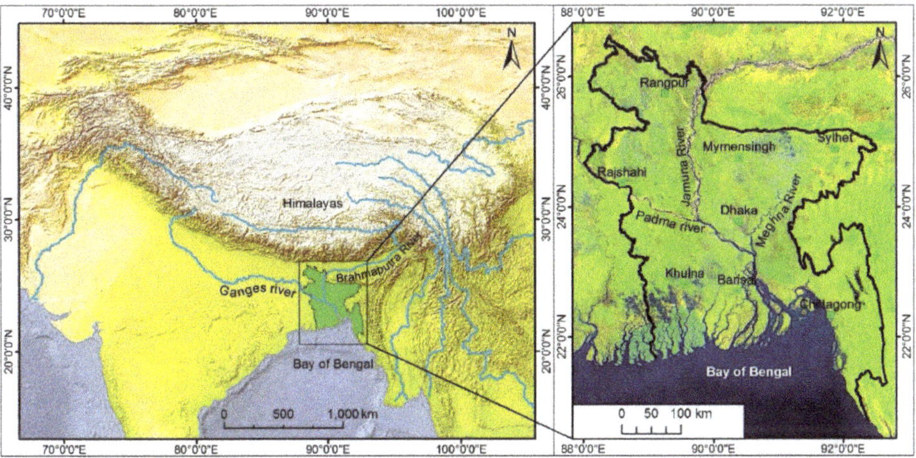

Figure 1. Study area of the entire country of Bangladesh showing three large river systems: The Ganges (Padma), Brahmaputra (Jamuna), and the Meghna forming the largest riverine delta in the world.

Approximately 50% of the country is within 7 m of mean sea level and most of the country is on a delta plain under the influence of the Padma, Jamuna, and Meghna rivers. Most of the plain lands is used for crop production and approximately 87% of rural households rely on agriculture for at least part of their income. Rice production in Bangladesh is a crucial part of the national economy. Recently,

Bangladesh has become one of the top 5 rice-producing countries in the world [46]. If no flood occurs in the coming year, Bangladesh will remain in the top 5 rice-producing countries. Although Bangladesh is a land of 6 seasons, the 3 most distinct seasons are the pre-monsoon hot season from March to May, the rainy monsoon season lasting from June through October, and a cool, dry winter season from November through February. The country has an average of 136 wet days per year, and approximately 80% of yearly rainfall occurs from June to September [47]. In a normal year, about one-third area of the country gets inundated by flood water [48]. The planet's highest rainfall occurs in Cherrapunji, which is located just a few kilometres away from the north-eastern border of Bangladesh. The high annual rainfall combined with the mountain terrain causes rivers from the north eastern border to flow with a very high current due to high gradient topography. When this water reaches Bangladesh territory, it spreads over a large area and regularly causes different levels of flood incidents in Bangladesh. Sometimes, locally concentrated prolonged heavy rainfall worsens the flood situations.

2.2. Materials

To enable comprehensive inundation mapping across Bangladesh, approximately 11 frames of Sentinel-1 C-band interferometric wide swath (IW) frames with a 250 km swath width were required (Figure 2).

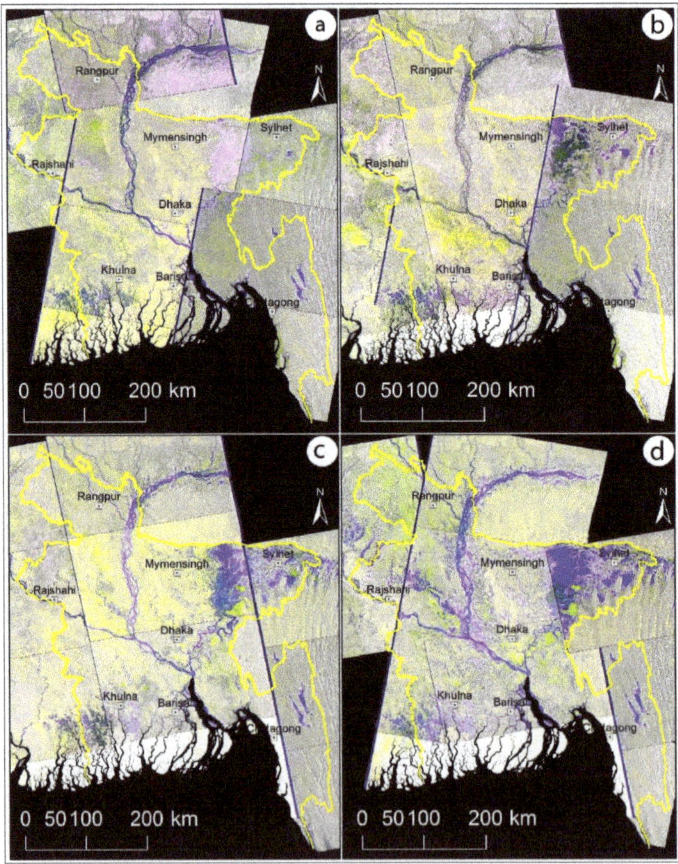

Figure 2. Sentinel-1 composites VH (red), VH (green), and VH/VV backscatter ratio (blue) images used for inundation mapping of (**a**) March, (**b**) April, (**c**) June, and (**d**) August 2017.

For the pre-flood (March) and post-flood (April, June, and August) inundation mapping, a total of 44 dual-polarization Sentinel-1 level-1 Ground Range Detected (GRD) products were used. The GRD products consist of focused SAR data that was detected, multi-looked, and projected to the ground range using the WGS-84 Earth ellipsoid model. The ellipsoid projection of the GRD products was corrected using the terrain height specified in the product's general annotation. Both the like-polarized (vertical transmit and vertical receive (VV)) and cross-polarized (vertical transmit and horizontal receive (VH)) channels were used in this study, and data were retrieved as Level-1 GRD products. Sentinel-1 SAR images were useful as the data were freely available within 3 h of acquisition for near real-time (NRT) emergency response and within 24 h for systematically archived data. Table S1 shows the Sentinel-1 images that were used in this study, which were freely downloaded from the Copernicus open access hub data portal of the European Space Agency (ESA).

In addition to SAR and as part of the flood damage assessment, pre-flood cloud-free Landsat-8 image collections between 1 January and 30 June 2017 were used for land use/land cover mapping in the Google Earth Engine. To support land cover classification, additional ancillary data used in this study included a 30-m resolution shuttle radar topography mission (SRTM) digital elevation models (DEM) [49], which were retrieved from the United States Geological Survey (USGS) archived data portal, as well as road network information from OpenStreetMap [50] and administrative boundary data from the database of Global Administrative Areas (GADM) [51].

2.3. Methods

The methods used for this study are presented in Figure 3. Specifically, Sentinel-1 image classification for flood mapping, with initial pre-processing carried out to mitigate the SAR-typical speckle noise signatures from the images. During the pre-processing step, we rectified the radiometric and geometric distortions due to the characteristics of the imaging system and imaging conditions and performed radiometric corrections to improve visualization and interpretation for flood mapping.

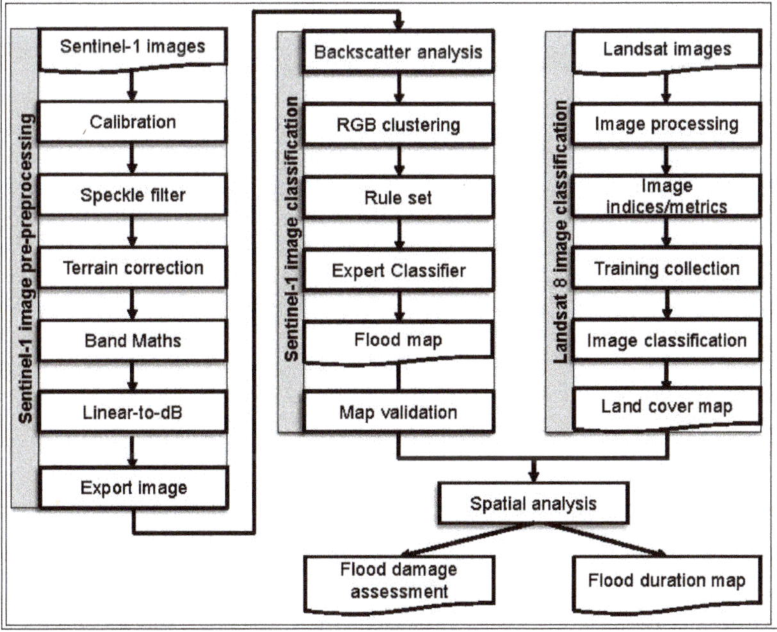

Figure 3. Overall methodological framework for the flood inundation mapping and damage assessment using multi-temporal Sentinel-1 and Landsat-8 satellite images.

The pre-processing steps, including data import, radiometric calibration, speckle filtering, radiometric terrain correction [52,53], linear-to-backscattering coefficient decibel scaling (dB) transformation, and data export, were implemented using ESA's Sentinel Application Platform (SNAP). The open-access SNAP toolbox is capable of reading, pre-processing, and visualizing Sentinel-1 SAR images.

During the Sentinel-1 pre-processing, Level-1 images were first imported into the SNAP Desktop tool. Secondly, Sentinel-1 images were radiometrically corrected by applying annotated image calibration constants to arrive at physically meaningful radar backscatter pixel values. Thirdly, speckle filters were applied to reduce the granular noise characteristic to SAR data. Fourthly, multi-look processing was carried out to reduce the speckle further and improve image interpretability. Fifthly, geometric distortions present in the SAR images were corrected by transforming the coordinates to a standard reference frame. The ratio band of VH/VV was generated dividing the VH by the VV band. Finally, a radiometric conversion from a linear scale to a dB scale was conducted using the following expressions:

$$\sigma^0_{dB} = 10 \cdot \log_{10} \sigma^0 \tag{1}$$

where, σ^0 (dB)—backscattering image in dB, σ^0—Sigma nought image.

The preprocessed images were exported for classification. For automation, all processing steps were assembled and connected through Graph Builder, which is available in SNAP and was run in batch processing mode. The pre-processed stack of Sentinel-1 SAR images was imported into ERDAS Imagine for knowledge-based image analysis. According to Janssen and Middelkoop [54], knowledge-based classifications contain the following 5 characteristics: Aim, ancillary data, domain knowledge, knowledge presentation, and inference. The knowledge engineer provides the interface for an expert with first-hand knowledge of the data to develop an algorithm into a hierarchical decision tree using logic or rules [55]. While performing expert classification, the Sentinel-1 images were clustered to create a thematic raster layer by RGB clustering functions in ERDAS Imagine. In geospatial applications, the unsupervised classification used to be known as the iso-clustering or migrating means technique that helps to group the same type of features into homogeneous and diverse features into heterogeneous clusters [56–58]. The RGB clustering is the most common technique for data compression and iso-clustering works better on an optimal number of classes usually unknown [59,60]. The RGB clustering functions are a simple classification algorithm that quickly compresses a three-band image into a single-band pseudo-colour image without necessarily classifying any particular features and without a signature file and decision rule. The RGB clustering provides greater control over the parameters used to partition the pixels into similar classes [61,62]. During Sentinel 1 image clustering, the VH band is designated as a red band, the VV band as a green band, and the VH/VV band as the blue band. Secondly, the RGB clustered image was processed to generate a clamped image for converting thematic class values into uniquely numbered "polygons", representing contiguous groups of the original class values. Thirdly, within the clamped image, mean radar backscatter (dB) values of the VH and VV bands were generated and used as input for an expert-guided classification of flood and non-flood. VH and VV-band backscatter (dB) statistics for flood and non-flood samples are presented in Figure S1. For this analysis, only two major classes, a "waterbodies" class and an "others" class, were considered. Box plots were used to show the statistical distribution of the data. A distinct separable backscatter value for water and other classes was used for image classification [44,63].

To understand the quality of the produced thematic maps, a validation process was required for the classification results [39,64]. During the flood disaster, it is challenging to conduct fieldwork for flood map validation [65]. Cloud-free Landsat-8 images (Landsat Surface Reflectance Level-2) available during the flood period (from 22 August 2017) was used for cross comparison. The Landsat-8 image acquired on 22 August 2017 was classified for flood mapping in eCognition developer using an object-based image analysis (OBIA) method called the geographic object-based image analysis (GEOBIA). The detailed methodology used to prepare the flood maps is described in [43]. Briefly, eCognition Developer software was used to divide the image into segments. The GEOBIA method segments remote sensing imagery into meaningful image objects based on the spatial, spectral, and temporal characteristics of image pixels. During the Landsat based flood mapping, few procedures were applied in terms of selected attributes using indices such as the normalized difference water index (NDWI), normalized difference vegetation index (NDVI), a and the land and water mask (LWM) derived from spectral values of the image, together with land band information. Finally, developed rule sets exploring the image object mean value were used to generate Landsat-based flood maps. The Landsat-based flood maps are shown in Figure 4c and were used to see omission and commission errors with Sentinel-1-based flood maps. A comparison was made between Landsat and Sentinel-1-derived inundation areas. Figure 4 shows the comparison of optical data and SAR-based flood inundation maps.

In addition, 4500 reference points were collected from the Landsat-8 image of 22 August to validate the flood map of 29 August that was the closest available flood map from 22 August. The expert-guided classification scheme achieved an overall classification accuracy of 96.44%.

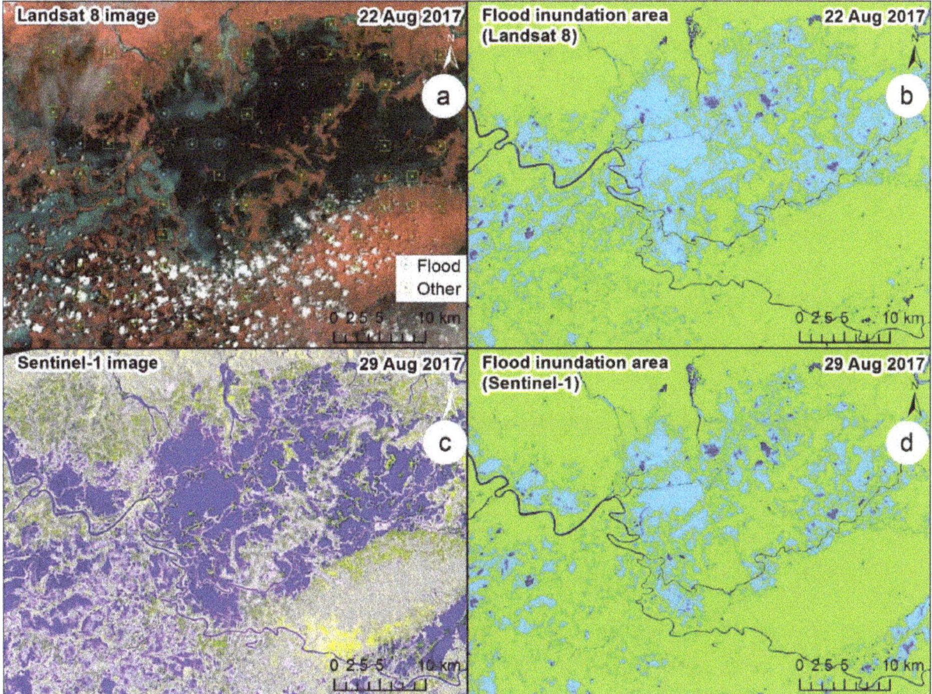

Figure 4. Comparison of optical data and synthetic aperture radar—(SAR) based flood inundation: (**a**) False color composite Landsat-8 image from 22 August 2017; (**b**) classification result based on Landsat-8 (dark blue: Perennial water; light blue: Flood inundation areas; green: Other areas); (**c**) false color composite Sentinel-1 image from 29 August 2017, showing water bodies in blue; (**d**) classification result based on Sentinel-1 data (same color assignment).

As the initially detected flood extent includes both permanent water bodies and flooded water, pre-flood water extent must be removed from the classified map of flooded area [66]. Pre-flood waterbodies classified from Landsat images acquired before 27 March 2017 was considered as perennial waterbodies because no news reports about floods were received before that date [22]. Overlaying the pre-flood waterbodies with the April, June, and August flood maps, the flood inundation area was separated from perennial water bodies. Further analysis of the derived flood information was carried out in the ArcGIS environment to perform damage assessment as well as associated statistical analysis at a 30-m resolution (flood area estimation, affected area estimation).

As part of the flood damage assessment, pre-flood land cover maps for the year 2017 were prepared in GEE using Landsat images acquired between January and April of that year. For land cover mapping, optical satellite images were used, e.g., Landsat are most common, and data were freely accessible and are often used to explore unique spectral characteristics of different land cover using image indices. The Google Earth Engine environment is a powerful computational fast analysis processing platform that can handle huge volumes of remote sensing imagery [67,68]. GEE provides online access to archived pre-processed Landsat imagery [69]. In the GEE, a sequence of processing steps was followed (Figure 2) for image analysis and land use/land cover mapping. All the 2017 pre-flood Landsat data for the entirety of Bangladesh were processed to derive the pre-flood water extent (https://code.earthengine.google.com/). First, as part of the atmospheric image correction, a cloud-free Landsat-8 image composites were prepared using partially cloudy images available for the period of January and April 2017. Using the derived cloud free image composites, the normalized difference vegetation index (NDVI), normalized difference water index (NDWI), normalized difference moisture index (NDMI), bare soil index (BSI), normalized pigment chlorophyll ratio index (NPCRI), and land and water mask (LWM) index were created and used for land cover mapping.

A supervised classification scheme was used to generate 2017 land cover maps from the set of Landsat-8-based index layers. To facilitate the 2017 land cover classification, training data were acquired using Google Earth High-resolution images as a basis. A total of 5484 training sets were collected for areas under tree coverage (Madhupur forest, hill forest, and mangrove forest, as well as rural settlement and homestead orchard), 8 were acquired for grassland, 346 for the barren area, 1039 for cropland, 3142 for waterbodies, and 86 for built-up areas. All the land cover legends were developed based on the land cover classification system (LCCS), which was developed by FAO to provide a consistent framework for the classification and mapping of land cover [70,71]. All training data were imported into the Google Earth Engine environment in the form of a fusion table. Finally, a classification and regression tree (CART) land cover classifier was applied using the imported training sets and the Landsat-related raster layers. The CART is a decision tree (DT) based machine-learning method for constructing prediction models from a set of training data using the concept of information entropy that shows the strongly improved performance of classification [72,73]. In addition to the Landsat bands "B1", "B2", "B3", "B4", "B5", "B6", "B7", "B8", and "B9", the indices "NDVI", "NDWI", "NDMI", "BSI", "NPCRI", "LWM", and the SRTM DEM were also used in the classification. As with any digitally classified land cover product, there can be reasons for misclassification related to the environmental conditions at the time of image acquisition (clouds, fog, etc.), variations in local forest types or limitations in computational algorithms. Finally, the derived land cover maps were validated using another set of samples collected from field data and independent training data from high-resolution images of Google Earth. Publicly open earth observation data and online map tools like Google Map, Google Earth, Collect Earth Online and OpenStreetMap allow the accuracy assessment of a national level land cover based on very high-resolution satellite images [38,39,74–76]. The accuracy of the 2017 Landsat-derived land cover map was assessed using (10 km × 10 km) 1400 reference points from Google Earth and 65 points from the ground. These were compared with the land cover map to calculate the error matrix, and an overall accuracy of 87.51% was found.

3. Results

Developed national level pre and post-flood inundation maps for the entire country, during March, April, June, and August 2017, based on the Sentinel-1 images, are presented in Figure 5. The results show the presence of perennial waterbodies in March 2017 covering an area of 5.03% in Bangladesh. In April 2017, a total flood-inundated area was 2.01%, with most inundation was occurring in cropland (1.51%), followed by rural settlement and homestead orchard areas (0.21%), and other areas (0.29%).

Similarly, more areas were inundated during the catastrophic June and August 2017 months, with inundation covering 4.53% and 7.01%, respectively.

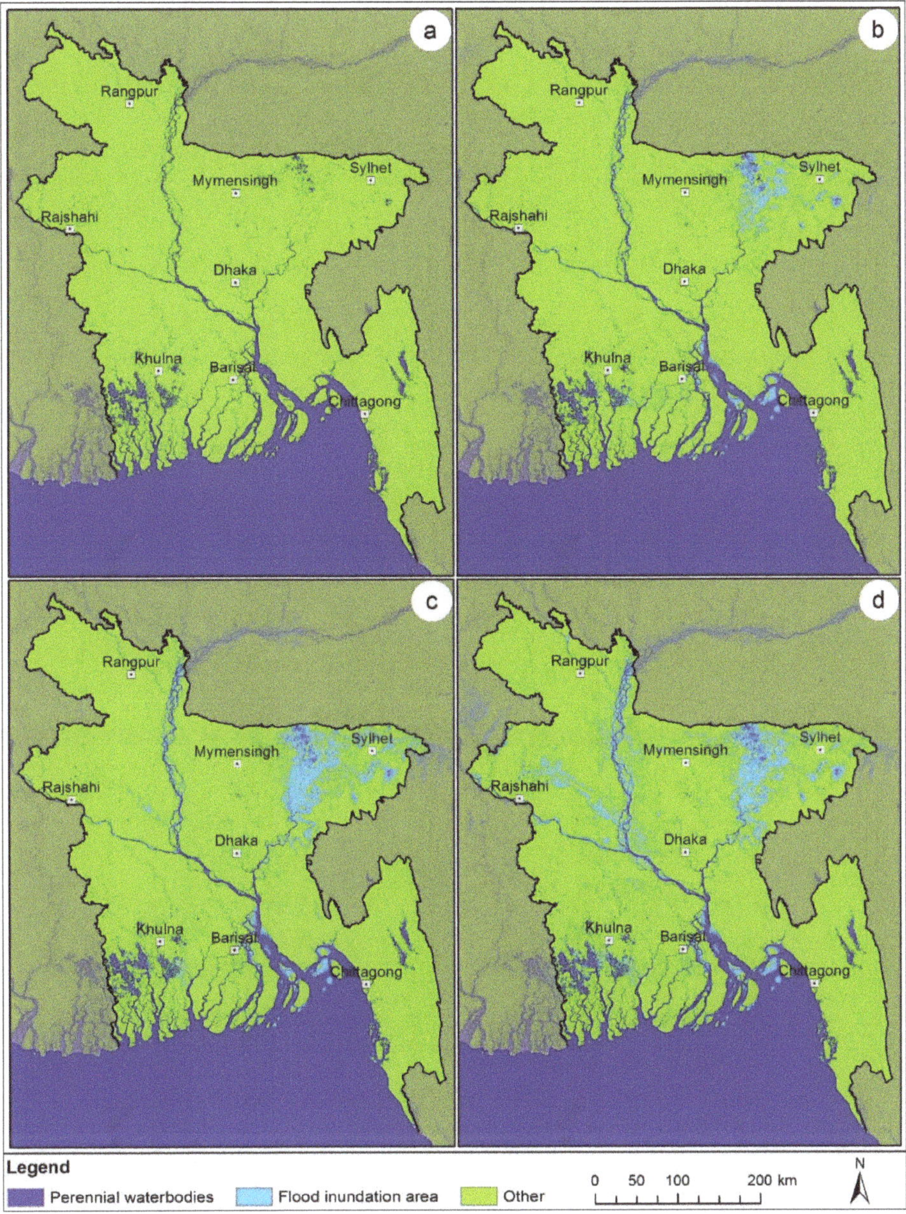

Figure 5. Comprehensive flood inundation map of Bangladesh for the months of (**a**) March, (**b**) April, (**c**) June, and (**d**) August 2017.

The 2017 flood of Bangladesh caused significant inundation of cropland, rural settlement, and homestead orchard, and other land use areas. Considering all crop-related land use and land cover types (Figure 6), the percentage of inundated cropland was found to be 1.51%, 3.46%, and 5.30% in April, June, and August respectively. Significant inundation also occurred to residential property, public infrastructure, and fish farming ponds. A total of 0.21% of the rural settlement and homestead orchard areas were inundated in April, increasing to 0.47% in June, and 0.65% in August (Figure 7).

The time series flood data in Figure 8 shows that for the April to August 2017 time frame, some of the areas experienced continuous inundation, while some areas were progressively inundated, and some recovered from the flood waters as time progressed. Within the April and June 2017 time frame, an inundated area of 257,729 ha was common for both months, while 410,853 ha were newly flooded, and 38,776 ha recovered from the flood inundation. From June to August 2017, 532,173 ha were common for both months, while 502,927 ha were newly inundated, and 136,406 ha recovered from the floods (Figure 8).

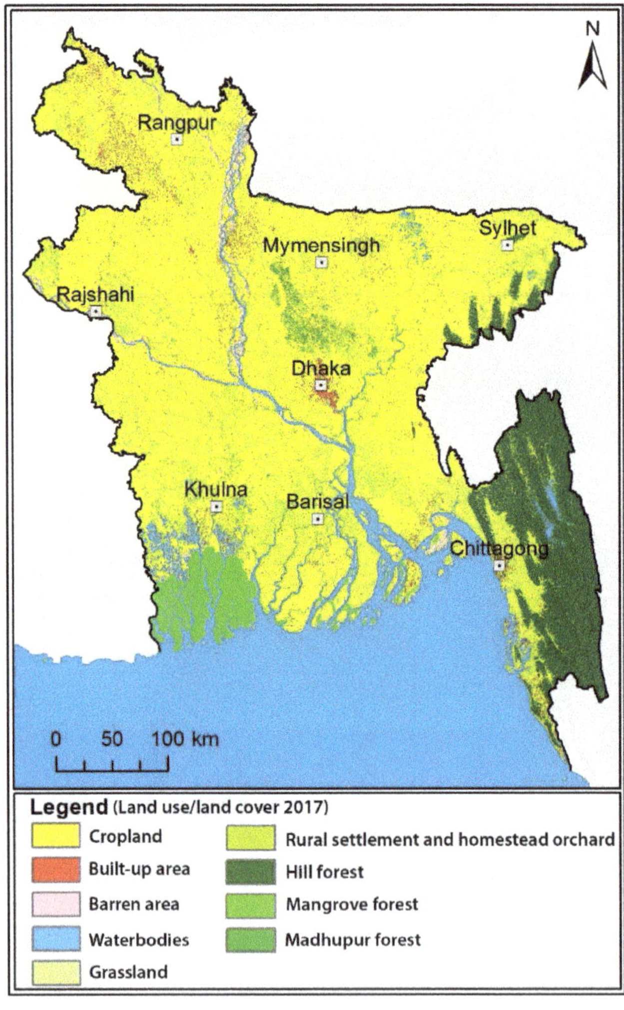

Figure 6. 2017 land cover map of Bangladesh developed using Landsat 8 and Google Earth Engine cloud computing.

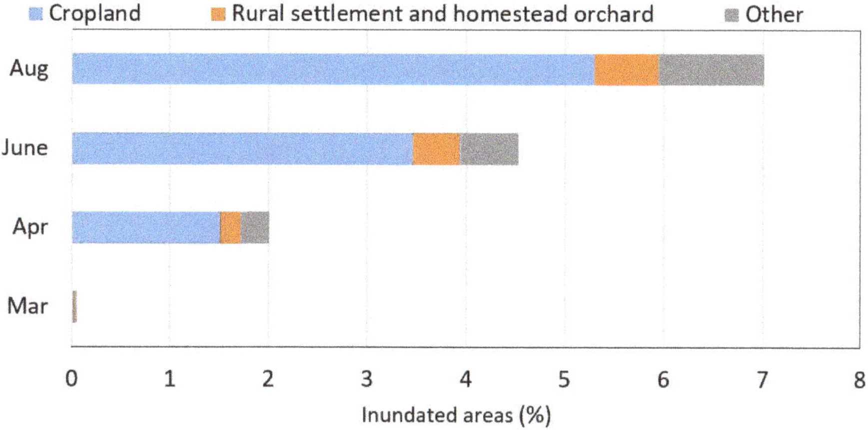

Figure 7. Flood affected major land cover categories. Flood extent was derived from Sentinel-1 SAR data while land cover classes were extracted from cloud-free Landsat-8 imagery.

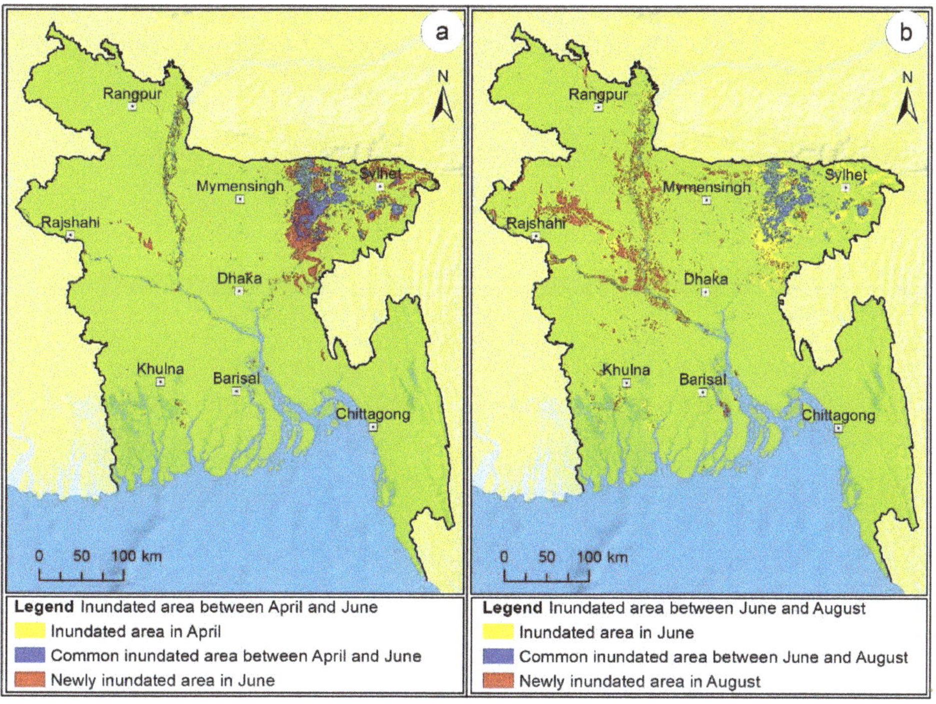

Figure 8. Flood recession and rise areas of Bangladesh between (**a**) April and June, (**b**) June and August.

A land cover map for the same year was also derived from Landsat-8 data for potential flood damage assessment. The derived map was produced from a cloud-free Landsat-8 composite between 1 January and 1 April 2017 and is shown in Figure 6. The map consists of nine classes, namely, tree cover (Madhupur forest, hill forest, and mangrove forest, as well as rural settlement and homestead orchard), grassland, cropland, barren area, built-up area, and waterbodies.

Through the possibility of an automatic SAR-based processing chain, flood inundation mapping based on Sentinel-1 images have the potential to rapidly provide flood information for flood management. During flood mapping, a significant difference in backscatter values for waterbodies and non-water areas enables a separation between inundated areas and other land areas. Derived optimal backscatter ranges for flood inundated areas in Sentinel-1 images shows a clear distinction from other classes, as presented in Figure S1. The backscatter response in VH polarizations for inundated areas were between −24.25 dB and −17.4 dB for this area of interest. In VV polarizations, water bodies showed a range between −22.4 dB and −12.9 dB. The backscatter response in VH polarizations for non-water areas was determined between −16 dB and −9.6 dB, and in VV polarizations it ranged between −10.8 dB and −1 dB. Identified optimal ranges of backscatter values can be applied for the automation of flood mapping using Sentinel-1 images to produce flood inundation maps for areas of similar topography.

For an accuracy assessment, the Sentinel-1 classification result for 29 August 2017, was evaluated with the waterbodies map derived from Landsat-8 (22 August 2017) using 4500 reference points collected from the Landsat-8 classification map. The overall accuracy of the 2017 August flood inundation map was found to be 96.44%, with a kappa value of 0.81, standard error kappa of 0.02, and a 95% confidence interval between 0.770 to 0.850 (Table 1). The evaluation of the flood map from SAR data compared to the optical image-based inundation map (Table S4) shows that for a particular cloud free window, the Landsat- (22 August 2017) based map produced an inundated area of 70% while the SAR-based map showed 59% of the areas inundated. Within the 70% Landsat-based inundated maps, a 53% inundated area was common, 17% of the area was an omission, and 6% of the area differed from the Sentinel-1-based study. A visual comparison of the Landsat-based (22 August 2017) and Sentinel-1-based (29 August 2017) flood extent is also shown in Figure 4.

Table 1. Error matrix for the land cover map of 2017.

Class Name	Flood	Other	Total	Accuracy
Flood	1335	123	1458	91.56
Other	37	3005	3042	98.78
Total	1372	3128	4500	n.a.
Producer's accuracy (%)	97.30	96.07	n.a.	n.a.

The accuracy of the 2017 Landsat-derived land cover map was found to be 87.51%, with a kappa value of 0.81, a standard error kappa of 0.02, a 95% confidence interval between 0.770 and 0.850, and a 0.906 maximum possible unweighted kappa, given the observed marginal frequencies (Table 2).

Table 2. Error matrix for the land cover map of 2017.

Land Cover	Hill Forest	Madhupur Forest	Mangrove Forest	Rural Settlement and Homestead Orchard	Grassland	Cropland	Barren Area	Built-Up Area	Waterbodies	Total	Users Accuracy (%)
Hill forest	155	1		1		8			1	166	93.37
Madhupur forest		7		1		1				9	77.78
Mangrove forest			43	1						44	97.73
Rural settlement and homestead orchard	1	1	1	214		60			2	279	76.70
Grassland			1		3		1		1	6	50.00
Cropland				59	1	722	3		12	797	90.59
Barren area						1	19	1	1	22	86.36
Built-up area				3		2		6		11	54.55
Waterbodies						18			113	131	86.26
Total	156	9	45	279	4	812	23	7	130	1465	n.a.
Producer's accuracy (%)	99.36	77.78	95.56	76.70	75.00	88.92	82.61	85.71	86.92	n.a.	n.a.

4. Discussion

It is well known that Bangladesh has a long history of natural disasters. Between 1980 and 2008, it experienced 219 natural disasters that caused over USD $16 billion in total damage [77]. Due to the flat topography and climatic features, more than 80% of the population is potentially exposed to floods. Following the devastating disaster issues, Bangladesh has made significant efforts to reduce its disaster vulnerability primarily through post-disaster management. The Government of Bangladesh (GOB) recently constructed a good number of flood shelters, built flood protection embankments, sluice gates and regulators on different rivers, and have been dredging the drainage channels and canals [78]. Compared to previous efforts, Bangladesh is now much safer from disasters due to these post-disaster actions. Another survey of published literature concluded that considerably less research has been conducted for operational flood mapping using earth observation data for emergency response. Space based earth observation (EO) data can be used to deliver information on the extent of hazards during response operations so as to mitigate damage [79].

In the past, vital research was conducted for the 1988 flood mapping to support relief operations [7]. After that, few flood mapping studies have been conducted by academic researchers related to flood issues in Bangladesh without focusing on emergency response [6,14,21,33,35,48]. Relatively, cloud-free satellite images showed that during the last three weeks of September, areas of inundation were 31% to 42% of Bangladesh. The actual flooded area was more from satellite image estimates which differed from officially reported areas [80]. Although real-time flood monitoring plays a vital role in relief operations [81,82], flood maps also play an important role in decision-making, planning, and implementing flood management options [83]. Most of the studies have published their flood mapping results ten years after a flood event [32]. Sometimes the flood duration is quite long, although many researchers have mapped flooding areas for a single month. Furthermore, no dissemination systems were implemented to support sharing the inundated area maps during the crisis or after publishing the research article [84]. A number of monitoring systems have the potential for flood management in Bangladesh, however limited access to discharge data from the upstream, and lack of timely acquisition of geospatial data has an impact on the operational suitability of these systems. In many cases, model-based inundation mapping does not provide good results on plain land area [18]. Cloudy weather also prevents optical systems to provide coherent image coverage to use for flood inundation mapping. Due to the high level of cloud contamination during the monsoon time, cloud-free Landsat images identification was difficult for flood mapping [85]. For Bangladesh, one of the best opportunities for operational flood mapping comes from Sentinel-1 imaging as it is publically available [63]. In Bangladesh, the agriculture sector is greatly impacted by floods, despite conventional flood management systems paying very limited attention to this sector [63].

Under the circumstances that Bangladesh is under, the present study provides rapid flood inundation maps for March, April, June, and August 2017 using publicly available Sentinel-1 data using a replicable methodology. The derived flood maps provide spatial and temporal dynamics of the flooding area across the county. The utilization of the GEE for image processing facility in this study helped to promptly develop an operational land cover map for damage assessment. The advanced image processing tools of GEE enables the generation of accurate land coverage for large areas without downloading bulk data and prolonged desktop processing [86]. The method can be used for monitoring land cover regularly as the analysis can easily be re-run while new Landsat data is injected in GEE. The derived maps provide crucial information for local disaster management agencies, which helps to prioritize relief and rescue operations. At the same time, automatically generated land cover maps derived from pre-flood Landsat-8 data support the assessment of economic loss and help in the prioritization of financial compensation due to crop damage. The use of SAR images provides significant advantages, as their cloud-free and all-weather capabilities enable the production of regularly sampled flood extent information in a near real-time delivery manner [7,87]. Before the launch of Sentinel-1B, the revisit time of Sentinel-1A was 12 days, which was challenging for a rapid disaster response. After the launch of Sentinel-1B and the completion of the two satellite

Sentinel-1 constellation, revisit times improved to six days over the area of interest. In the future, a constellation of three Sentinel satellites would be much more useful for responding to flooding events in near real-time. As part of this study, the team provided georeferenced GIS layers in addition to jpg-formatted maps to encourage the broad use of data and to ensure that the data are fully compliant with the GIS environments used by the response agencies [88].

For this study, the use of Sentinel-1 dual-polarization radar images showed a high potential for flood mapping due to their free-of-charge nature. Adopting the image pre-processing techniques and knowledge-based classification methods used in this study helped in developing scene-specific standards and knowledge and achieved better classification accuracy on flood maps [89,90]. For this exercise, the developed flood mapping system was run on a desktop-type computer. Hence, the availability of sufficient computing hardware and internet bandwidth were the main limitations discovered in this study. For future implementations, we will consider the utilization of cloud computing resources, such as those offered by the Google Earth Engine, the ESA Thematic Exploitation Platforms, or Amazon Web Services-based environments offered by NASA's Alaska Satellite Facility (ASF) DAAC with all Sentinel images. At the same time, utilizing recently released public-domain Landsat-8 datasets via the Google Earth Engine or similar platforms may enable the establishment of a framework for rapid land cover monitoring on a national level [40].

The present study produced a flood inundation map with optimum accuracy for the entirety of Bangladesh. However, there were some uncertainties of flood maps due to the floating vegetated areas [91]. On the Sentinel-1 images, sometimes cultivated land for rice plantation was appeared as inundated areas. To overcome those uncertainties, local knowledge is crucial. Alternatively, unmanned aerial vehicles (UAVs) are emerging tools for the monitoring of real time floods in disaster management [92]. Due to flight endurance and payload capacity, UAVs would not be functional if large areas were inundated [69]. In general, SNAP tools take a longer time for preprocessing Sentinel-1 images. There is also a lag time between the image availability in ESA Sentinel hub and in GEE. If lag time is reduced in the future, the method could be implemented in GEE environment for a more rapid production of flood extent.

Bangladesh is a flood-prone country and thus under constant threat of flooding. Every year, floods destroy lives, livestock, and infrastructure, bringing an enormous financial toll. During disasters, obtaining reliable information is crucial. As part of the operational methods, flood inundation was validated using Landsat-8 images. Unfortunately, due to the dominance of cloudy conditions, no (cloud-free) Landsat images were available during the peak flood period. Apart from wall-to-wall comparisons with optical or field-based studies, a determination of the extent of flooding in Bangladesh could serve disaster management purposes such as relief operations.

5. Conclusions

Based on the results of the study, we can conclude that earth observation and geospatial technologies provide prompt information for effective decisions for comprehensive flood disaster management for Bangladesh. Due to the predominance of severe weather conditions during flooding time, freely available and regularly sampled Sentinel-1 SAR earth observation data has great potential in producing flood information with high accuracy and high spatial resolution in a six day interval. The method was based on publicly available free-of-charge data, particularly useful for less developed countries. Cloud-based computation environments, such as the GEE platform, proved to be particularly valuable for operational users in planning a flood-related emergency response and for understanding flood damage by land cover mapping. Natural flood disasters are common and cannot be stopped. However, efficient tools for flood inundation mapping and flood damage assessment can be useful for emergency response and disaster management.

Supplementary Materials: The following are available online at http://www.mdpi.com/2072-4292/11/13/1581/s1, Figure S1. Box plots of Sentinel-1 VH and VV backscatter for flood and non-flood samples, Table S1. List of Sentinel-1 images used for the flood inundation mapping, Table S2. Landsat 8 operational land imager (OLI) spectral bandwidth spatial resolution, Table S3. Land cover classification system (LCCS) along with adopted code, Table S4. Omission and commission (%) errors between Landsat (22 August 2017) and Sentinel images (29 August 2017).

Author Contributions: K.U. and M.A.M. conceptualized overall research design; K.U. performed image analysis, prepared the flood map, land cover maps, flood damage assessment, and change assessment, and drafted the first version of the manuscript; M.A.M. and F.J.M. reviewed and provided feedback; finally all the authors read, edited, critiqued the manuscript, and approved the final version.

Acknowledgments: We express our gratitude to David Molden, Director General and Eklabya Sharma, Deputy Director General of ICIMOD for the overall guidance. This study was partially supported by core funds of ICIMOD contributed by the governments of Afghanistan, Australia, Austria, Bangladesh, Bhutan, China, India, Myanmar, Nepal, Norway, Pakistan, Switzerland, and the United Kingdom. This paper has been prepared under the SERVIR-Himalaya initiative funded by the NASA and USAID. Our special gratitude goes to Birendra Bajracharya, Chief of Party, SERVIR HKH for their encouragement and very active support in bringing out this work. Finally, we would like to thank those who reviewed the articles incomprehensibly and provided valuable suggestions to improve the manuscript.

References

1. Ozaki, M. *Disaster Risk Financing in Bangladesh*; ADB South Asia Working Paper Series No. 46; Asian Development Bank: Metro Manila, Philippines, 2016.

2. Chowdhury, R.B.; Moore, G.A. Floating agriculture: a potential cleaner production technique for climate change adaptation and sustainable community development in Bangladesh. *J. Clean. Prod.* **2017**, *150*, 371–389. [CrossRef]

3. Dasgupta, S.; Huq, M.; Khan, Z.H.; Sohel Masud, M.; Ahmed, M.M.Z.; Mukherjee, N.; Pandey, K. Climate proofing infrastructure in Bangladesh: The incremental cost of limiting future flood damage. *J. Environ. Dev.* **2011**, *20*, 167–190. [CrossRef]

4. Biswas, A.K. Management of Ganges-Brahmaputra-Meghna System: Way Forward. In *Management of Transboundary Rivers and Lakes*; Varis, O., Biswas, A.K., Tortajada, C., Eds.; Springer: Berlin/Heidelberg, Germany, 2008; pp. 143–164. [CrossRef]

5. Ahmad, Q.; Ahmed, A.U. Regional cooperation in flood management in the Ganges-Brahmaputra-Meghna region: Bangladesh perspective. In *Flood Problem and Management in South Asia*; Springer: Berlin/Heidelberg, Germany, 2003; pp. 181–198.

6. Banerjee, L. Effects of flood on agricultural productivity in Bangladesh. *Oxf. Dev. Stud.* **2010**, *38*, 339–356. [CrossRef]

7. Rasid, H.; Pramanik, M. Areal extent of the 1988 flood in Bangladesh: How much did the satellite imagery show? *Nat. Hazards* **1993**, *8*, 189–200. [CrossRef]

8. Mirza, M.M.Q. Climate change, flooding in South Asia and implications. *Reg. Environ. Chang.* **2011**, *11*, 95–107. [CrossRef]

9. Bangladesh Bureau of Statistics. *Banglaxesh Disaster Related Statistics 2015*; BBS: Dhaka, Bangladesh, 2016.

10. Mirza, M.M.Q. Global warming and changes in the probability of occurrence of floods in Bangladesh and implications. *Glob. Environ. Chang.* **2002**, *12*, 127–138. [CrossRef]

11. Kundzewicz, Z.W.; Kanae, S.; Seneviratne, S.I.; Handmer, J.; Nicholls, N.; Peduzzi, P.; Mechler, R.; Bouwer, L.M.; Arnell, N.; Mach, K.; et al. Flood risk and climate change: global and regional perspectives. *Hydrol. Sci. J.* **2014**, *59*, 1–28. [CrossRef]

12. Dash, J.; Paul, R. *Worst Monsoon Floods in Years Kill More Than 1200 across South Asia*; Reuters: London, UK, 2017.

13. Lin, L.; Di, L.; Tang, J.; Yu, E.; Zhang, C.; Rahman, M.; Shrestha, R.; Kang, L. Improvement and Validation of NASA/MODIS NRT Global Flood Mapping. *Remote Sens.* **2019**, *11*, 205. [CrossRef]

14. Chowdhury, E.H.; Hassan, Q.K. Use of remote sensing data in comprehending an extremely unusual flooding event over southwest Bangladesh. *Nat. Hazards* **2017**, *88*, 1805–1823. [CrossRef]

15. Amarnath, G.; Rajah, A. An evaluation of flood inundation mapping from MODIS and ALOS satellites for Pakistan. *Geomat. Nat. Hazards Risk* **2016**, *7*, 1526–1537. [CrossRef]

16. Shen, X.; Wang, D.; Mao, K.; Anagnostou, E.; Hong, Y. Inundation Extent Mapping by Synthetic Aperture Radar: A Review. *Remote Sens.* **2019**, *11*, 879. [CrossRef]

17. Bates, P.D. Remote sensing and flood inundation modelling. *Hydrol. Process.* **2004**, *18*, 2593–2597. [CrossRef]

18. Jung, Y.; Kim, D.; Kim, D.; Kim, M.; Lee, S. Simplified flood inundation mapping based on flood elevation-discharge rating curves using satellite images in gauged watersheds. *Water* **2014**, *6*, 1280–1299. [CrossRef]

19. Rahman, M.S.; Di, L. The state of the art of spaceborne remote sensing in flood management. *Nat. Hazards* **2017**, *85*, 1223–1248. [CrossRef]

20. Chowdhury, M.R. Consensus seasonal Flood Forecasts and Warning Response System (FFWRS): An alternate for nonstructural flood management in Bangladesh. *Environ. Manag.* **2005**, *35*, 716–725. [CrossRef] [PubMed]

21. Islam, A.S.; Bala, S.K.; Haque, M. Flood inundation map of Bangladesh using MODIS time-series images. *J. Flood Risk Manag.* **2010**, *3*, 210–222. [CrossRef]

22. Ahmed, M.R.; Rahaman, K.R.; Kok, A.; Hassan, Q.K. Remote Sensing-Based Quantification of the Impact of Flash Flooding on the Rice Production: A Case Study over Northeastern Bangladesh. *Sensors* **2017**, *17*, 2347. [CrossRef]

23. Rahman, M.S.; Di, L.; Shrestha, R.; Eugene, G.Y.; Lin, L.; Zhang, C.; Hu, L.; Tang, J.; Yang, Z. Agriculture flood mapping with Soil Moisture Active Passive (SMAP) data: A case of 2016 Louisiana flood. In Proceedings of the 2017 6th International Conference on Agro-Geoinformatics, Fairfax, VA, USA, 7–10 August 2017; pp. 1–6.

24. Joyce, K.E.; Belliss, S.E.; Samsonov, S.V.; McNeill, S.J.; Glassey, P.J. A review of the status of satellite remote sensing and image processing techniques for mapping natural hazards and disasters. *Prog. Phys. Geogr.* **2009**, *33*, 183–207. [CrossRef]

25. Nandi, I.; Srivastava, P.K.; Shah, K. Floodplain Mapping through Support Vector Machine and Optical/Infrared Images from Landsat 8 OLI/TIRS Sensors: Case Study from Varanasi. *Water Resour. Manag.* **2017**, *31*, 1157–1171. [CrossRef]

26. Doody, T.; Overton, I.; Pollock, D. Floodplain inundation mapping. In *Ecological Outcomes of Flow Regimes in the Murray–Darling Basin*; Report Prepared for the National Water Commission by CSIRO Water for a Healthy Country Flagship; Overton, I., Colloff, M.J., Doody, T.M., Henderson, B., Cuddy, S.M., Eds.; CSIRO: Canberra, Australia, 2009; pp. 289–308.

27. Wilson, A.M.; Jetz, W. Remotely sensed high-resolution global cloud dynamics for predicting ecosystem and biodiversity distributions. *PLoS Biol.* **2016**, *14*, e1002415. [CrossRef]

28. Greifeneder, F.; Wagner, W.; Sabel, D.; Naeimi, V. Suitability of SAR imagery for automatic flood mapping in the Lower Mekong Basin. *Int. J. Remote Sens.* **2014**, *35*, 2857–2874. [CrossRef]

29. Ajmar, A.; Boccardo, P.; Broglia, M.; Kucera, J.; Giulio-Tonolo, F.; Wania, A. Response to Flood Events: The Role of Satellite-Based Emergency Mapping and the Experience of the Copernicus Emergency Management Service. *Flood Damage Surv. Assess. New Insights Res. Pract.* **2017**, *228*, 213–228.

30. Ohki, M.; Watanabe, M.; Natsuaki, R.; Motohka, T.; Nagai, H.; Tadono, T.; Suzuki, S.; Ishii, K.; Itoh, T.; Yamanokuchi, T. Flood Area Detection Using ALOS-2 PALSAR-2 Data for the 2015Heavy Rainfall Disaster in the Kanto and Tohoku Area, Japan. *J. Remote Sens. Soc. Jpn.* **2016**, *36*, 348–359.

31. Voormansik, K.; Praks, J.; Antropov, O.; Jagomagi, J.; Zalite, K. Flood mapping with TerraSAR-X in forested regions in Estonia. *IEEE J. Sel. Top. Appl. Earth Obs. Remote Sens.* **2014**, *7*, 562–577. [CrossRef]

32. Hoque, R.; Nakayama, D.; Matsuyama, H.; Matsumoto, J. Flood monitoring, mapping and assessing capabilities using RADARSAT remote sensing, GIS and ground data for Bangladesh. *Nat. Hazards* **2011**, *57*, 525–548. [CrossRef]

33. Werle, D.; Martin, T.C.; Hasan, K. Flood and Coastal Zone Monitoring in Bangladesh with Radarsat ScanSAR: Technical Experience and Institutional Challenges. *Johns Hopkins APL Tech. Dig.* **2000**, *21*, 148–154.

34. Dewan, A.M.; Islam, M.M.; Kumamoto, T.; Nishigaki, M. Evaluating flood hazard for land-use planning in Greater Dhaka of Bangladesh using remote sensing and GIS techniques. *Water Resour. Manag.* **2007**, *21*, 1601–1612. [CrossRef]

35. Roy, S.K.; Sarker, S.C. Integration of Remote Sensing Data and GIS Tools for Accurate Mapping of Flooded Area of Kurigram, Bangladesh. *J. Geogr. Inf. Syst.* **2016**, *8*, 184. [CrossRef]

36. Uddin, K.; Guring, D.R. Land cover change in Bangladesh: A knowledge based classification approach. In Proceedings of the 10th International Symposium on Hill Mountain Remote Sensing Cartography, Kathmandu, Nepal, 8–11 September 2008.

37. Gilani, H.; Shrestha, H.L.; Murthy, M.; Phuntso, P.; Pradhan, S.; Bajracharya, B.; Shrestha, B. Decadal land cover change dynamics in Bhutan. *J. Environ. Manag.* **2015**, *148*, 91–100. [CrossRef]

38. Uddin, K.; Abdul Matin, M.; Maharjan, S. Assessment of Land Cover Change and Its Impact on Changes in Soil Erosion Risk in Nepal. *Sustainability* **2018**, *10*, 4715. [CrossRef]

39. Uddin, K.; Shrestha, H.L.; Murthy, M.; Bajracharya, B.; Shrestha, B.; Gilani, H.; Pradhan, S.; Dangol, B. Development of 2010 national land cover database for the Nepal. *J. Environ. Manag.* **2015**, *148*, 82–90. [CrossRef] [PubMed]

40. Huang, H.; Chen, Y.; Clinton, N.; Wang, J.; Wang, X.; Liu, C.; Gong, P.; Yang, J.; Bai, Y.; Zheng, Y. Mapping major land cover dynamics in Beijing using all Landsat images in Google Earth Engine. *Remote Sens. Environ.* **2017**, *202*, 166–176. [CrossRef]

41. Xiong, J.; Thenkabail, P.; Tilton, J.; Gumma, M.; Teluguntla, P.; Oliphant, A.; Congalton, R.; Yadav, K.; Gorelick, N. Nominal 30-m cropland extent map of continental Africa by integrating pixel-based and object-based algorithms using Sentinel-2 and Landsat-8 data on Google Earth Engine. *Remote Sens.* **2017**, *9*, 1065. [CrossRef]

42. Hansen, M.C.; Potapov, P.V.; Moore, R.; Hancher, M.; Turubanova, S.; Tyukavina, A.; Thau, D.; Stehman, S.; Goetz, S.; Loveland, T.R. High-resolution global maps of 21st-century forest cover change. *Science* **2013**, *342*, 850–853. [CrossRef] [PubMed]

43. Uddin, K.; Gurung, D.R.; Giriraj, A.; Shrestha, B. Application of Remote Sensing and GIS for Flood Hazard Management: A Case Study from Sindh Province, Pakistan. *Am. J. Geogr. Inf. Syst.* **2013**, *2*, 1–5.

44. Manjusree, P.; Kumar, L.P.; Bhatt, C.M.; Rao, G.S.; Bhanumurthy, V. Optimization of threshold ranges for rapid flood inundation mapping by evaluating backscatter profiles of high incidence angle SAR images. *Int. J. Disaster Risk Sci.* **2012**, *3*, 113–122. [CrossRef]

45. Syed, M.A.; Al Amin, M. Geospatial modeling for investigating spatial pattern and change trend of temperature and rainfall. *Climate* **2016**, *4*, 21. [CrossRef]

46. WorldAtlas. 10 Largest Rice Producing Countries. Available online: https://www.worldatlas.com/articles/the-countries-producing-the-most-rice-in-the-world.html (accessed on 25 June 2018).

47. Temps, C. Rainfall/Precipitation in Dhaka, Bangladesh. Available online: https://en.climate-data.org/asia/bangladesh/dhaka-division/dhaka-1062098/ (accessed on 17 June 2018).

48. Paul, B.K.; Rasid, H. Flood damage to rice crop in Bangladesh. *Geogr. Rev.* **1993**, *83*, 150–159. [CrossRef]

49. NASA. NASA Shuttle Radar Topography Mission Global 1 Arc Second. Available online: https://cgiarcsi.community/data/srtm-90m-digital-elevation-database-v4-1/ (accessed on 17 June 2018).

50. Haklay, M.; Weber, P. Openstreetmap: User-generated street maps. *IEEE Pervasive Comput.* **2008**, *7*, 12–18. [CrossRef]

51. Global Administrative Areas. GADM Database of Global Administrative Areas, Version 2.0. Available online: http://www.gadm.org (accessed on 25 June 2018).

52. Small, D. Flattening gamma: Radiometric terrain correction for SAR imagery. *IEEE Trans. Geosci. Remote Sens.* **2011**, *49*, 3081–3093. [CrossRef]

53. Ajadi, O.A.; Meyer, F.J.; Webley, P.W. Change detection in synthetic aperture radar images using a multiscale-driven approach. *Remote Sens.* **2016**, *8*, 482. [CrossRef]

54. Janssen, L.L.; Middelkoop, H. Knowledge-based crop classification of a Landsat Thematic Mapper image. *Int. J. Remote Sens.* **1992**, *13*, 2827–2837. [CrossRef]

55. Mwaniki, M.W.; Matthias, M.S.; Schellmann, G. Application of remote sensing technologies to map the structural geology of central Region of Kenya. *IEEE J. Sel. Top. Appl. Earth Obs. Remote Sens.* **2015**, *8*, 1855–1867. [CrossRef]

56. Gan, G.; Ma, C.; Wu, J. *Data Clustering: Theory, Algorithms, and Applications*; Society for Industrial and Applied Mathematics: Ontario, Canada, 2007; Volume 20.

57. Ghosh, A.; Mishra, N.S.; Ghosh, S. Fuzzy clustering algorithms for unsupervised change detection in remote sensing images. *Inf. Sci.* **2011**, *181*, 699–715. [CrossRef]

58. Mather, P.; Tso, B. *Classification Methods for Remotely Sensed Data*; CRC Press: Boca Raton, FL, USA, 2016.
59. Mohamed, I.N.; Verstraeten, G. Analyzing dune dynamics at the dune-field scale based on multi-temporal analysis of Landsat-TM images. *Remote Sens. Environ.* **2012**, *119*, 105–117. [CrossRef]
60. Komac, M. A landslide susceptibility model using the analytical hierarchy process method and multivariate statistics in perialpine Slovenia. *Geomorphology* **2006**, *74*, 17–28. [CrossRef]
61. Mahi, H.; Farhi, N.; Labed, K. Unsupervised classification of satellite images using K-Harmonic Means Algorithm and Cluster Validity Index. *EARSeL eProc.* **2016**, *15*, 10.
62. Kumar, G.; Sarthi, P.P.; Ranjan, P.; Rajesh, R. Performance of k-means based Satellite Image Clustering in RGB and HSV Color Space. In Proceedings of the 2016 International Conference on Recent Trends in Information Technology (ICRTIT), Chennai, India, 8–9 April 2016; pp. 1–5.
63. Rahman, M.; Di, L.; Yu, E.; Lin, L.; Zhang, C.; Tang, J. Rapid Flood Progress Monitoring in Cropland with NASA SMAP. *Remote Sens.* **2019**, *11*, 191. [CrossRef]
64. Khatami, R.; Mountrakis, G.; Stehman, S.V. Mapping per-pixel predicted accuracy of classified remote sensing images. *Remote Sens. Environ.* **2017**, *191*, 156–167. [CrossRef]
65. Wing, O.E.; Bates, P.D.; Sampson, C.C.; Smith, A.M.; Johnson, K.A.; Erickson, T.A. Validation of a 30 m resolution flood hazard model of the conterminous United States. *Water Resour. Res.* **2017**, *53*, 7968–7986. [CrossRef]
66. Jain, S.K.; Saraf, A.K.; Goswami, A.; Ahmad, T. Flood inundation mapping using NOAA AVHRR data. *Water Resour. Manag.* **2006**, *20*, 949–959. [CrossRef]
67. Firpi, O.A.A. Satellite Data for All? Review of Google Earth Engine for Archaeological Remote Sensing. *Internet Archaeol.* **2016**. [CrossRef]
68. Shelestov, A.; Lavreniuk, M.; Kussul, N.; Novikov, A.; Skakun, S. Exploring Google Earth Engine Platform for Big Data Processing: Classification of Multi-Temporal Satellite Imagery for Crop Mapping. *Front. Earth Sci.* **2017**, *5*, 17. [CrossRef]
69. Mesas-Carrascosa, F.-J.; Torres-Sánchez, J.; Clavero-Rumbao, I.; García-Ferrer, A.; Peña, J.-M.; Borra-Serrano, I.; López-Granados, F. Assessing optimal flight parameters for generating accurate multispectral orthomosaicks by UAV to support site-specific crop management. *Remote Sens.* **2015**, *7*, 12793–12814. [CrossRef]
70. Bajracharya, B.; Uddin, K.; Chettri, N.; Shrestha, B.; Siddiqui, S.A. Understanding Land Cover Change Using a Harmonized Classification System in the Himalaya. *Mt. Res. Dev.* **2010**, *30*, 143–156. [CrossRef]
71. Gregorio, A.D. *Land Cover Classification System Classification Concepts and User Manual Software Version (2)*; Food and Agriculture Organization of the United Nations: Rome, Italy, 2005.
72. Loh, W.Y. Classification and regression trees. *Wiley Interdiscip. Rev. Data Min. Knowl. Discov.* **2011**, *1*, 14–23. [CrossRef]
73. Goldblatt, R.; You, W.; Hanson, G.; Khandelwal, A.K. Detecting the boundaries of urban areas in India: A dataset for pixel-based image classification in Google Earth Engine. *Remote Sens.* **2016**, *8*, 634. [CrossRef]
74. Uddin, K.; Chaudhary, S.; Chettri, N.; Kotru, R.; Murthy, M.; Chaudhary, R.P.; Ning, W.; Shrestha, S.M.; Gautam, S.K. The changing land cover and fragmenting forest on the Roof of the World: A case study in Nepal's Kailash Sacred Landscape. *Landsc. Urban Plan.* **2015**, *141*, 1–10. [CrossRef]
75. Aguirre-Gutiérrez, J.; Seijmonsbergen, A.C.; Duivenvoorden, J.F. Optimizing land cover classification accuracy for change detection, a combined pixel-based and object-based approach in a mountainous area in Mexico. *Appl. Geogr.* **2012**, *34*, 29–37. [CrossRef]
76. Saah, D.; Johnson, G.; Ashmall, B.; Tondapu, G.; Tenneson, K.; Patterson, M.; Poortinga, A.; Markert, K.; Quyen, N.H.; San Aung, K. Collect Earth: An online tool for systematic reference data collection in land cover and use applications. *Environ. Model. Softw.* **2019**, *118*, 166–171. [CrossRef]
77. UNDP. Bangladesh: Disaster Risk Reduction as Development. Available online: https://www.undp.org/content/undp/en/home/librarypage/poverty-reduction/supporting_transformationalchange/Bangladesh-drr-casestudy-transformational-change.html (accessed on 29 June 2018).
78. Dewan, T.H. Societal impacts and vulnerability to floods in Bangladesh and Nepal. *Weather Clim. Extrem.* **2015**, *7*, 36–42. [CrossRef]
79. Zoka, M.; Psomiadis, E.; Dercas, N. The Complementary Use of Optical and SAR Data in Monitoring Flood Events and Their Effects. *Proceedings* **2018**, *2*, 644. [CrossRef]

80. Falter, D.; Schröter, K.; Dung, N.V.; Vorogushyn, S.; Kreibich, H.; Hundecha, Y.; Apel, H.; Merz, B. Spatially coherent flood risk assessment based on long-term continuous simulation with a coupled model chain. *J. Hydrol.* **2015**, *524*, 182–193. [CrossRef]

81. Borah, S.B.; Sivasankar, T.; Ramya, M.; Raju, P. Flood inundation mapping and monitoring in Kaziranga National Park, Assam using Sentinel-1 SAR data. *Environ. Monit. Assess.* **2018**, *190*, 520. [CrossRef] [PubMed]

82. Atijosan, A.; Salau, A.O.; Badru, R.A.; Alaga, T. Development of a Low Cost Community Based Real Time Flood Monitoring and Early Warning System. *Int. J. Sci. Res. Sci. Eng. Technol.* **2017**, *3*, 189–195.

83. World Meteorological Organization. *Integrated Flood Management Tools Series*; No. 20; WMO: Zurich, Switzerland, 2013; pp. 1–88.

84. Amir, M.; Khan, M.; Rasul, M.; Sharma, R.; Akram, F. Hydrologic and hydrodynamic modelling of extreme flood events to assess the impact of climate change in a large basin with limited data. *J. Flood Risk Manag.* **2018**, *11*, S147–S157. [CrossRef]

85. Klemas, V. Remote Sensing of Floods and Flood-Prone Areas: An Overview. *J. Coast. Res.* **2014**, *31*, 1005–1013. [CrossRef]

86. Zurqani, H.A.; Post, C.J.; Mikhailova, E.A.; Schlautman, M.A.; Sharp, J.L. Geospatial analysis of land use change in the Savannah River Basin using Google Earth Engine. *Int. J. Appl. Earth Obs. Geoinf.* **2018**, *69*, 175–185. [CrossRef]

87. Reiche, J.; Hamunyela, E.; Verbesselt, J.; Hoekman, D.; Herold, M. Improving near-real time deforestation monitoring in tropical dry forests by combining dense Sentinel-1 time series with Landsat and ALOS-2 PALSAR-2. *Remote Sens. Environ.* **2018**, *204*, 147–161. [CrossRef]

88. WFP. *Terai Flood 72 Hour Assessment*; Version 1; The United Nations World Food Programme, Ed.; WFP: Kathmandu, Nepal, 2017.

89. Lobry, S.; Denis, L.; Tupin, F.; Fjørtoft, R. Double MRF for water classification in SAR images by joint detection and reflectivity estimation. In Proceedings of the 2017 IEEE International Geoscience and Remote Sensing Symposium (IGARSS), Fort Worth, TX, USA, 23–28 July 2017.

90. Ban, Y.; Hu, H.; Rangel, I.M. Fusion of Quickbird MS and RADARSAT SAR data for urban land-cover mapping: Object-based and knowledge-based approach. *Int. J. Remote Sens.* **2010**, *31*, 1391–1410. [CrossRef]

91. Tsyganskaya, V.; Martinis, S.; Marzahn, P.; Ludwig, R. SAR-based detection of flooded vegetation—A review of characteristics and approaches. *Int. J. Remote Sens.* **2018**, *39*, 2255–2293. [CrossRef]

92. Casado, M.R.; Irvine, T.; Johnson, S.; Yu, D.; Butler, J. Drone watch: UAVs for flood extent mapping and damage assessment. In Proceedings of the Small Unmanned Aerial Systems for Environmental Research, Worcester, UK, 28–29 June 2018.

 *remote sensing*

Article

Urban Flood Detection with Sentinel-1 Multi-Temporal Synthetic Aperture Radar (SAR) Observations in a Bayesian Framework: A Case Study for Hurricane Matthew

Yunung Nina Lin [1,2,*], Sang-Ho Yun [3], Alok Bhardwaj [2] and Emma M. Hill [2,4]

1   Institute of Earth Sciences, Academia Sinica, Taipei City 115, Taiwan
2   Earth Observatory of Singapore, Nanyang Technological University, Singapore 639798, Singapore
3   Jet Propulsion Laboratory, California Institute of Technology, Pasadena, CA 91109, USA
4   Asian School of the Environment, Nanyang Technological University, Singapore 637459, Singapore
*   Correspondence: ninalin@earth.sinica.edu.tw

Received: 4 June 2019; Accepted: 24 July 2019; Published: 29 July 2019

Abstract: In this study we explored the application of synthetic aperture radar (SAR) intensity time series for urban flood detection. Our test case was the flood in Lumberton, North Carolina, USA, caused by the landfall of Hurricane Matthew on 8 October 2016, for which airborne imagery—taken on the same day as the SAR overpass—is available for validation of our technique. To map the flood, we first carried out normalization of the SAR intensity observations, based on the statistics from the time series, and then construct a Bayesian probability function for intensity decrease (due to specular reflection of the signal) and intensity increase (due to double bounce) cases separately. We then formed a flood probability map, which we used to create our preferred flood extent map using a global cutoff probability of 0.5. Our flood map in the urban area showed a complicated mosaicking pattern of pixels showing SAR intensity decrease, pixels showing intensity increase, and pixels without significant intensity changes. Our approach shows improved performance when compared with global thresholding on log intensity ratios, as the time series-based normalization has accounted for a certain level of spatial variation by considering the different history for each pixel. This resulted in improved performance for urban and vegetated regions. We identified smooth surfaces, like asphalt roads, and SAR shadows as the major sources of underprediction, and aquatic plants and soil moisture changes were the major sources of overprediction.

Keywords: SAR intensity time series; urban flood mapping; double bounce effect; Hurricane Matthew

---

1. Introduction

Flood extent maps based on synthetic aperture radar (SAR) have increasingly been used in recent emergency response operations. For example, in the Sentinel Asia consortium (https://sentinel.tksc.jaxa.jp), where national space agencies, research institutions, and end-user organizations work together on emergency observation requests in the Asia-Pacific region, the responses to 20 of 23 activated flood-related events in the year of 2017 used ALOS-2 SAR flood-mapping results [1]. These statistics demonstrate the need for radar's all-weather, day-and-night sensing capability, where in most cases cloud cover and rains persist for the duration of a flood.

Flood extent maps are primarily extracted from SAR intensity (the squared amplitude of the SAR return) information. The backscattering coefficient (σ^0, log intensity in dB) decreases due to the specular surface of open flood waters if radar energy is mostly forward-scattered. In other cases, σ^0 may increase if the radar wave bounces first off the water surface (away from the satellite) and then off a semivertical structure, such as a building wall, tree trunk, or even a car in the flood (towards the satellite); this is the "double bounce" effect. Compared with rural floods, urban flood mapping suffers more from layover and shadow effects due to SAR's side-looking nature. The layover zone is in front of buildings, toward the satellite direction, and the length is usually longer than the building height at incidence angles lower than 45° (layover length = building height × cot(incidence angle) [2]). Within this zone there is a good chance of seeing stronger backscattering due to the double bounce effect, and its strength is a function of the oblique angle between the flight direction and the building orientation. The double-bounce intensity increase can be larger than 10 dB at 0° and remains high up to 5°; at an angle larger than 10° the increase will drop to a constant level [3,4]. The shadow zone, on the other hand, has a smaller length (shadow length = building height × tan(incidence angle) [2]), and the backscattering stays low at all times. The low intensity within the shadow may lead to false flood detection if one uses a during-flood SAR image alone.

Several studies have tried to improve the accuracy of urban flood mapping by addressing layover and shadow effects. For example, Mason et al. and Giustarini et al. [2,5,6], for a case study using TerraSAR-X data, masked out the layover and shadow zones by using a SAR simulator. The accuracy of open water flood mapping increases after applying the mask, but flooded pixels that show increased brightness due to double-bounce scattering are left out from the flood map. In another example, Pulvirenti et al. [7] developed an algorithm that adopted the double-bounce intensity values from electromagnetic modeling as initial fuzzy thresholds, and they used fuzzy logic to map out the urban flood with intensity increase in COSMO-SkyMed images. Mason et al. [8] also demonstrated the use of an electromagnetic scattering model and high-resolution elevation data to simulate the double-bounce effects. The simulated double-bounce scattering strength agrees with the observed data; hence, it can provide information about the statistical distribution of pixels showing double bounce for better thresholding decisions. Both model-based approaches have the advantage of being independent from user biases when handling double-bounce effects, although the auxiliary data of a high-resolution building model (or digital surface model) is not always available. Tanguy et al. [9] proposed to use ancillary hydraulic data to refine the flood detected with RADARSAT-2 so as to overcome the SAR limitations associated with viewing geometry. The availability of such high-resolution hydraulic data is not, however, guaranteed. Recently, some studies have found that using interferometric coherence in conjunction with intensity will improve the detection accuracy particularly associated with double bounce [4,10–12]. The efficacy of augmenting moderate resolution interferometric coherence (from Sentinel-1 SAR, for example) in urban flood detection still awaits systematic, quantitative validation [10].

In parallel to the developments in urban flood mapping, multitemporal SAR data analyses are also gradually being adopted by scientists to study flood extents. Hostache et al. [13] suggested to select the most appropriate pre-flood image for change detection from a time series perspective. Schlaffer et al. [14,15] carried out harmonic analysis on seven years of ENVISAT ASAR data and identified floods by looking at anomalies in the time series. D'Addabbo et al. [16] utilized Bayesian networks to jointly consider SAR intensity and InSAR coherence time series of COSMO-SkyMed data, as well as ancillary information including several hydraulic parameters, to study flood extent variation over time. Clement et al. [17] utilized Sentinel-1 SAR intensity difference (change detection) time series to identify the evolution of a flood with time. Ouled Sghaier et al. [18] utilized texture analysis on multitemporal RADARSAT-2 and Sentinel-1 intensity data to study flood history. Among these studies, some of them focus on flood evolution and treat each epoch independently, while others emphasize the use of pixel history to improve the accuracy of flood detection on particular events.

The approach that we propose in this study focuses on both the urban and time series dimensions. We would like to obtain the statistics of each pixel from the time domain and use them in a Bayesian flood probability function. The flood probability is calculated based on the historical pixel intensity values. The test data are Sentinel-1 SAR, whose open access and short repeat time make them a priority choice in many emergency response cases. We test our approach on the Hurricane Matthew flood in early October 2016 at the town of Lumberton, in North Carolina, USA. The Lumberton flood offers a great opportunity in which nearly concurrent airborne optical imagery and spaceborne SAR observations coexist for the flooding epoch. It is also an ideal target to examine the efficacy of urban flood mapping algorithms, as more than half of the flooded areas are in urban settings.

2. Study Area and Data

2.1. Study Area and Weather Event

Lumberton is located on the Coastal Plains of North Carolina, with an average elevation of 40 m and the Lumber River flowing through the town center (Figure 1). The test area was right around the town center, approximately 5.5 × 12 km in dimension, with ~30% of the area being an urban environment.

On 8 October 2016, Hurricane Matthew made landfall on the southeast coast of South Carolina and slowly moved northwards into North Carolina. Even though it had significantly weakened from a Category 5 to Category 1 hurricane at the time of landing, and was an extratropical cyclone by the time it arrived at North Carolina [19], torrential rains still brought >200 mm cumulative precipitation depths on 9 October (according to the National Oceanic and Atmospheric Administration (NOAA)'s ground weather station USC00315177; see Figure 1 for location) and caused record-breaking flood levels along the Lumber River. A levee broke, resulting in four flood-related deaths and more than 1500 people evacuated. Most of the region was still in knee-deep water two days after the hurricane passed [19].

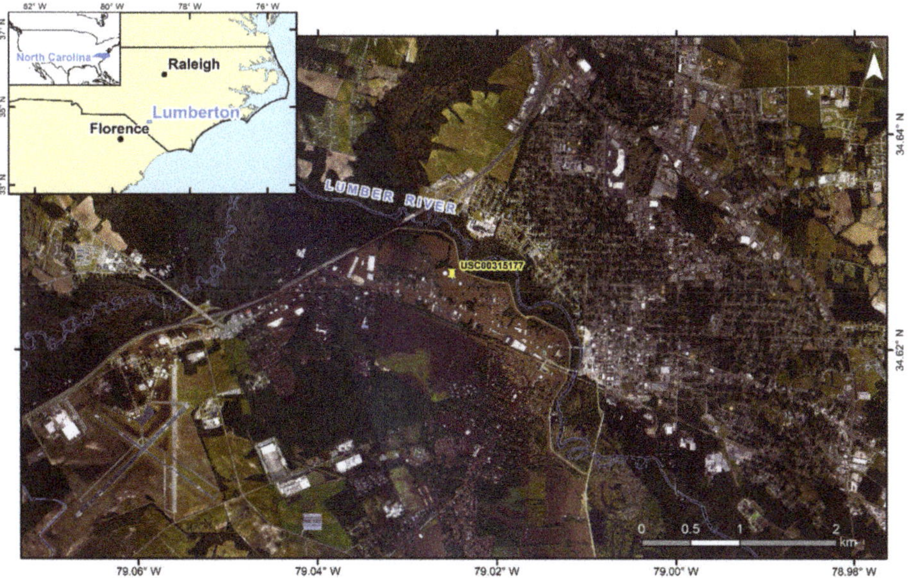

Figure 1. Lumberton in Robeson County, North Carolina, and the during-event airborne optical imagery taken on 11 October 2016. The Lumber River flows right through the middle of the town. USC00315177 is National Oceanic and Atmospheric Administration (NOAA)'s ground weather station.

2.2. Optical and Synthetic Aperture Radar (SAR) Imagery

Because the Carolinas experienced such serious flooding during the hurricane, the National Oceanic and Atmospheric Administration (NOAA) Remote Sensing Division acquired inland aerial photos between 11–16 October as rapid response imagery using a nadir-looking camera mounted on NOAA's King Air 350ER aircraft [20]. The Lumberton swath was taken during the daytime (around noon local time, from 15:53 to 16:35 UTC) on 11 October at a ground resolution of ~30 cm (Figure 1). The same-day Sentinel-1A interferometric wide swath SAR image was acquired at 19:13 EDT (23:13

UTC). Given the seven-hour difference in the acquisition time, there will be some small but minimal differences between the flood extents mapped from the two datasets, with potentially smaller flood areas recorded in the SAR image.

Besides the aerial photos taken by NOAA, we also obtained a SPOT-6 image acquired on 6 September 2016 in order to study the pre-flood water bodies. This image, together with other historical WorldView/QuickBird images available on Google Earth, showed that the location and area of pre-existing water bodies along the Lumber River remained fairly constant through time.

For the SAR imagery, we obtained all scenes available in ascending track 77 that covered Lumberton since the beginning of the Sentinel-1 mission until the end of August 2018. There is a total of 63 images, including the during-flood scene on 11 October 2016. Descending track 84 that covers this region only contains 2 images throughout the whole time; therefore, we excluded the descending track from our analysis. We treated all scenes except for the 11 October scene as nonflood imagery in our multitemporal analysis, assuming a stationarity in land cover for all times except when the flood occurred. All SAR images were processed and geocoded at 15 × 15 m ground pixel spacing using a SRTMv3 DEM [21] oversampled by a factor of two.

3. Data Processing and Analysis

3.1. Validation Dataset

Radar intensity data contain more complicated responses than a simple binary dissection between wet and dry areas. We therefore decided to create a validation dataset that also honored possible SAR σ^0 responses based on different land cover types. We manually classified the during-flood aerial image into the following six classes (Table 1 and Figure 2):

Flood: This class was the open standing water surface as directly seen in the aerial image. It was considered as an area of flooding associated with Hurricane Matthew. Intensity may decrease or increase in the during-event epoch depending on the incidence angle of the SAR image and the proximity of the pixel to any nearby vertical structures, such as trees or buildings.

Permanent Water: This class was identified based on the water bodies in the pre-flood SPOT-6 image. The SAR σ^0 tended to stay low both in the non- and during-event images given the constant specular reflection surface at all times.

Flooded Vegetation: This class represented small to medium-size patches of trees surrounded by flood water from Hurricane Matthew. They mainly appeared in the urban area of Lumberton. Whether the radar wave penetrated the tree crowns was subject to factors like radar wavelength, polarization, and the density and structure of vegetation. If it did, double bounce and enhanced backward scattering may occur, resulting in an increase in the during-event σ^0. If the tree crown was too dense to penetrate, SAR σ^0 stayed relatively stable with potential seasonal variations.

Dry Vegetation: This class stood for small to medium-size patches of trees standing on dry land. The SAR σ^0 should also stay at a relatively constant level with potential seasonal variations.

Dry: The dry class represented buildings, roads, dry bare land or lawns. Any pixels that appeared to be dry without ambiguity of underlying flood water were included in this class. The pixel values may vary with time depending on the detailed land cover type (paved road, lawn, bare soil, etc.). We did not expect to see significant intensity anomalies in buildings and roads on the during-event epoch, although intensity changes associated with soil moisture may be observed on the bare land or lawns.

Uncertain: A large area of the Lumberton region is covered by dense forest where the Lumber River flows through. Careful investigation of the spatial context showed that many of these large forest patches were surrounded by flood water on all sides. What remains unclear is whether if there was also flood water under the tree canopy. To honor this unidentifiable condition based on the high-resolution optical imagery, we classified these large forest patches as uncertain. The SAR σ^0 may stay at a relatively high level due to the dense tree tops, but the actual level and the during-event response may vary within the same forest patch. We did not include pixels in this class in the final confusion matrix calculation.

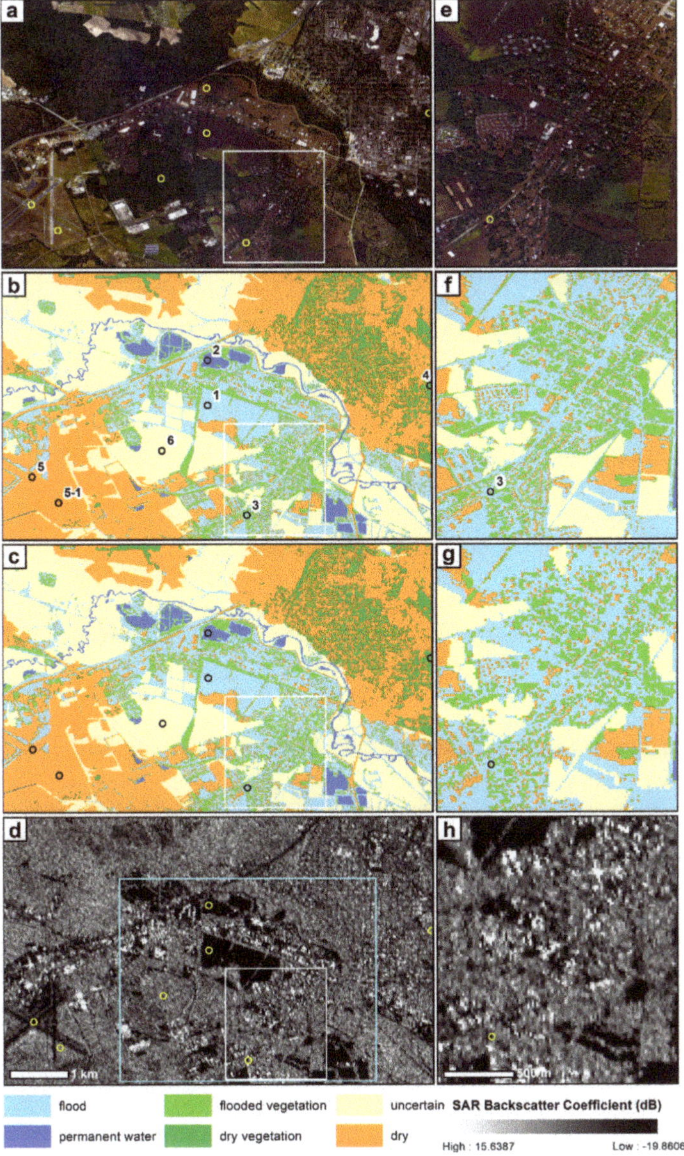

Figure 2. (**a**) The during-event aerial image and (**b**) the validation vector image with 30 cm resolution. (**c**) The rasterized validation image at 15 m resolution. (**d**) The during-event Sentinel-1 synthetic aperture radar (SAR) image at 15 m posting. The cyan box indicates the region used for Gaussian curve fitting. (**e**,**f**) The magnified views of (**a**–**d**) for the urban area. The open circles labeled 1 to 6 in (**b**) are the sample pixels of the time series (Figure 4) for each class.

After manual classification was done, we applied a nearest-neighbor sampling to convert the validation vector data of 30 cm resolution (Figure 2b) into a raster image of 15 m resolution (Figure 2c).

We took cautious steps to ensure that each pixel in the converted raster image registered to the pixels in the SAR image. Note that there was potential change or loss of information during the down-sampling. We also tested converting the vector file into a 30 m raster image, and the result showed significant loss of spatial context, especially in urban regions. For example, buildings were concatenated together or coalesced with trees; and smaller buildings could fully disappear. Our conclusion is that 15 m ground resolution is minimally needed to undertake proper flood mapping in urban regions.

Table 1. Classes in the validation dataset and their backscattering responses.

| | Responses in During-Event Scene | | | |
	Intensity Drop	**Intensity Increase**	**Intensity Stays Low**	**Intensity Stays High**
Flooded	Flood	Flood		
		Flooded Vegetation		
Nonflooded			Permanent Water	Dry Vegetation
			Dry	Dry
Not Determined	Uncertain (not considered in evaluation metric)			

3.2. Sentinel-1 SAR Data Processing

We developed our amplitude stack processing pipeline using the NASA Jet Propulsion Laboratory's InSAR Scientific Computing Environment (ISCE) version 2 (ISCE2 is now open-sourced at GitHub: https://github.com/isce-framework/isce2). In this pipeline (Figure 3), we started from SLC files in VV polarization mode, and incorporated Sentinel-1 radiometric calibration and thermal noise calibration [22] on a burst-by-burst basis. Then we co-registered all slave images to a single master image before merging the bursts. In this study, to ensure proper comparison with the validation dataset, we produced a stack of intensity images in georeferenced coordinates. For an actual operational system, we would switch to stack processing in radar coordinates.

After producing the geocoded amplitude stack of 15 × 15 m posting, we applied a 5 × 5 Lee filter [23,24] to reduce speckle noise. We also tested 3 × 3 and 7 × 7 Lee filters, and comparison with the validation dataset shows that the 5 × 5 window gave the optimal result at a ground pixel spacing of 15 × 15 m. Since the incidence angle variation was small across the study area, from 35.7° at near range to 36.4° at far range, and the Lumberton region is relatively flat, we skipped the local incidence angle correction proposed by [25]. Finally, we computed the backscattering coefficients (σ^0) in decibels (dB) by the following definition:

$$\sigma^0 = 10log_{10}\left(A^2\right) \tag{1}$$

where A stands for the amplitude of the SAR image in the complex domain after radiometric calibration and thermal noise calibration [22]. We then looked into these σ^0 values on a pixel-by-pixel basis in the following multitemporal analysis.

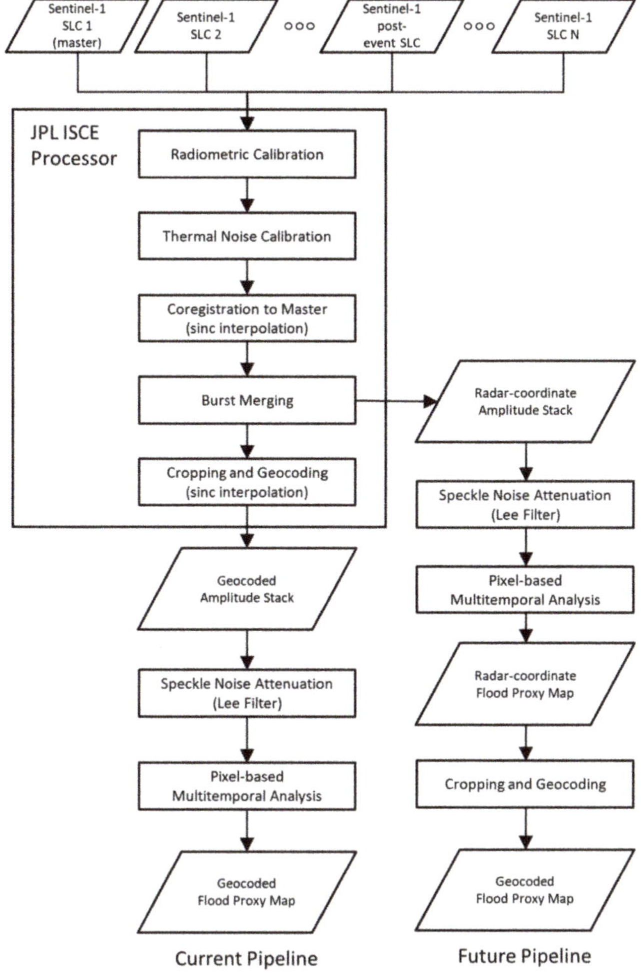

Figure 3. The processing flow chart for the intensity stack and flood proxy map. JPL ISCE, Jet Propulsion Laboratory's InSAR Scientific Computing Environment.

3.3. SAR Intensity Time Series

We first tried to observe if there was any characteristic temporal pattern associated with each of the six classes defined in Section 3.1. For the Flood class (Figure 4a), we picked a pixel sitting in the middle of open standing water (sample pixel 1 in Figure 2), and we observed a significant intensity decrease (~15 dB) for the during-event epoch, while the background value remained between −5 and −12 dB. It is worth pointing out that the histogram for this pixel (Figure 4b) looked similar to the Dry class (Figure 4f), reflecting the nature that this pixel was dry under ordinary conditions. The flood epoch was, therefore, an outlier (vertical line in Figure 4b) from the dry epochs, and this formed the core of our flood detection approach (Section 4).

We picked the sample pixel for the Permanent Water class in the middle of a pond. Its σ^0 slightly varied with time but mostly stayed low, around −20 dB (Figure 4a; sample pixel 2 in Figure 2). In some epochs the value may go 5 dB higher, close to some of the nonflooded epochs in the flood class. This natural variation may be associated with the changes of floating aquatic plants in the pond, which can be observed in the WorldView/QuickBird images in Google Earth.

For the Flooded Vegetation and Dry Vegetation classes, we selected one sample each from the flooded neighborhood in the south (sample pixel 3) and from the dry neighborhood in the north (sample pixel 4). They both showed σ^0 values between −5 and −8 dB in general, with the Dry Vegetation class higher on average than the other (Figure 4b). This difference may simply represent different vegetation density or structures between two sample pixels. The more important difference was the intensity increase (~3 dB) in the time series of the Flooded Vegetation class on the during-event epoch, revealing possible double-bounce backscattering between the tree and flood water surfaces.

The σ^0 values for Dry class (sample pixel 5) stayed at a constant low level around −15 dB, even more stable than the Permanent Water sample pixel (Figure 4c). This pattern reflects the characteristics of paved road (Figure 2a,b), with the asphalt layer showing low backscattering intensity. We also examined another pixel in the Dry class (sample pixel 5-1), and the values fluctuated within a wider range, possibly due to the changes between land cover type (bare ground vs. lawn) and/or changes in soil moisture. Regardless of different temporal patterns in these two sample pixels, neither of them showed any intensity anomalies for the during-event epoch.

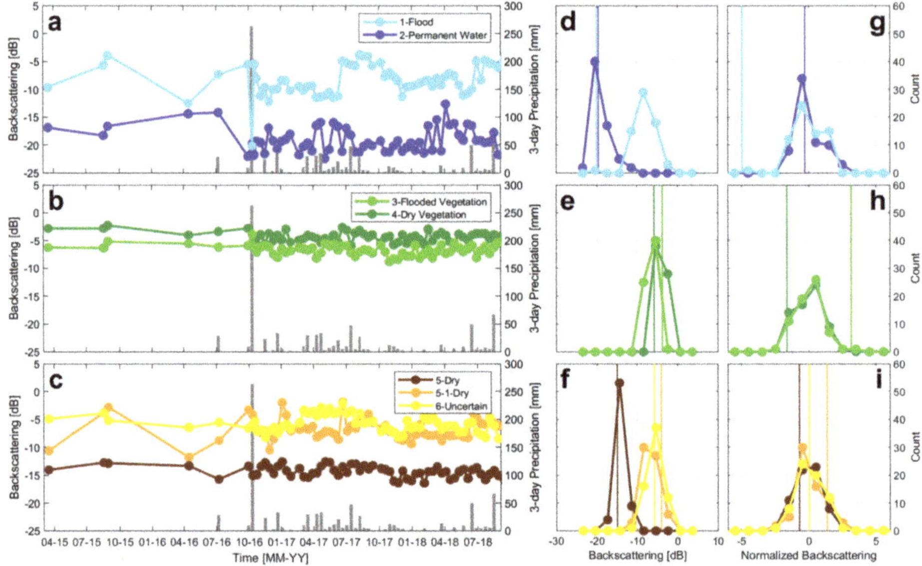

Figure 4. (**a–c**) σ^0 time series of selected pixels for the 6 classes in the validation dataset. The location of each sample pixel is marked in Figure 2, with the same ID number as shown in the legend (e.g., "1-Flood" in the legend of Figure 4a is from the sample pixel 1 in Figure 2). The grey bars in the background are 3-day cumulative precipitation from Global Precipitation Measurement daily solutions [26]. The during-event epoch is marked by the precipitation record of >250 mm. (**d–f**) Histogram of the time series. The vertical lines stand for the σ^0 values on the during-event epoch. (**g–i**) Histogram of normalized time series. Vertical lines stand for the during-event σ^0 values after normalization.

Compared to the Dry class, the σ^0 values for the Uncertain class seemed to show low-frequency seasonal variations between the years of 2017 and 2018. As the pixel was located in the middle of a dense forest (sample pixel 6), the undulations in the time series may reflect seasonal changes in the forest.

We computed the histograms of σ^0 for each of the time series. In Figure 4d, we can see that the histogram for the Flood and Permanent Water classes looked very different, but their during-event

σ^0 values were almost identical. For the Flooded Vegetation and Dry Vegetation classes (Figure 4e), however, the histograms were more similar, with also small differences in the σ^0 values on the during-event epoch. Figure 4d,e together demonstrated that, in general, it was easier to map out the open-water flood, whereas the double-bounce effect associated with flooded vegetation cannot be easily identified. Given the high-frequency variation of σ^0 values in almost every class, plus a much smaller during-event σ^0 change in the Flooded Vegetation class than that in the Flood class, identifying the double bounce effect in the SAR image will never be an easy task.

4. Methods

4.1. Probabilistic Thesholding on Normalized Intensity Time Series

One way to address the information in the time series is to estimate the best-fitting model of the time series—similar to what geodesists do with GPS or InSAR displacement time series (see [27] for example). In a complete suite of parameterized time series analyses, one needs to decide what functional terms to use in the modeling, such as a linear function (long-term rate), sinusoidal function (seasonal patterns), step function (sudden and nonrecoverable changes), delta function (sudden and recoverable changes), and even integrated B-splines (for transient changes). For SAR intensity time series, Schlaffer et al. [14,15] used harmonic (sinusoidal) modeling on 7 years of ENVISAT intensity data to account for seasonal variations before carrying out thresholding on the model-observation residuals. In our study, we decided not to carry out this harmonic modeling for two reasons. The first reason, also the most critical one, was that the number of total epochs may not always satisfy rigorous time series modeling. In actual emergency responses we may have even fewer epochs than we have in this study. Second, it may require time-consuming quality checks on the results of the model fitting, as the existence of secular or transient signals may bias the fitting of sinusoidal patterns [28], which in return may bias the model-observation residuals and, hence, the thresholding results. As the determination of functional terms needed for each land cover type is a nontrivial process, we chose not to pursue this direction in this study.

Here we propose a hybrid approach that combines the distribution normalization (so called z-score) and Bayesian probability, with the latter based on the flood probability estimation and classification [15,29]. We named this approach p50-ts for reference later in the discussion. The procedure is as follows. On each pixel,

(1)   Compute the mean (μ_{ts}) and standard deviation (S_{ts}) of σ^0 from nonflood epochs in the time series by excluding the during-event (k^{th}) epoch:

$$\mu_{ts} = \frac{\sum_{i=1}^{n} \sigma_i^0}{n} \quad S_{ts}^2 = \frac{\sum_{i=0}^{n}\left(\sigma_i^0 - \mu_{ts}\right)^2}{n-1} \quad where\ i = 1, 2, \ldots, n\ and\ i \neq k \tag{2}$$

(2)   Normalize the whole time series, including the during-event k^{th} epoch:

$$\widetilde{\sigma}_i^0 = \frac{\sigma_i^0 - \mu_{ts}}{S_{ts}} \tag{3}$$

This was the most critical step in our approach. The normalized intensities are read as the deviation from their ordinary state (the historical mean) on the same scale (after being divided by the time series standard deviation). The normalized during-event intensity can indicate how anomalous it is from all the other pre-event epochs after considering the natural variations. When we look at the histogram after normalization, the curve will be centered at zero (Figure 4g–i), with the normalized during-event intensity being at either the left or right far end of the distribution due to the presence of specular reflection or double bounce. As mentioned in Section 3.3, the histogram actually reflects the probability of mainly the nonflooded condition. To honor the fact that the flooded condition should also have its own probability, next we tried to incorporate the Bayesian approach into our method.

(3)   With the normalized time series, for each pixel, construct the conditional probability of an epoch to be flooded, using the following equation:

$$p\left(F\middle|\widetilde{\sigma_i^0}\right) = \frac{p(\widetilde{\sigma_i^0}|F)p(F)}{p\left(\widetilde{\sigma_i^0}\middle|F\right)p(F) + p\left(\widetilde{\sigma_i^0}\middle|\overline{F}\right)p(\overline{F})} \tag{4}$$

$p(F)$ and $p\left(\overline{F}\right)$ are the prior probabilities for the flooded and nonflooded epochs. Here we assumed the direction of intensity change on the same pixel, should it be flooded, remained unchanged throughout the short time series. With this assumption, we can adopt the noninformative priors of $p(F) = p(\overline{F}) = 0.5$ for simplicity [29], on both cases of flooded with intensity decrease and intensity increase. For the likelihood functions $p(\widetilde{\sigma_i^0}|F)$ and $p(\widetilde{\sigma_i^0}|\overline{F})$, we assumed Gaussian distributions for both, and the equations are as follows:

$$p\left(\widetilde{\sigma_i^0}\middle|F\right) = \frac{1}{\sqrt{2\pi}s_F}exp\left[-\frac{1}{2}\frac{\left(\widetilde{\sigma_i^0} - m_F\right)^2}{s_F^2}\right] \tag{5}$$

$$p\left(\widetilde{\sigma_i^0}\middle|\overline{F}\right) = \frac{1}{\sqrt{2\pi}s_{\overline{F}}}exp\left[-\frac{1}{2}\frac{\left(\widetilde{\sigma_i^0} - m_{\overline{F}}\right)^2}{s_{\overline{F}}^2}\right] \tag{6}$$

$m_{\overline{F}}$ and $s_{\overline{F}}$ are the mean and standard deviation for the nonflooded epochs in the normalized backscattering $(\widetilde{\sigma_k^0})$ time series; therefore, their values are 0 and 1, respectively. The other set of statistical descriptors, m_F and s_F, are for the flooded epochs. Since we only had 1 flooded epoch in the whole time series, we proposed to use the statistics from the spatial domain, which we can estimate from histogram fitting of the normalized during-event backscattering $(\widetilde{\sigma_k^0})$ using the Levenberg–Marquardt algorithm [30] as suggested in [29,31]. We first chose a region that covered central Lumberton, where intensity decreases and increases due to flood should both existed and their components were more likely to be statistically meaningful (Figure 3d). Then, instead of two Gaussians, we fit the histogram with the sum of three Gaussian curves:

$$h(y) = G_1 + G_2 + G_3 = A_1exp\left[-\frac{1}{2}\frac{[y-m_1]^2}{s_1^2}\right] + A_2exp\left[-\frac{1}{2}\frac{[y-m_2]^2}{s_2^2}\right] + A_3exp\left[-\frac{1}{2}\frac{[y-m_3]^2}{s_3^2}\right] \tag{7}$$

The third Gaussian curve fits the bulging part at the high end of the histogram and, hence, gives lower root-mean-square errors compared with the two-Gaussian model (Figure 5a,b). We used the parameters for the Gaussian curves on the left and on the right (Figure 5a), (m_1, s_1) and (m_3, s_3), to approximate the m_F and s_F in the likelihood functions for the pixels of intensity decrease $(p_D(\widetilde{\sigma_i^0}|\overline{F}))$ and intensity increase $(p_U(\widetilde{\sigma_i^0}|\overline{F}))$ respectively (Figure 5c). From here we can construct the conditional probability function for each case separately (denoted as p_D and p_U in Figure 5d using (4).

(4)   Generate the flood probability map for the during-event epoch (k^{th}) by putting $\widetilde{\sigma_k^0}$ in p_D and p_U. We can also define a probability cutoff value and form a binary flood map. In this case, we adopted $p = 0.5$, which has been identified to be associated with the transition zone [29]. Next we will describe the validation process using this binary flood map.

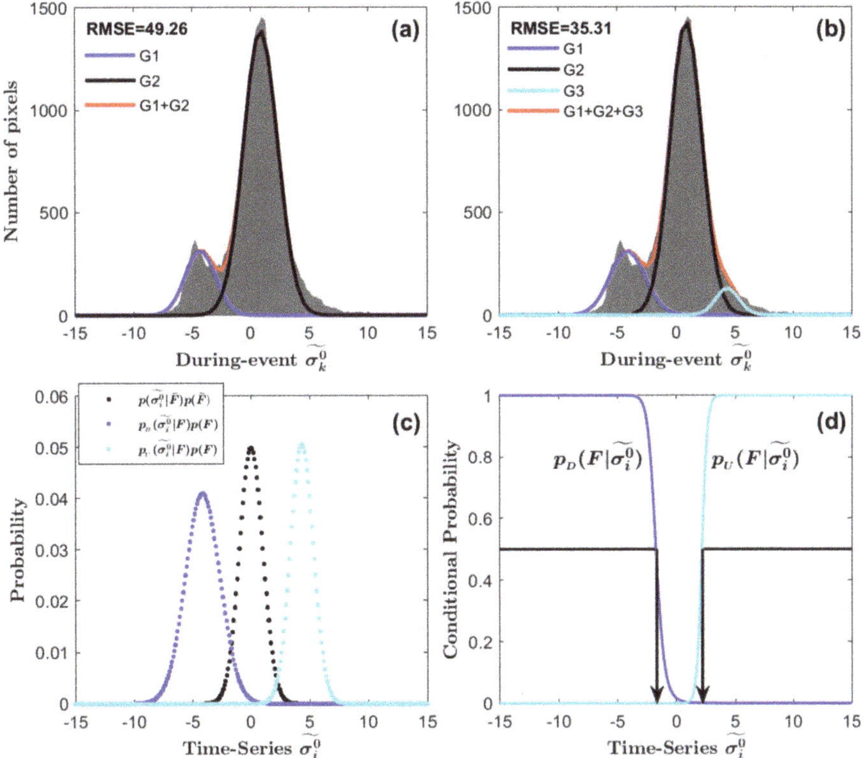

Figure 5. (**a**) The histogram and 2-Gaussian curve fitting for the during-event epoch in the normalized time series. (**b**) The curve fitting for 3-Gaussian model. RMSE = root-mean-square error. (**c**) The probability for the nonflooded, the flooded with intensity decrease, and the flooded cases, with intensity increase for the normalized backscattering in the time series. (**d**) The conditional probability of the normalized backscattering in the time series being flooded with intensity decrease (blue curve) or intensity increase (cyan curve). Black line with arrow indicates the cutoff threshold of $p = 0.5$ for a binary flood map.

4.2. Validation Approach

As the flood probability maps for the during-event intensity decrease and increase cases are constructed separately, we created the maximum flood probability map of p_F by taking the maximum value from the maps of p_D and p_U. We can interpret it as the probability of during-event intensity changes, with $p_F = 0$ for no change and $p_F = 1$ for significant change (either intensity increase or decrease). We derived the reliability diagram by comparing the observed flood frequency p_o with the flood probability p_F. For each *i-th* of the 10 bins with equal p_F interval: $[0.0, 0.1]$, $(0.1, 0.2]$, ... ,$(0.9, 1.0]$, the corresponding p_o is derived by taking the ratio between the number of pixels in the Flood/Flooded Vegetation class and the total number of pixels in the probability bin (n_i). p_o can also be written in contingency terms:

$$p_{o_i} = \frac{P_i}{P_i + N_i} = \frac{TP_i + FP_i}{TP_i + TN_i + FP_i + FN_i} \tag{8}$$

where T, F, P, and N stand for true, false, positive, and negative in the contingency table and the combination thereof.

In addition to the reliability diagram, in which the proximity to the 1:1 line represents the goodness of model prediction, we can also compute the weighted root-mean-square error for the probability map by [15,29,32]:

$$\varepsilon = \sqrt{\frac{\sum_{i=1}^{10} n_i \left(p_{F_i} - p_{o_i}\right)^2}{\sum_{i=1}^{10} n_i}} \tag{9}$$

This metric can be considered as the mean error of the probability map and, hence, is a metric for reliability.

In addition to the reliability metric, we also looked at other metrics based on the contingency matrix. We chose the critical success index (CSI), which is defined as

$$\text{CSI} = \frac{TP}{TP + FP + FN} \tag{10}$$

As CSI removes the influence of the nonflooded fraction within the area of interest (AOI), this metric was considered to better estimate the binary model accuracy [33,34]. In addition, we also obtained the values of producer's accuracy PA = TP/(TP + FN), user's accuracy UA = TP/(TP + FP), and overall accuracy OA = (TP + TN)/(P + N). TP, FP, TN, FN stands for true positive, false positive, true negative and false negative in the contingency matrix. Higher PA and UA values stand for smaller numbers of underpredicted pixels (FN) and overpredicted pixels (FP), respectively. OA is sensitive to the proportion of dry pixels in the AOI, but since it is still widely used when discussing the performance of flood mapping algorithms, we still included this metric for reference.

We tried to compare our mapping result with the best-possible (highest CSI) result obtained through grid search in the threshold space. This way we could judge whether if the $p = 0.5$ criterion served as a good threshold for flood mapping. We also conducted the same search on the log intensity ratio between the during-event image and the epoch right before. By doing so we could better understand the improvement made by considering the temporal statistics of each pixel.

5. Results

Figure 6a shows the mapping result at the cutoff probability of $p_D = 0.5$ (black pixels) and $p_U = 0.5$ (grey pixels). The largest open-water flood body (with a w-shape) near the central part of the AOI was well depicted. In the urban area (Figure 6b), we saw that a large proportion of the urban floods were mapped by the criterion of $p_U \geq 0.5$, indicating that these pixels experienced an intensity increase during the event epoch as compared with other epochs in their own history, and the increase was significant enough such that the probability of being flooded was over 50%. However, we could still observe that a visible portion of the urban flood could not be mapped by this p50-ts method (Figure 6b). The unmapped pixels represented flooded areas without significant changes of during-event backscattering in the time series. The mosaicking pattern of intensity decrease, intensity increase, and even unchanged intensity signified the complexity of radar backscattering patterns of floods in urban areas.

Next, we looked at the probability maps (Figure 6c–f). The transition zone of intermediate probability (yellow color) was narrow, with the majority of the mapped flood at the high probability end ($p > 0.9$; red color). The effect of underprediction was also demonstrated by having low probability values in the pixels within the Flood and Flooded Vegetation classes (Figure 6b vs. Figure 6d,f). Their spatial distributions were more clearly shown by the contingency map (Figure 7), where the mapped urban flood was mainly surrounded by FN pixels. Overprediction (FP pixels), on the other hand, was not as common within the urban area.

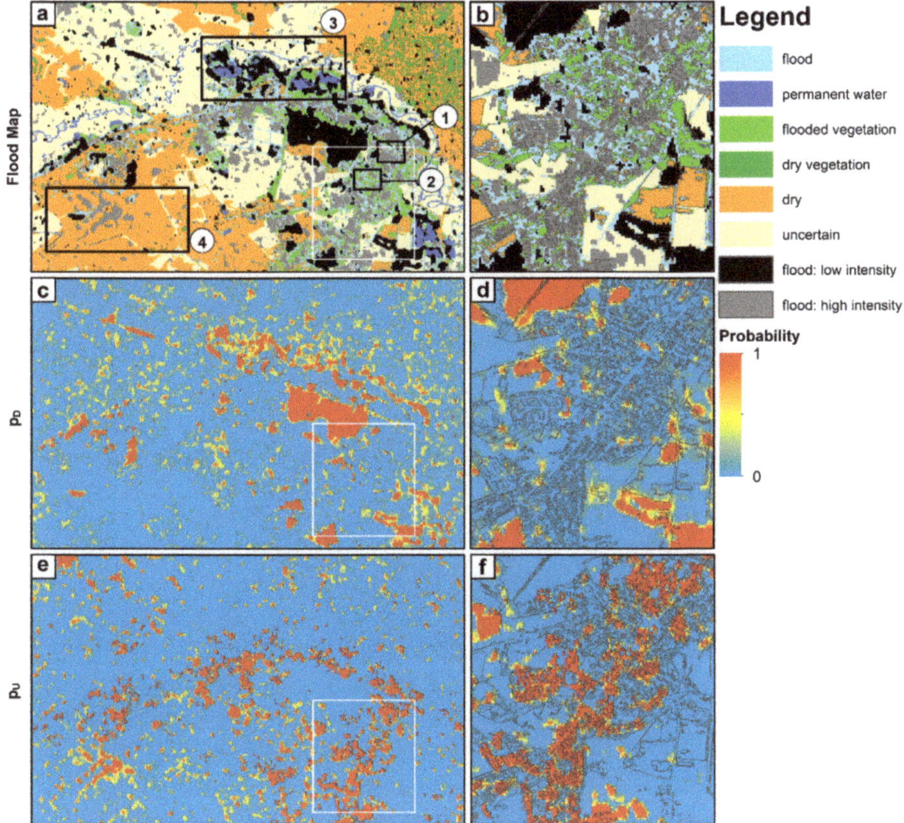

Figure 6. Flood mapped by the probability threshold of $p = 0.5$ for (**a**) the whole area of interest (AOI) and (**b**) the urban area. The second and third panels are for the probability maps of (**c–d**) intensity decrease (p_D) and (**e–f**) intensity increase (p_U). Numbers in white circles are the patch IDs used in discussion.

On the reliability diagram (Figure 8), we saw that the plots were mostly below the 1:1 line, with the flood probability larger than the observed frequency, except for the first and second bin ($p_F = 0 \sim 0.2$). In the first bin, p_o was mainly determined by FN/(FN+TN), so the ~5% higher in p_o than p_F came from the larger number of FN pixels than what p_F predicted. In the last bin, p_o was mainly determined by TP/(TP+FP), so the ~14% lower value in p_o than p_F came from the larger number of FP pixels than that predicted by p_F (overprediction). It would be wrong to interpret the numbers as indicating that the mapping results suffered more overprediction than underprediction because the sample numbers varied greatly in each bin. The best way to view reliability is by looking at the weighted root-mean-square error between the plot and the 1:1 line (ε in (9)), which was around 13%. This can be interpreted as the average error in the probability map.

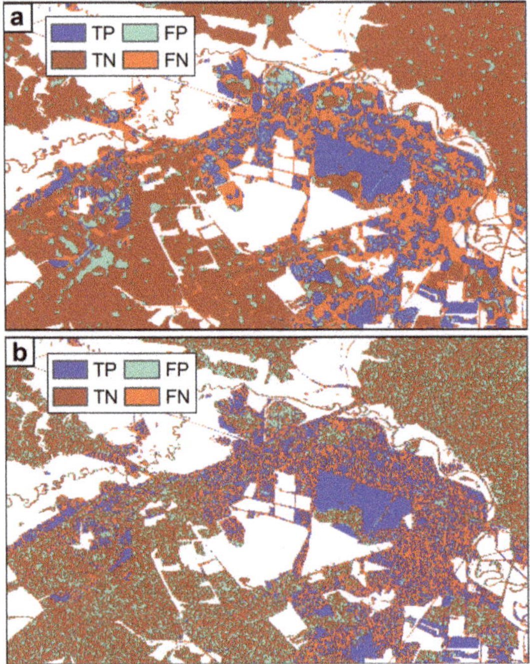

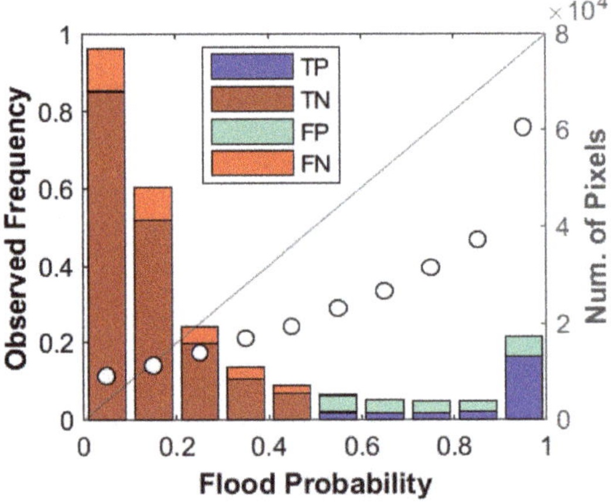

Figure 7. Comparison between (**a**) the contingency map obtained by $p = 0.5$ cutoff threshold and (**b**) the contingency map from the best result of grid search on log intensity ratio of during- and pre-event image (i.e., nontemporal analysis). Pixels in the Uncertain class are masked out from the map.

Figure 8. The reliability diagram between the flood probability p_F and the observed frequency p_o (white circles). The number of pixels for each probability bin is shown in vertical bars color-coded by contingency types. The deviation of the circles from the 1:1 line represents the error for each probability bin. The pixel counts in different contingency types shows that the error comes from FN (underprediction) for probability bins below 0.5, and FP (overprediction) for probability bins above 0.5.

The evaluation metrics echoed what we found in the maps and plots (Table 2). The comparison with the best grid search result of the same normalized dataset showed that the p50-ts method could reach identical performance as the best choice of thresholds. When comparing with the best grid search result of log intensity ratio, we could see that the p50-ts method had higher overall CSI (34% vs. 24%) and OA (81% vs. 64%). There was a slight decrease in PA (~6%), but UA went from 30% for the log intensity ratio method to 57% for our technique, suggesting significant improvement in reducing the number of overpredicted (FP) pixels. In the urban area, the metrics from our method and the log intensity ratio method were similar, but the resulting flood maps showed different flood patterns (Figure 7). The difference was due to the fact that although global thresholds were used in both approaches, the time series normalization accounted for a certain level of spatial difference by considering the different history for each pixel; hence, the corresponding σ^0 thresholds spatially varied.

Table 2. Evaluation Metrics.

Type	Overall				Urban			
	CSI$^+$ [%]	OA$^+$ [%]	PA$^+$ [%]	UA$^+$ [%]	CSI [%]	OA [%]	PA [%]	UA [%]
p50-ts method	34.4	80.8	45.5	56.7	40.2	42.8	43.4	83.3
Best result of grid search* on normalized σ^0	34.1	80.7	46.4	56.2	40.1	42.9	43.0	83.8
Best result of grid search* on log intensity ratio	24.0	64.2	51.9	30.5	41.3	44.6	44.3	85.2
p50-ts method (treating Flooded Veg. as nonflooded)	33.0	82.7	50.0	48.7	36.0	49.4	41.2	72.2

* Using overall critical success index (CSI) as the criterion. For reference purposes only. $^+$CSI = critical success index; OA = overall accuracy; PA = producer's accuracy; UA = user's accuracy

In summary, the p50-ts method improved flood mapping in the case of Hurricane Matthew flood within the Lumberton area. The result is close to the best of what can be done with the optimal uniform thresholding on a pair of SAR images. However, the result still suffers obvious underprediction and overprediction. Next, we discuss the potential sources for the false predictions.

6. Discussion

The time series normalization allows us to identify the flood-related double bounce pixels and specular reflection pixels of statistical significance. With that we can study the statistical distribution for these two effects. We sorted out the pixels that were in the Flood class and mapped as flooded due to intensity increase, and we interpreted these as due to double-bounce effect. We compared their histogram with histograms for (1) pixels in the Flood class mapped as flooded due to intensity decrease, interpreted as open water flood that saw specular reflection, and (2) pixels in the Dry class mapped as nonflooded. The plots in Figure 9 indicate that, from the histogram perspective, the normalized during-event intensity (Figure 9c) could better separate the pixels that saw double-bounce scattering than the log intensity ratio (Figure 9b). For the open water flood, the normalized during-event intensity and the log intensity ratio had similar efficacies in separating them from the nonflooded. The single during-event intensity (Figure 9a) had the worst histogram separation among all three. The better separation capability for double-bounce scattering marks the value of this method in studying urban floods.

We also looked at the intensity changes with respect to the p50 threshold in the six validation classes individually (Figure 10). In the Flood class, only 50% of the pixels were detected as flood, among which 32% were identified with σ^0 decrease and 18% with σ^0 increase. In the remaining 50% there was no clear σ^0 anomaly based on the thresholds given. In the Flooded Vegetation class, we saw more pixels with σ^0 increase (18%) than those with σ^0 decrease (12%). However, about 70% of the pixels in the Flooded Vegetation class did not see significant σ^0 changes. In the Permanent Water class, 22% of the pixels were identified as flood by σ^0 decrease. As for the Dry and Dry Vegetation classes, we saw a small fraction of false positives (6%–10%). Next, we would like to address the potential sources that cause the underprediction in the Flood and Flooded Vegetation class as well as the overprediction in the Permanent Water and Dry class.

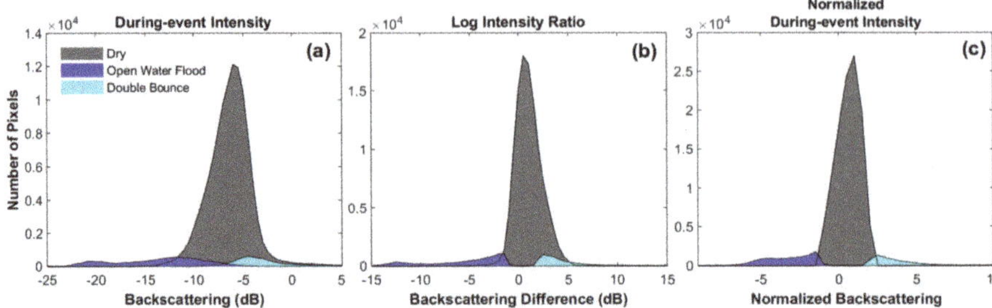

Figure 9. The histogram comparison between the pixels in the Dry class and mapped as nonflooded (grey), Flood class and mapped as flooded with intensity decrease (cyan, interpreted as open water flood), and Flood class and mapped as flood with intensity increase (cyan, interpreted as double bounce scattering). (**a**) Histogram for the pixel during-event intensity. (**b**) Histogram for the pixel log intensity ratio. (**c**) Histogram for the pixel during-event intensity after time-series normalization.

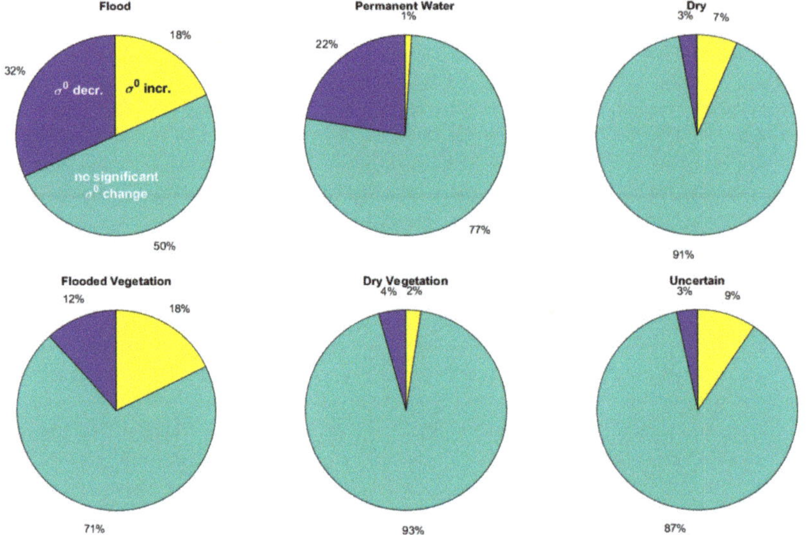

Figure 10. Pie charts showing, for each of the 6 classes, the proportions of pixels detected as flooded with either σ^0 decrease (blue) or σ^0 increase (yellow), or as nonflooded with insignificant σ^0 change at the p50 thresholds.

6.1. Uncertainties in the Validation Dataset

One possible source of error is uncertainties in the validation dataset. We generated the validation data by rasterizing the validation vectors of ~30 cm resolution into 15 × 15 m pixels. At a regional scale, the rasterized image agreed in general with the vector file (Figure 2). However, when we zoomed in to the local scale, we observed discrepancies between the two (Figure 10a,b and Figure 11a,b). The discrepancies were associated with the sizes and orientations of the objects as well as their relative portion within a pixel. The difference was particularly obvious in a setting where the objects were densely distributed but isolated from one another. Thus, although the number of pixels falsely validated may only account for a small portion in the AOI, we should always be aware of the existence of such uncertainties, especially when utilizing moderate-resolution SAR images in flood mapping.

6.2. Source of Underprediction

Underprediction (FN) accounted for the majority of false predictions. We first used patch 1 (Figure 6a) to discuss the possible sources of underprediction. In Figure 11d, there was a clear pattern of underprediction, with four elongated zones of FN subparallel to each other. According to the pre-flood optical image, these zones were asphalt roads (Figure 11c). Several studies have pointed out that asphalt will show up as dark pixels in radar images due to its smooth surface and low subsurface soil moisture content [32–34]. What we saw in this study was that the backscattering from asphalt surfaces could be as low as that from a water surface and, hence, indistinguishable from the during-event flood in the time series. This agrees with the point made by [6], that smooth surfaces such as tarmac, paved road, and parking lots may serve as water surface-like radar response areas. In the cases where they are actually flooded, there may not be any intensity anomaly compared to dry conditions. This is one of the major sources of underprediction.

In patch 2, we saw another source of underprediction. This patch was inside a region with trees, houses, and parallel small roads. These densely packed houses and trees may cause serious shadow effects, leading to undetectable zones of constant low intensity (Figure 12c,d). Therefore, shadow was an effect that the time series normalization would not be able to deal with.

One thing worth pointing out is that shadow and smooth surfaces usually cause overprediction in the flood mapping with a single intensity image, while it causes underprediction in the log intensity ratio method as well as our time series normalization method.

There is one more phenomenon that drew our attention: only a small portion of the Flooded Vegetation was mapped as flooded (Figures 11e and 12e). Some of the tree patches showed significant during-event intensity changes while others did not (Figure 11e). The logic behind the Flooded Vegetation class is that since the small tree patches were surrounded by open water flood, backscattering could be stronger when the radar wave penetrated the tree crown and reflected at the tree trunks and at the water surface, or vice versa. This expected behavior was, however, only seen in 18% of the Flooded Vegetation class, with another 12% seeing intensity decrease that could not be explained by the double-bounce mechanism. Therefore, despite the higher probability of observing intensity increase as compared with the Dry Vegetation class (Figure 10), it is likely that the backscattering for this class was the combination of changes in trees and flood. There is no easy way to separate the contribution from these two mechanisms. We had quickly tested to re-categorize this class as nonflooded, and this gave better overall OA (+2%) and PA (+4.5%) but lower CSI (−1.4%) and UA (−8%) (Table 2). As tree penetration capability is also a function of radar wavelength and polarization [35], there may not be a single answer to what would be the most appropriate way to validate the flood mapping in this class.

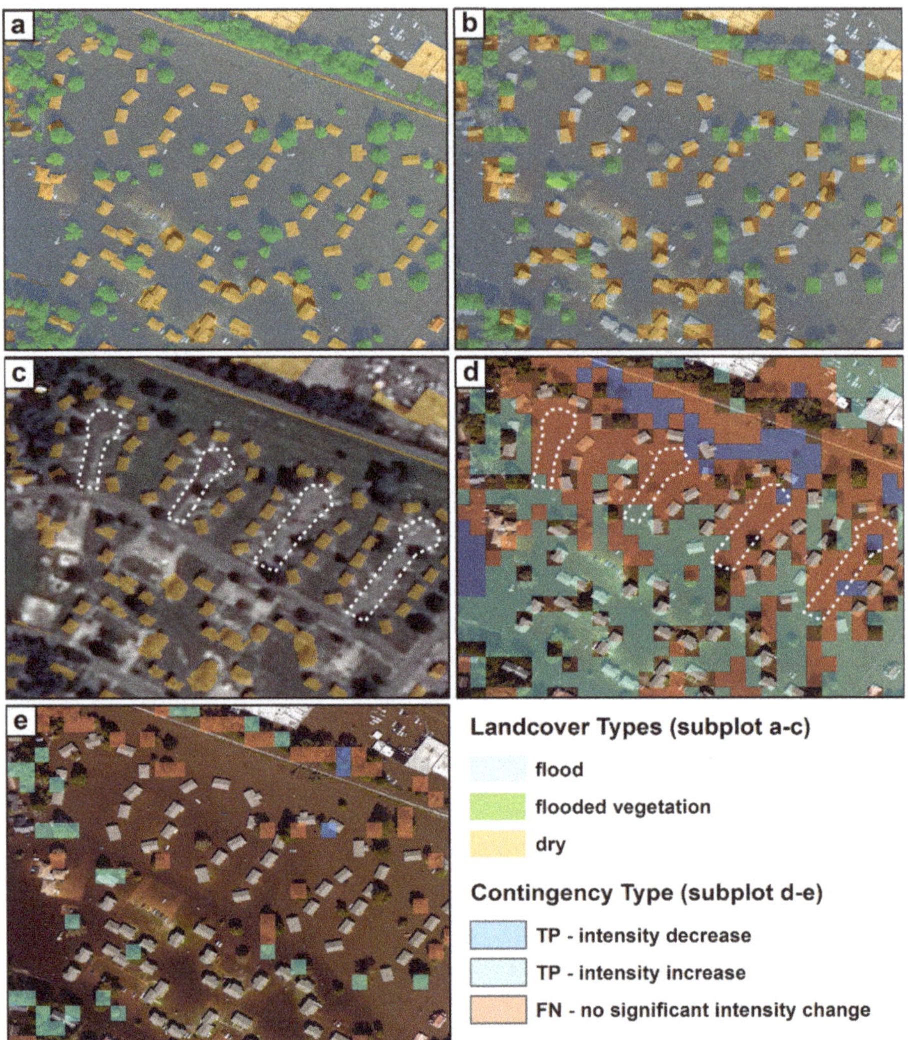

Figure 11. The during-event aerial photo for patch 1 (see Figure 6a for location), overlaid with (**a**) the land cover classes in vector format, and (**b**) the rasterized land cover map in 15 × 15m resolution. (**c**) The pre-event satellite image overlaid with the Dry class for reference. The white dashed lines represent asphalt driveways. (**d**) The contingency map for the Flood class. (**e**) The contingency map for the Flooded Vegetation class.

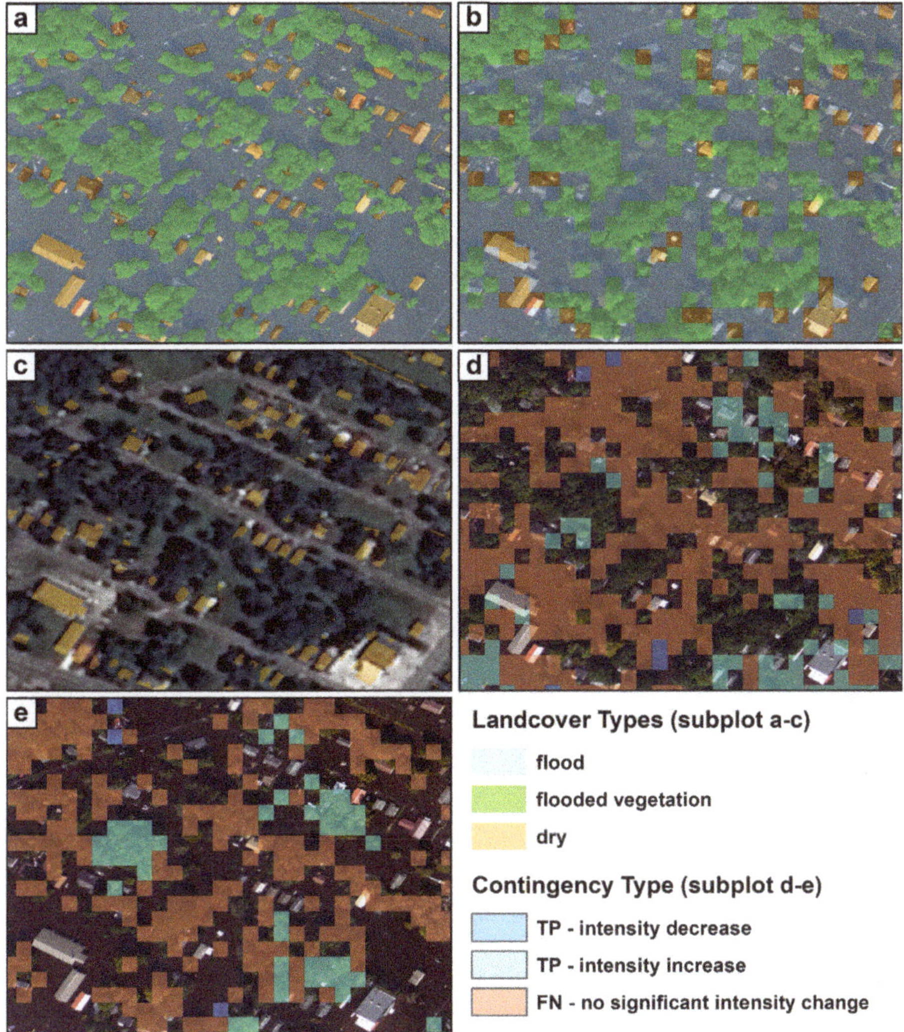

Figure 12. Same caption as Figure 10 but for patch 2. See Figure 6a for location.

6.3. Source of Overprediction

A large proportion of overprediction (FP) was seen in the Permanent Water class (Figure 10). One example is in patch 3, where part of the pond water was mapped as flooded (Figure 13a-c). When we compared the pre-event and during-event optical images, we observed that some of the ponds were possibly vegetated with aquatic plants. As aquatic plants such as macrophytes are known to cause high backscattering intensity in C-band SAR [35], it is likely that we would see a during-event intensity decrease when the pond water level increased higher than the plants. Vegetation in the permanent waterbody is, therefore, one potential source of overprediction.

Another source of overprediction comes from the during-event intensity increase in the Dry class (Figure 11). One example is shown in patch 4, in which the soil underneath the sparse low meadow appeared darker in the during-event aerial image (Figure 12e) than the pre-event satellite image (Figure 12d); However, there was no water visible on the surface. This is a known effect of soil moisture, where the rise in soil moisture will also give rise to a radar backscattering intensity increase [36,37]. This effect is more obvious in bare soil or sparse meadow land cover types and, hence, affects more the prediction of rural floods than urban floods.

Finally, the increase in backscattering intensity from buildings, possibly due to cumulated water on the roof, also accounted for some overprediction in the Dry class, although the proportion was relatively small in this case study.

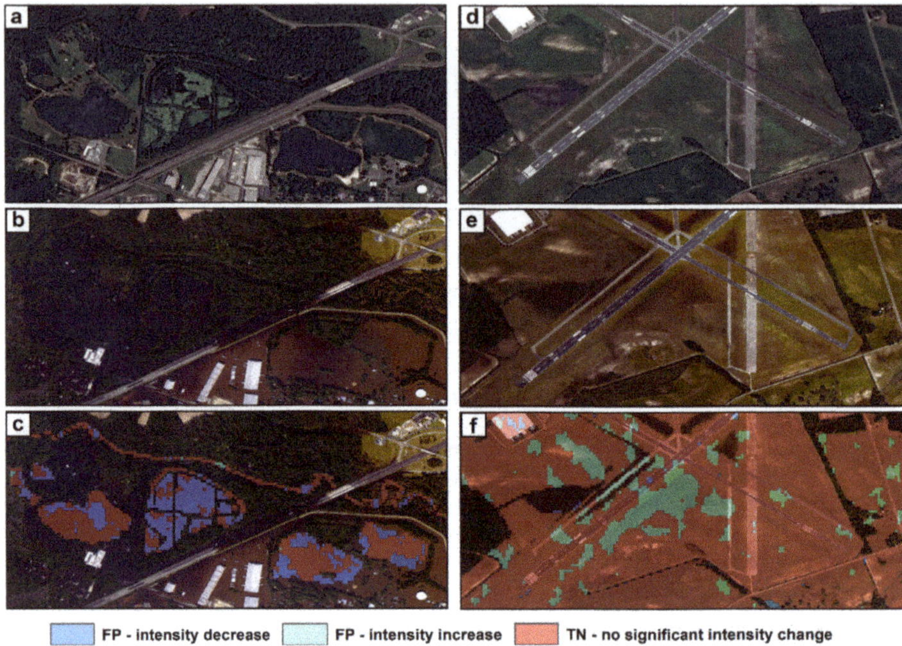

Figure 13. The (**a**) pre-event optical image and (**b**) during-event aerial photo for patch 3. See Figure 6a for location. (**c**) The contingency map for the Flood class. (**d–e**) Same as (**a–b**) for patch 4. (**f**) The contingency map for the Dry class.

7. Conclusions

In this paper we presented an approach to utilize multitemporal SAR intensity information in a Bayesian probability framework for mapping floods in Lumberton, North Carolina, caused by the 2016 Hurricane Matthew. We normalized during-event SAR intensity observations with statistics from the SAR intensity time series, and we computed the flood probability with prior and likelihood functions. Flood detections based on a cutoff probability of 0.5 showed improved performance when compared with results from an approach that used the optimal uniform threshold in pre- and during-SAR intensity pair analyses. The mapping result showed that a high percentage of the urban flood was associated with SAR intensity increase (double-bounce effect), and the urban flood in Lumberton was a complicated mosaic of pixels, with during-event intensity increase and intensity decrease as well as pixels without significant intensity changes. Underprediction was as high as 50%, which we interpreted to be mainly associated with asphalt surface cover and shadow effects. Overprediction is possibly related to vegetation in permanent water bodies and local soil moisture increase.

Author Contributions: Y.N.L. and S.-H.Y. carried out the algorithm development and experiments together, while S.-H.Y. provided the conceptual model and idea about the study area. A.B. produced the validation dataset. E.M.H. was heavily involved in all the discussions and manuscript/proposal writing.

Remote Sens. **2019**, *11*, 1778

Funding: This research was supported by the Earth Observatory of Singapore via its funding from the National Research Foundation Singapore, the Singapore Ministry of Education under the Research Centres of Excellence initiative, and by Singapore National Research Foundation Investigatorship Award No. NRF-NRFI05-2019-0009. Part of the research was carried out at the Jet Propulsion Laboratory, California Institute of Technology, under a contract with the National Aeronautics and Space Administration (NASA Program NNH16ZDA001N-GEO).

Acknowledgments: We sincerely thank NOAA National Geodetic Survey Remote Sensing Division for acquiring and sharing the airborne optical imagery in a timely manner. We thank student interns Tan Fang Yi, Lisa Lim Yan Xin, Jana Lim Shu Xian, and Skye Lee Wee Ru at Nanyang Technological University for their tireless efforts to construct the validation dataset. We also thank Jungkyo Jung for valuable discussions throughout the course of this study. We thank the reviewers for their detail and constructive comments and suggestions. This is EOS paper number 248.

Conflicts of Interest: The authors declare no conflict of interest. The funders had no role in the design of the study; in the collection, analyses, or interpretation of data; in the writing of the manuscript, or in the decision to publish the results.

References

1. Sentinel Asia. *Sentinel Asia Sentinel Asia Annual Report 2017*; Japan Aerospace Exploration Agency: Tokyo, Japan, 2019; p. 48.

2. Mason, D.C.; Speck, R.; Devereux, B.; Schumann, G.J.P.; Neal, J.C.; Bates, P.D. Flood detection in urban areas using TerraSAR-X. *IEEE Trans. Geosci. Remote Sens.* **2010**, *48*, 882–894. [CrossRef]

3. Ferro, A.; Brunner, D.; Bruzzone, L.; Lemoine, G. On the Relationship between Double Bounce and the Orientation of Buildings in VHR SAR Images. *IEEE Geosci. Remote Sens. Lett.* **2011**, *8*, 612–616. [CrossRef]

4. Pulvirenti, L.; Chini, M.; Pierdicca, N.; Boni, G. Use of SAR data for detecting floodwater in urban and agricultural areas: The role of the interferometric coherence. *IEEE Trans. Geosci. Remote Sens.* **2016**, *54*, 1532–1544. [CrossRef]

5. Mason, D.C.; Davenport, I.J.; Neal, J.C.; Schumann, G.J.P.; Bates, P.D. Near real-time flood detection in urban and rural areas using high-resolution synthetic aperture radar images. *IEEE Trans. Geosci. Remote Sens.* **2012**, *50*, 3041–3052. [CrossRef]

6. Giustarini, L.; Hostache, R.; Matgen, P.; Schumann, G.J.P.; Bates, P.D.; Mason, D.C. A change detection approach to flood mapping in urban areas using TerraSAR-X. *IEEE Trans. Geosci. Remote Sens.* **2013**, *51*, 2417–2430. [CrossRef]

7. Pulvirenti, L.; Pierdicca, N.; Chini, M.; Guerriero, L. An algorithm for operational flood mapping from synthetic aperture radar (SAR) data using fuzzy logic. *Nat. Hazards Earth Syst. Sci.* **2011**, *11*, 529–540. [CrossRef]

8. Mason, D.C.; Giustarini, L.; Garcia-Pintado, J.; Cloke, H.L. Detection of flooded urban areas in high resolution Synthetic Aperture Radar images using double scattering. *Int. J. Appl. Earth Obs. Geoinf.* **2014**, *28*, 150–159. [CrossRef]

9. Tanguy, M.; Chokmani, K.; Bernier, M.; Poulin, J.; Raymond, S. River flood mapping in urban areas combining RADARSAT-2 data and flood return period data. *Remote Sens. Environ.* **2017**, *198*, 442–459. [CrossRef]

10. Chini, M.; Pelich, R.; Pulvirenti, L.; Pierdicca, N.; Hostache, R.; Matgen, P. Sentinel-1 InSAR Coherence to Detect Floodwater in Urban Areas: Houston and Hurricane Harvey as a Test Case. *Remote Sens.* **2019**, *11*, 107. [CrossRef]

11. Li, Y.; Martinis, S.; Wieland, M. Urban flood mapping with an active self-learning convolutional neural network based on TerraSAR-X intensity and interferometric coherence. *ISPRS J. Photogramm. Remote Sens.* **2019**, *152*, 178–191. [CrossRef]

12. Chaabani, C.; Chini, M.; Abdelfattah, R.; Hostache, R.; Chokmani, K. Flood Mapping in a Complex Environment Using Bistatic TanDEM-X/TerraSAR-X InSAR Coherence. *Remote Sens.* **2018**, *10*, 1873. [CrossRef]

13. Hostache, R.; Matgen, P.; Wagner, W. Change detection approaches for flood extent mapping: How to select the most adequate reference image from online archives? *Int. J. Appl. Earth Obs. Geoinf.* **2012**, *19*, 205–213. [CrossRef]

14. Schlaffer, S.; Matgen, P.; Hollaus, M.; Wagner, W. Flood detection from multi-temporal SAR data using harmonic analysis and change detection. *Int. J. Appl. Earth Obs. Geoinf.* **2015**, *38*, 15–24. [CrossRef]

15. Schlaffer, S.; Chini, M.; Giustarini, L.; Matgen, P. Probabilistic mapping of flood-induced backscatter changes in SAR time series. *Int. J. Appl. Earth Obs. Geoinf.* **2017**, *56*, 77–87. [CrossRef]

16. Addabbo, A.D.; Refice, A.; Pasquariello, G.; Lovergine, F.P.; Capolongo, D.; Manfreda, S. A Bayesian network for flood detection combining SAR imagery and ancillary data. *IEEE Trans. Geosci. Remote Sens.* **2016**, *54*, 3612–3625. [CrossRef]

17. Clement, M.A.; Kilsby, C.G.; Moore, P. Multi-temporal synthetic aperture radar flood mapping using change detection. *J. Flood Risk Manag.* **2018**, *11*, 152–168. [CrossRef]

18. Ouled Sghaier, M.; Hammami, I.; Foucher, S.; Lepage, R. Flood extent mapping from time-series sar images based on texture analysis and data fusion. *Remote Sens.* **2018**, *10*, 237. [CrossRef]

19. Stewart, S.R. *Tropical Cyclone Report: Hurricane Matthew*; National Hurricane Center: Miami, FL, USA, 2017; p. 96.

20. National Geodetic Survey, Remote Sensing Division (Data Producer). *Hurricane Matthew: Rapid Response Imagery of the Surrounding Regions*; OAA's Ocean Service: Silver Spring, MD, USA, 2016.

21. Farr, T.G.; Rosen, P.A.; Caro, E.; Crippen, R.; Duren, R.; Hensley, S.; Kobrick, M.; Paller, M.; Rodriguez, E.; Roth, L.; et al. The Shuttle Radar Topography Mission. *Rev. Geophys.* **2007**, *45*. [CrossRef]

22. Miranda, N.; Meadows, P.J. *Radiometric Calibration of S-1 Level-1 Products Generated by the S-1 IPF*; ESA-EOPG-CSCOP-TN-0002; European Space Agency: Paris, France, 2015.

23. Jong-Sen, L.; Papathanassiou, K.P.; Ainsworth, T.L.; Grunes, M.R.; Reigber, A. A new technique for noise filtering of SAR interferometric phase images. *IEEE Trans. Geosci. Remote Sens.* **1998**, *36*, 1456–1465. [CrossRef]

24. Jong-Sen, L.; Grunes, M.R.; Grandi, G.D. Polarimetric SAR speckle filtering and its implication for classification. *IEEE Trans. Geosci. Remote Sens.* **1999**, *37*, 2363–2373. [CrossRef]

25. Pathe, C.; Wagner, W.; Sabel, D.; Doubkova, M.; Basara, J.B. Using ENVISAT ASAR global mode data for surface soil moisture retrieval over Oklahoma, USA. *IEEE Trans. Geosci. Remote Sens.* **2009**, *47*, 468–480. [CrossRef]

26. Huffman, G.; Bolvin, D.; Braithwaite, D.; Hsu, K.; Joyce, R.; Xie, P. *Integrated Multi-Satellite Retrievals for GPM (IMERG), Version 4.4*; NASA's Precipitation Processing Center: Washington, DC, USA, 2014.

27. Hetland, E.A.; Musé, P.; Simons, M.; Lin, Y.N.; Agram, P.S.; DiCaprio, C.J. Multiscale InSAR Time Series (MInTS) analysis of surface deformation. *J. Geophys. Res.* **2012**, 117. [CrossRef]

28. Riel, B.; Simons, M.; Agram, P.; Zhan, Z. Detecting transient signals in geodetic time series using sparse estimation techniques. *J. Geophys. Res.* **2014**, *119*, 5140–5160. [CrossRef]

29. Giustarini, L.; Hostache, R.; Kavetski, D.; Chini, M.; Corato, G.; Schlaffer, S.; Matgen, P. Probabilistic Flood Mapping Using Synthetic Aperture Radar Data. *IEEE Trans. Geosci. Remote Sens.* **2016**, *54*, 6958–6969. [CrossRef]

30. Marquardt, D. An Algorithm for Least-Squares Estimation of Nonlinear Parameters. *J. Soc. Ind. Appl. Math.* **1963**, *11*, 431–441. [CrossRef]

31. Chini, M.; Hostache, R.; Giustarini, L.; Matgen, P. A hierarchical split-based approach for parametric thresholding of SAR images: Flood inundation as a test case. *IEEE Trans. Geosci. Remote Sens.* **2017**, *55*, 6975–6988. [CrossRef]

32. Landuyt, L.; Wesemael, A.V.; Schumann, G.J.; Hostache, R.; Verhoest, N.E.C.; Coillie, F.M.B.V. Flood Mapping Based on Synthetic Aperture Radar: An Assessment of Established Approaches. *IEEE Trans. Geosci. Remote Sens.* **2019**, *57*, 722–739. [CrossRef]

33. Aronica, G.; Bates, P.D.; Horritt, M.S. Assessing the uncertainty in distributed model predictions using observed binary pattern information within GLUE. *Hydrol. Process.* **2002**, *16*, 2001–2016. [CrossRef]

34. Wilks, D.S. *Statistical Metods in the Atmospheric Sciences*, 3rd ed.; Elsevier, Academic Press: Oxford, UK, 2011.

35. Hess, L.L.; Melack, J.M.; Filoso, S.; Wang, Y. Delineation of inundated area and vegetation along the Amazon floodplain with the SIR-C synthetic aperture rada. *IEEE Trans. Geosci. Remote Sens.* **1995**, *33*, 896–904. [CrossRef]

36. Paloscia, S.; Pettinato, S.; Santi, E.; Notarnicola, C.; Pasolli, L.; Reppucci, A. Soil moisture mapping using Sentinel-1 images: Algorithm and preliminary validation. *Remote Sens. Environ.* **2013**, *134*, 234–248. [CrossRef]

37. Susan Moran, M.; Hymer, D.C.; Qi, J.; Sano, E.E. Soil moisture evaluation using multi-temporal synthetic aperture radar (SAR) in semiarid rangeland. *Agric. For. Meteorol.* **2000**, *105*, 69–80. [CrossRef]

Article

Performance Evaluation of a Potential Component of an Early Flood Warning System—A Case Study of the 2012 Flood, Lower Niger River Basin, Nigeria

Dorcas Idowu and Wendy Zhou *

Department of Geology and Geological Engineering, Colorado School of Mines, Golden, CO 80401, USA
* Correspondence: wzhou@mines.edu; Tel.: +1-303-384-2181

Received: 8 July 2019; Accepted: 15 August 2019; Published: 21 August 2019

Abstract: Floods frequently occur in Nigeria. The catastrophic 2012 flood in Nigeria claimed 363 lives and affected about seven million people. A total loss of about 2.29 trillion Naira (7.2 billion US Dollars) was estimated. The effect of flooding in the country has been devastating because of sparse to no flood monitoring, and a lack of an effective early flood warning system in the country. Here, we evaluated the efficacy of using the Gravity Recovery and Climate Experiment (GRACE) terrestrial water storage anomaly (TWSA) to evaluate the hydrological conditions of the Lower Niger River Basin (LNRB) in Nigeria in terms of precipitation and antecedent terrestrial water storage prior to the 2012 flood event. Furthermore, we accessed the potential of the GRACE-based flood potential index (*FPI*) at correctly predicting previous floods, especially the devastating 2012 flood event. For validation, we compared the GRACE terrestrial water storage capacity (TWSC) quantitatively and qualitatively to the water budget of TWSC and Dartmouth Flood Observatory (DFO) respectively. Furthermore, we derived a water budget-based *FPI* using Reager's methodology and compared it to the GRACE-derived *FPI* quantitatively. Generally, the GRACE TWSC estimates showed seasonal consistency with the water budget TWSC estimates with a correlation coefficient of 0.8. The comparison between the GRACE-derived *FPI* and water budget-derived *FPI* gave a correlation coefficient of 0.9 and also agreed well with the flood reported by the DFO. Also, the *FPI* showed a marked increase with precipitation which implies that rainfall is the main cause of flooding in the study area. Additionally, the computed GRACE-based storage deficit revealed that there was a decrease in water storage prior to the flooding month while the *FPI* increased. Hence, the GRACE-based *FPI* and storage deficit when supplemented with water budget-based *FPI* could suggest a potential for flood prediction and water storage monitoring respectively.

Keywords: flood; *FPI*; GRACE; terrestrial water storage anomaly; storage deficit

---

1. Introduction

Flooding is a major disaster in Nigeria, especially along the Niger and Benue Rivers. In Nigeria, it occurs in three main forms: Coastal floods which occur in mangrove and delta coastlines; river floods which occur on the flood plains of large rivers; and flash floods which are short-lived events developing in less than 6 hours from rainfall to the onset of flooding [1,2].

In 2012, heavy rainfall during the wet season combined with the release of water from Ladgo Dam in Cameroon led to a catastrophic flooding event that affected 30 states out of the 36 states of the country. The flooding which was described by the Nigeria National Emergency Management Agency as the worst in 40 years, claiming 363 lives, affecting about 7 million people, while a total loss of about 2.29 trillion Naira (7.2 billion US Dollars) was estimated.

The developed countries are still affected by disasters resulting from floods but have flood alert systems (such as the European Flood Alert System [3] and the US National Weather Service Automated

Flood Warning System [4]) in place which are effective in providing a monitoring and warnings service. A significant portion of the economic losses caused by floods occur in developing countries where ground flood monitoring and management programs are still inefficient, and the costs of building water control infrastructure such as dams, weirs, embankments and gauging stations can be prohibitive [5]. Also, ground-based methods used to monitor floods are based on hydro-meteorological data such as discharge and precipitation which are time-consuming in terms of collection and processing and are also affected by varying weather conditions. Furthermore, worthy of mentioning is the recent problem of security in Nigeria which may also inhibit the installation of these systems.

Over the years, there have been novel advances on remote sensing for forecasting and monitoring hydrological extremes such as floods and droughts. Victor [6] studied the use of satellite data for flood delineation, monitoring and prediction. Nasreddine et al. [7] developed a new flood forecasting approach for flood disaster management in poorly or totally ungauged watersheds using precipitation measurements. Sheffield et al. [8] and Zhang et al. [9] applied satellite data to monitoring and forecasting drought.

In recent decades, satellite data availability has improved dramatically and to complement the ground-based observations, flood monitoring has increasingly relied on the products obtained with space-borne sensors such as National Aeronautics and Space Administration (NASA) advanced microwave scanning radiometer for EOS (AMSR-E) [10] and moderate resolution imaging spectroradiometer (MODIS) [11]. Zhan et al. [12] explored the marginal benefit of incorporating space-borne soil moisture measurements into a hydrologic model for improved streamflow and flood prediction. They incorporated the surface soil moisture data from the AMSR-E into the Noah land surface model within the land information system (LIS). Their findings suggested the potential for improving flood forecasting through the assimilation of remotely sensed soil moisture data into a hydrologic model. Among the remote sensing products that have been used for flood monitoring, prediction and forecasting, data from the Gravity Recovery and Climate Experiment (GRACE) [13,14] are unique in that the changes in the amount of terrestrial water can be directly measured.

The GRACE satellite mission was launched in March 2002. It presents a means to observe monthly variations in total/terrestrial water storage within large (>200,000 km^2) river basins based on measurements of changes in Earth's gravity field [15]. These changes result when the amount of water stored in a region increases or decreases, which produces a ripple effect leading to the gravity signal in that region increasing or decreasing proportionately. The predictive ability of a GRACE-based flood potential has been compared to flood prediction models that use traditional input data sources such as river heights, snow amounts and the wetness of surface soils [16]. The method of GRACE storage deficit estimates could be used in combination with traditional remote sensing methods of precipitation forecasting to help assess the likelihood for flooding [16]. However, their reliability and efficacy for applications in developing countries need to be assessed due to the sparse availability of ground measurement data.

Reager and Famiglietti [17] proposed the flood potential index (*FPI*) to estimate flood risks worldwide based on GRACE terrestrial water storage anomaly (TWSA) and precipitation records. A qualitative comparison of *FPI* with a record of observed floods from the Dartmouth Flood Observatory (DFO) data set suggested that the proposed *FPI* product is useful for flood risk assessment in most regions [17].

Molodtsova et al. [18] tested the *FPI* in the United States where a dense network of flood gauges has been established, and reported that, potentially, a greater use of this method is in developing countries, where due to inadequate monitoring capability, floods tend to cause significant damage and the most loss of life. Additionally, they reported that floods in African countries, as found through the DFO database, are mainly caused by heavy rainfall events, for which the *FPI* seems to perform well in predicting flood potential.

Molodtsova et al. [18] went further to study the Juba–Shabelle River Basin, a 783,000 km^2 watershed shared between Somalia and Ethiopia, and found an increasing *FPI* was in the watershed one month prior to the flood and during the month of the flood, where both predictions agreed well with the actual flood extent area reported by the DFO. Based on their analysis, they inferred that developing countries with sparse or inadequate flood monitoring networks are potential beneficiaries of this approach.

Sun et al. [19] evaluated the GRACE *FPI* over the Yangtze River Basin (YRB) in China and suggested that estimates of terrestrial water storage based on GRACE, measured as FPI, are critical for understanding and predicting flooding. Thus, they concluded that GRACE data can be effectively used for monitoring and examining large floods in the YRB and elsewhere.

For our study, we chose the Lower Niger River Basin, Nigeria (Figure 1) as our area of interest, and the flooding event in 2012 as our case study (Table 1) because it was the worst in 40 years. We investigated the capacity of the GRACE TWSA (terrestrial water storage capacity, TWSC) in accurately capturing and predicting the 2012 flood event, and other flood years within the basin. We also evaluated the hydrological condition of the basin in terms of the available storage and predisposition to flooding. The GRACE-derived *FPI* was validated using the DFO report and compared to a water budget-derived *FPI*.

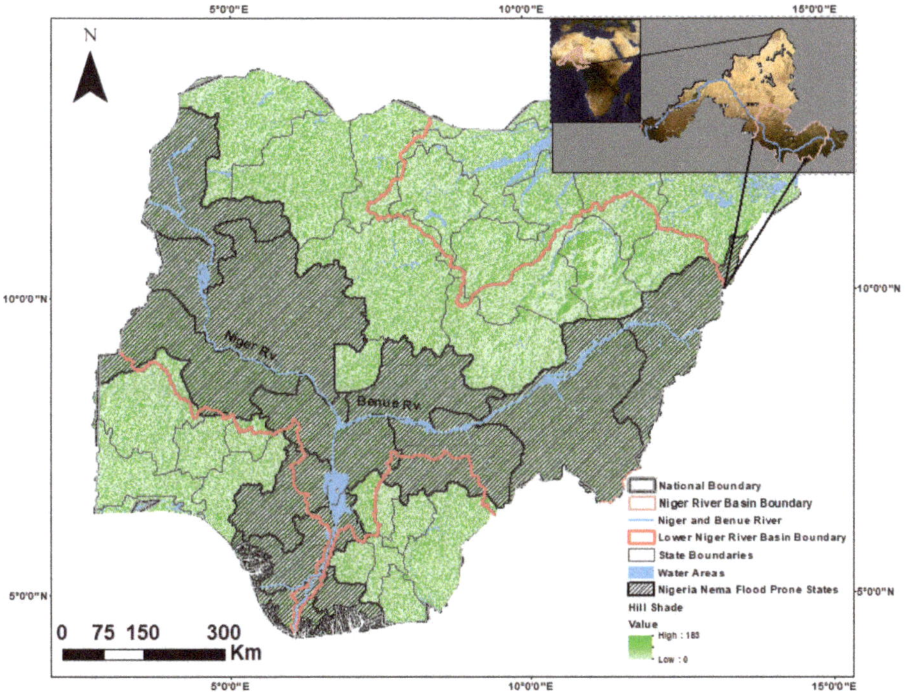

Figure 1. Map showing the boundaries of the Lower Niger River Basin in Nigeria with the states prone to Flooding (Nigeria National Emergency Management Agency).

Table 1. Statistics for the most devastating flood disasters in 2012 (compiled by authors based on [20]).

Country	Date	Death Toll	Number of People Affected	Number of People Displaced	Cause(s)	Cost of Damage
Pakistan	September	455	>5,000,000	350,000	Heavy monsoon rains	N/A
Nigeria	July–October	363	7,000,000	2,100,000	Heavy rains and water release from Dam	US $7.2 billion
North Korea	July–September	330	N/A	241,547	Torrential rains and tropical storm Khanun	N/A
Russia	July	171	30,000	13,000	Heavy rainfall	N/A
Philippines	August	95	1,230,000	15,134	Torrential rains	US $14.31 million
China	July	79	>1,600,000	56,933	Heavy rainfall	US $1.6 billion
India	August	35	>12,000	N/A	Monsoon rainfall	US $89 million
Nepal	May	26	N/A	N/A	Flooding from the outburst of a landslide dam	N/A

2. Data and Methods

2.1. Study Area

The Lower Niger River Basin (LNRB) is so termed because of its location within the Niger River Basin (NRB). Located in West Africa, the NRB covers 7.5% of the continent and cuts across ten countries. With a total area of approximately 2.2 million km^2 and a total length of 4100 km, the NRB is the third-longest river in Africa. It is divided into four parts, the Upper Niger River System, the Inner Delta, the Middle Niger River System and the LNRB. The major river within the basin is the Niger River which starts in the highlands of Guinea (upstream) threading eastwards mainly through Mali, Niger and Nigeria (downstream) before entering the Gulf of Guinea to the Atlantic Ocean. Its unusual crescent shape takes it inland towards the Sahara before turning south-west to the Gulf. Along its route, the river hydrology changes from its rain-fed headwaters, it loses flow and volume as it nears the Sahara where it forms an inland delta. The inland delta is an area of high evaporation that is composed of a number of slow-moving channels. Only after the Benue River joins the river in Nigeria does it become a large river once more. The Benue River (Figure 1) which is the major tributary that feeds the Niger River Basin in Nigeria meets the Niger River (Figure 1) to form a confluence in Lokoja (Figure 2), Nigeria. Rivers Niger and Benue (Figures 1 and 2) are the two largest rivers in West Africa. The water in the Niger River is partially regulated through dams.

In September, the Benue reaches its flood level. It begins to fall in October and falls rapidly in November, continuing slowly over the next three months to reach its lowest level in March and April. Annually, these rivers experience flooding as a result of the annual heavy rainfall which coincides with the wet season in Nigeria [21] and because of poor urban planning, settlements located within the floodplains and in the proximity of the river get flooded [22] (Figure 2).

Figure 2. NASA's Terra (Moderate Resolution Imaging Spectroradiometer) satellite images showing the pre flood (normal river geometry) and post flood river geometry of the Benue River and Niger and Benue confluence point. Post – flood image was captured in 2012. (Modified after [23])

In 2012, the flooding was devastating and in spite of the increasing awareness in combating flood hazard in along the rivers, the menace had recurred. This is because past flood control strategies have not achieved the desired result due to a lack of understanding of the hydrological variables that influence the persistence of these floods. The Nigerian National Emergency Management Agency (NEMA), in 2012 listed the flood-prone states in the country. Most of these states are within the LNRB and along the Niger and Benue Rivers (Figure 1).

2.2. Datasets

2.2.1. GRACE Terrestrial Water Storage Anomaly Products

The three official solutions (spherical harmonics solutions), the JPL (Jet Propulsion Laboratory), GFZ (GeoforschungsZentrum Potsdam) and CSR (Center for Space Research at University of Texas, Austin) of the GRACE RL05 TWSA product [24] were downloaded (http://grace.jpl.nasa.gov), from January 2004 to December 2012. The workflow in Figure 3 was applied to the datasets to derive the TWSA for the baseline of the study. The scaling factor suggested by the GRACE Tellus data portal [24] was applied to the GRACE data to account for the attenuation of small-scale surface mass variations [25]. For some years within our baseline of our study, some monthly TWSA data were missing. This is because, since early 2011, the GRACE instruments were periodically turned off due to active battery management. Those months were not considered in our study.

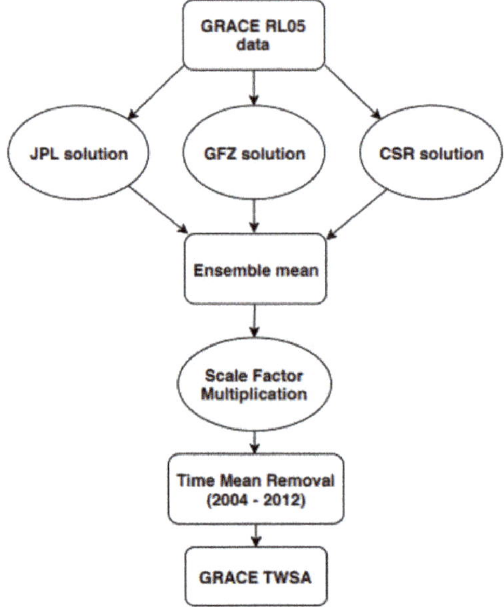

Figure 3. The basic workflow for the gravity recovery and climate experiment (GRACE) RL05 processing.

2.2.2. Evaluation of GRACE and Water Budget Terrestrial Water Storage Change (TWSC)

The GRACE TWSA was evaluated against the traditional water balance estimates before being used in generating the *FPI*. First, we calculated the TWSC from GRACE TWSA, then from the traditional water budget equation. The following water balance equation was used:

$$\frac{ds}{dt} = P - R - ET - SM - GW \tag{1}$$

where $\frac{ds}{dt}$ is the monthly change in terrestrial water storage, P is monthly precipitation, R is a monthly runoff, ET is monthly evapotranspiration, SM is soil moisture and GW is groundwater.

The change was calculated for our time steps using Equation (2).

$$ds/dt = TWSC(t) - TWSC(t-1)/t \tag{2}$$

For this study, the water balance data (P, R, ET, SM and GW) were obtained from the eartH2Observe water cycle integrator (WCI) [26].

2.2.3. Global Precipitation Climatology Centre (GPCC)

The 1×1 GPCC precipitation data, provided by the NOAA/OAR/ESRL PSD, Boulder, Colorado, USA [27] were used for deriving the *FPI*. Monthly data from January 2004 to December 2012 were downloaded while the datasets corresponding to the missing datasets in GRACE TWSA were removed to create consistency in data comparison.

2.2.4. Dartmouth Flood Observatory

Since ground flood monitoring data range from sparse to not available in the study area, the DFO data were used as validation of the performance of the DFO data beginning in 1985 and is based on flood reports from news and governmental sources and therefore mainly refers to large floods in densely populated regions. It also classifies a large flood event by the significant damage to structures, agricultural land, loss to human life and/or long duration [18]. The DFO data was downloaded as a GIS shapefile set providing catalog numbers and area affected map outlines, with much of the tabular attribute data (e.g., dates, duration and fatalities) also included. It is worthy of note that DFO data are mainly based on media reports which are expected to be biased towards the more densely populated regions and/or regions of interest [18].

2.3. Methods

2.3.1. GRACE-Derived Flood Potential Index

We followed the methodology proposed by Reager and Famiglietti [17] to compute monthly 2004–2012 *FPI* for the study area using the GRACE TWSA product. For each grid, we defined and computed the maximum water storage capacity (*Smax*) and storage deficit (*Sdef*). *Smax* is the historic maximum water storage capacity of the soil within a region [17] which for our study area we estimated to be the maximum of GRACE TWSA for LNRB from 2004 to 2012. The *Sdef*, which represents the available water on land before obtaining *Smax*, was calculated for each grid and month:

$$Sdef(t) = Smax - TWSA(t-1) \tag{3}$$

where $TWSA(t-1)$ is the saturation condition of the soil from the previous month [18]. The storage deficit shows how much more water the soil within an area can store before achieving the maximum capacity and was computed using the data from the previous month thus establishing a potential for forecasting. It is, however, expected that *Sdef* is low during wetter parts of the year and high during the drier part of the year. For visualization, *Sdef* for the study area was normalized to display the hydrological state of the basin in terms of available water.

GPCC monthly precipitation anomalies (*P*) were multiplied by the length of each month to estimate the amount of rainfall (in cm) that fell in the averaging interval:

$$Pmon(t) = P(t)dt. \tag{4}$$

Example of GRACE TWSC, *Sdef* and *Pmon* are shown in Figure 7 for 9.5°N, 12.5°E. The flood potential (*F*) for the month (*t*) was computed by:

$$F(t) = Pmon(t) - Sdef(t) \tag{5}$$

where *Pmon* is monthly precipitation. Flood potential, *F* (*t*) is the quantity of incoming water that cannot be stored based on the basin exceeding its maximum storage capacity. A high probability of flooding in the current month would mean a low storage deficit and high precipitation for the previous month [17]. The flood potential was further normalized to derive the Reager's flood potential index (*FPI*):

$$FPI(t) = F\frac{t}{\max[F[t]]}.$$ (6)

The values of *FPI* vary from $-\infty$ to 1 with positive values indicating that water input from precipitation is above the mean water storage and should be interpreted as a potential risk for flooding [17]. When normalized *FPI* nears 1, it indicates an abnormally high difference between precipitation and regional storage ability and therefore high flood likelihood [17]. The derived *FPI* was qualitatively validated against the DFO observational flood datasets.

2.3.2. Water Budget-Derived Flood Potential Index

The methodology in Section 2.2.1 was used in estimating the water budget *FPI*. The water budget *Smax* was estimated from the time series of the water budget TWSC. Then we calculated the water budget-based *Sdef* using Equation (3) and flood potential using Equation (5). For consistency, we used the GPCC precipitation in the equation for calculating the water budget flood potential. Using Equation (6), we derived the water budget-based flood potential index.

3. Results

3.1. Analysis of GRACE TWSA and Validation

We compared the three official GRACE TWSA data to each other and generated the time series for the three solutions which gave a correlation coefficient of approximately 0.99 (Figure 4) showing similar accuracy although processed using different solutions. As a result, we used the ensemble mean [28,29] from the three solutions in our analysis.

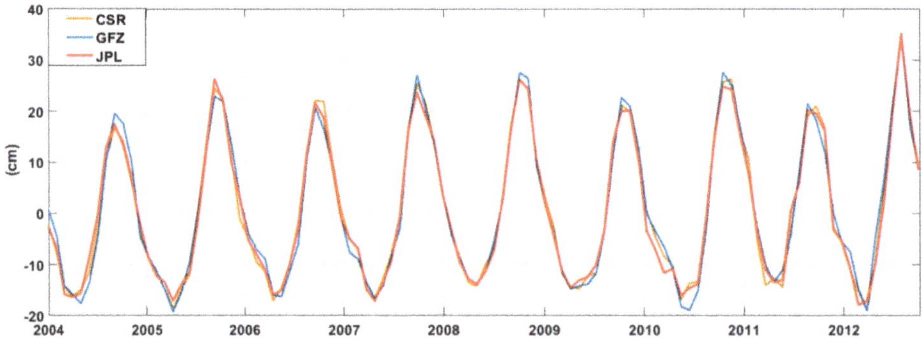

Figure 4. GRACE terrestrial water storage anomaly (TWSA) time series for the three solutions: CSR, GFZ and JPL for 9.5°N, 12.5°E from 2004 to 2012.

Furthermore, the time series of the GRACE-based TWSC and the derived water budget-based TWSC estimates show a considerable consistency with a correlation coefficient of 0.8. Additionally, they both are generally negative during the dry months (November to March) and positive during the wet months (April to October). Figures 5 and 6 display the graphical relationship between GRACE TWSC and water budget TWSC in the LNRB.

3.2. Hydrological State of the LNRB

Figures 7 and 8 graphically explain the relationship among the variables; GRACE-based TWSC, water budget-based TWSC and their respective *Sdef* and *Pmon*. It also shows how an increasing GRACE and water budget TWSC increases with precipitation, while there is a decrease in available storage relative to the other variables. The months during which these three variables intersect implies a potential for flooding to occur. In the LNRB, 2005, 2007, 2008, 2009, 2010, 2011 and 2012 were flooding years [22,30,31] which is consistent with Figures 7 and 8. Figure 7 also reveals that the 2012 flood was the worst event among all.

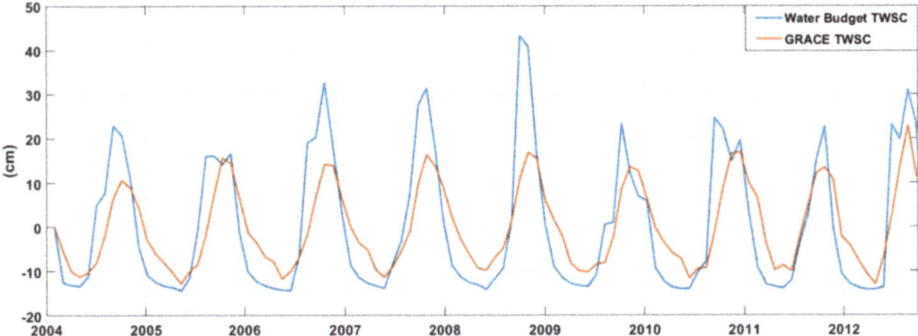

Figure 5. Comparison between GRACE terrestrial water storage capacity (TWSC) and derived water balance TWSC from 2004 to 2012 at the location of longitude 12.5 and latitude 9.5.

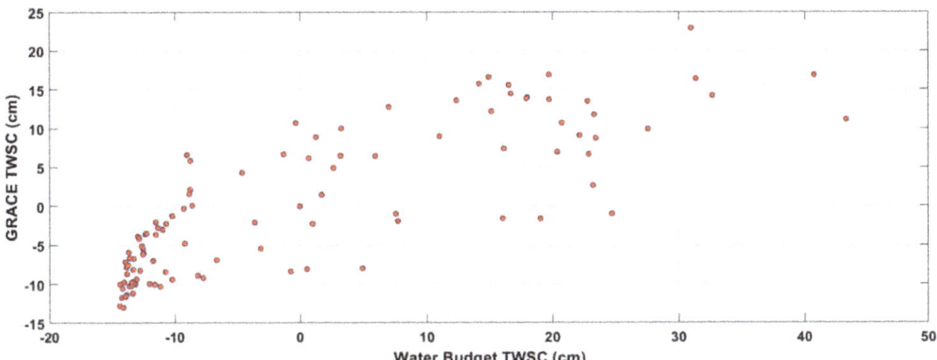

Figure 6. Scatterplot for GRACE TWSC and water budget-derived TWSC.

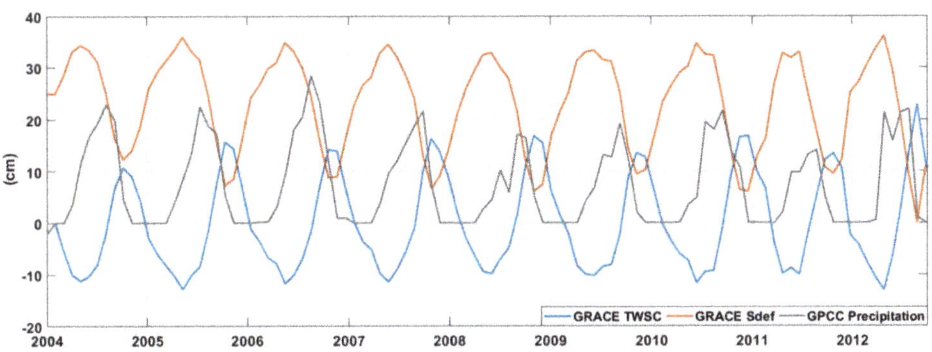

Figure 7. Variations in time series of monthly GRACE TWSC, storage deficit and precipitation for longitude 12.5 and latitude 9.5 from 2004 to 2012.

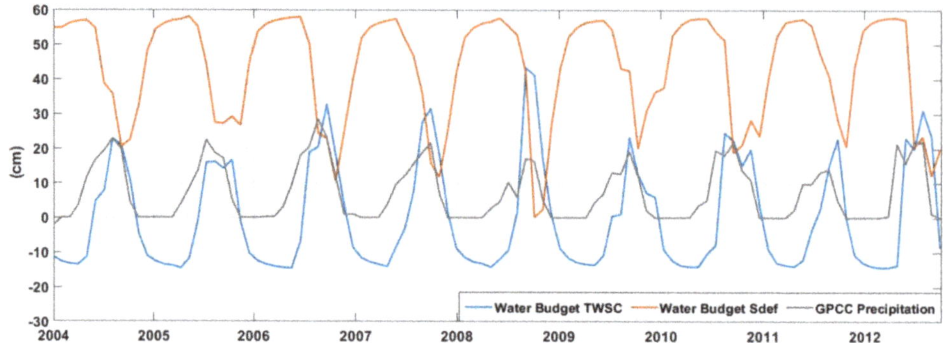

Figure 8. Variations in time series of monthly water budget TWSC, storage deficit and precipitation at the location of longitude 12.5 and latitude 9.5 from 2004 to 2012.

3.2.1. Precipitation within the LNRB

Rainy season varies for different geopolitical zones in Nigeria which by extension is applicable to the LNRB. The GPCC precipitation data corresponds to the wet season within the basin (Figure 8). The rainy season is between April and October. The driest months are January and December which implies 0 precipitation in both January and December. The peak month in the north is August and September in the south, which agrees well with our results as shown in Figure 9.

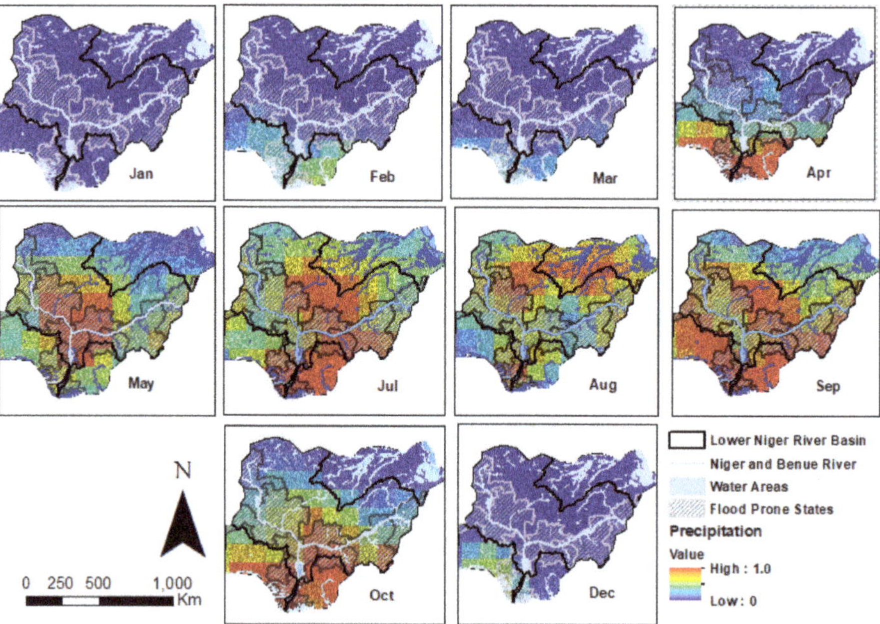

Figure 9. Spatiotemporal distribution of precipitation in the Lower Niger River Basin (LNRB) in 2012 from January to December. June and November were not displayed so as to show consistency when compared to GRACE data.

3.2.2. GRACE-Based Storage Deficit within the LNRB

Zooming in to the 2012 flood year in the LNRB, Figure 10 visually depicts how the available storage changes from surplus to deficit; because the storage deficit was derived using GRACE TWSA from the previous month, the amount of available storage for the coming month was approximated. For example, as shown in Figure 10, available storage for the month of September 2012 was derived using the GRACE TWSA from August and the result shows a low *Sdef* for September 2012. With this, we estimate the amount of hydrological input necessary to cause the system to flood, hence, establishing the potential for flood prediction. Comparing Figures 9 and 10, we can infer that the available storage began to decline from the rainfall peak months, August and September.

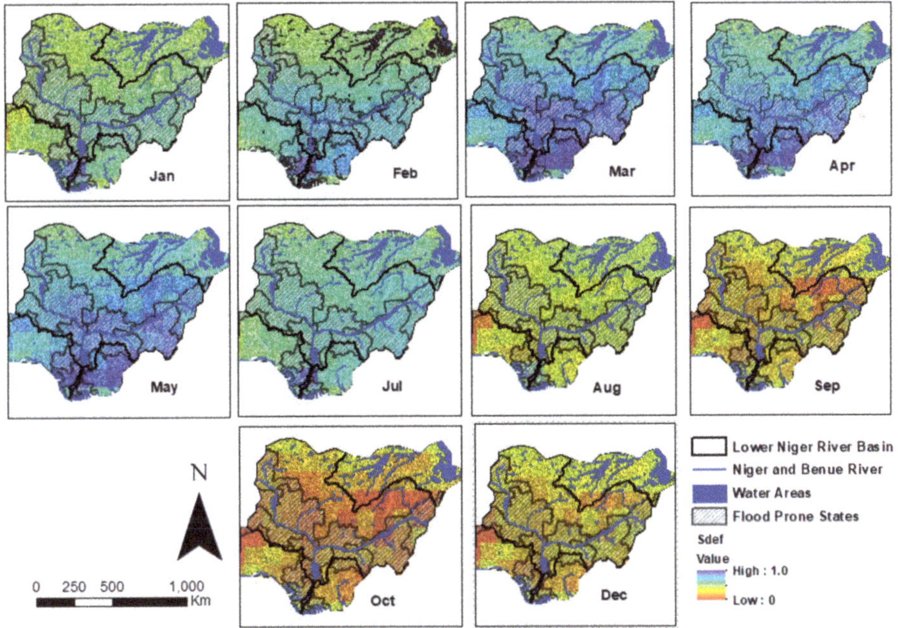

Figure 10. Spatiotemporal display of storage deficit (*Sdef*) (normalized) in the LNRB from January to December 2012. May and October 2012 are missing months in the GRACE TWSA time series due to battery management. Hence, no *Sdef* and flood potential index (*FPI*) for June and November. The red color shows areas with low *Sdef* while the blue areas represent areas with high *Sdef*.

3.2.3. GRACE Flood Potential Index (*FPI*)

The *FPI* for 2012 in the LNRB is shown in Figure 10. When compared to Figure 8, one could see how sensitive the index is to precipitation. Different areas within the basin experiences flooding at different rainfall peak times as shown in Figure 10. However, according to the Nigerian NEMA, in September 2012, 30 out of 36 states in Nigeria was affected by flooding which corresponds to the prediction by the *FPI*.

3.2.4. GRACE-Based RFPI Validation

We validated the GRACE-based *FPI* both quantitatively and qualitatively. The derived GRACE-based *FPI* was quantitatively compared to the water budget-derived *FPI* from 2004 to 2012 (Figure 11) and qualitatively validated against the DFO flood data for September 2005, 2007, 2009 and 2012 (Figure 11). We found a good agreement between the *FPI* derived from GRACE and water budget estimates as shown in Table 2, Figures 11 and 12. Figure 13 shows that both GRACE-based *FPI* and water budget-based *FPI* trend in the same direction. Table 2 further shows the similarities between the *FPI* from GRACE and water budget.

Statistical Test for GRACE and Water Budget Flood Potential Index

For statistical significance, we posed a question of whether there is a significant difference in the flood potential derived using GRACE TWSC and water budget TWSC. We used a Wilcoxon rank-sum test to test the null hypothesis that there is no difference between the *FPI* estimates. We also tested an alternative hypothesis, that there is a difference between the *FPI* estimates. Using an Alpha level of 0.05, our result showed that we can accept the null hypothesis, and at P > 0.05, we have no reason to reject the null hypothesis that there is no difference in the flood potential estimates.

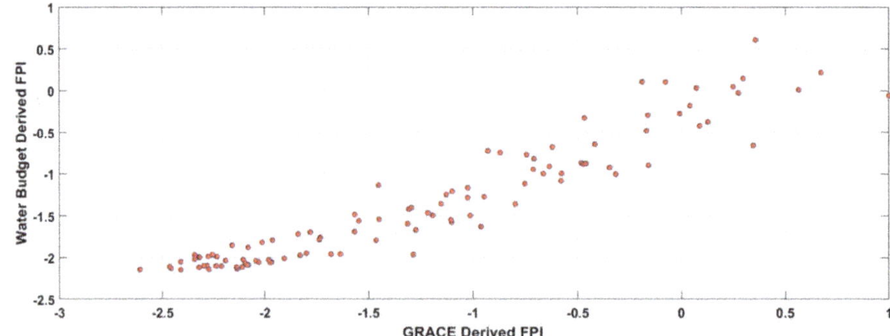

Figure 11. Scatterplot for GRACE-derived *FPI* and water budget-derived *FPI*.

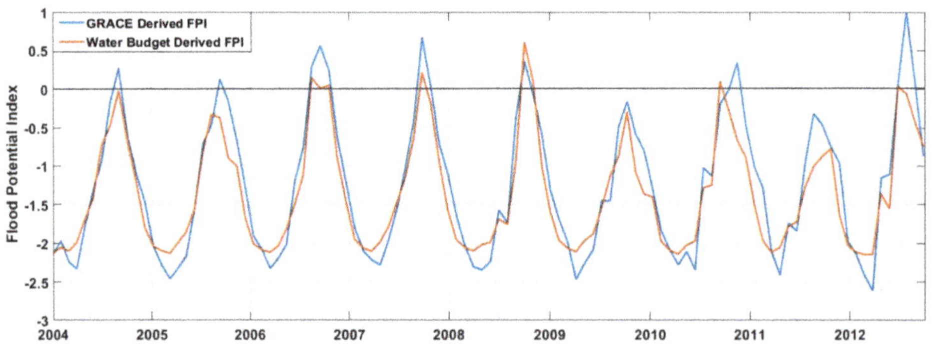

Figure 12. Graphical comparison and validation of *FPI* from GRACE using the *FPI* from water budget estimates.

Table 2. Tabular representation of the comparison between *FPI* from GRACE and water budget estimates.

Years	Flood Potential Index	
September	GRACE-Based *FPI*	Water Budget-Based *FPI*
2005	0.2	−0.3
2006	0.6	0.2
2007	0.7	0.3
2009	−0.3	−0.2
2010	0.4	0.2
2012	0.9	0.1

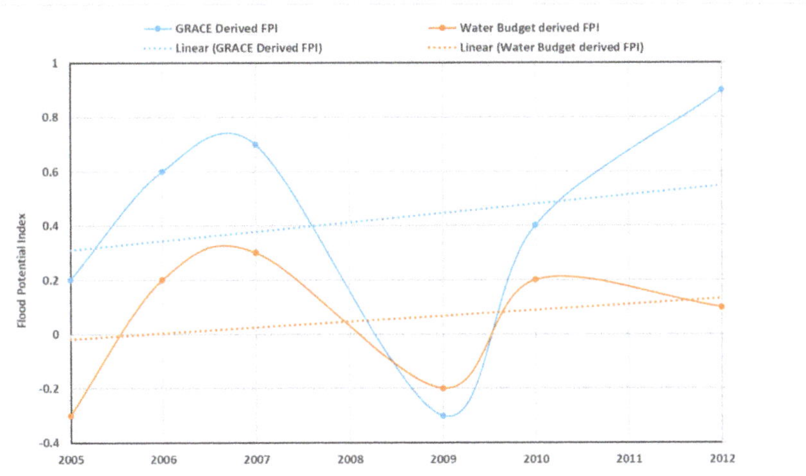

Figure 13. Trend plots for GRACE-derived and water budget-derived *FPIs*.

We chose big flood events that got publicity within our period of study so as to see how well the GRACE-based *FPI* compares to the DFO reported floods. However, because the DFO flood report is based on news, it is more biased towards urban flood events. These flood events reported in our study area in September 2005, 2007, 2009 and 2012 were also predicted by the GRACE-based *FPI* (Figure 14). Also, the *FPI* values from GRACE (Table 3) predicted flooding for the flood-prone/worst-hit state in September 2012 as reported by the Nigerian NEMA.

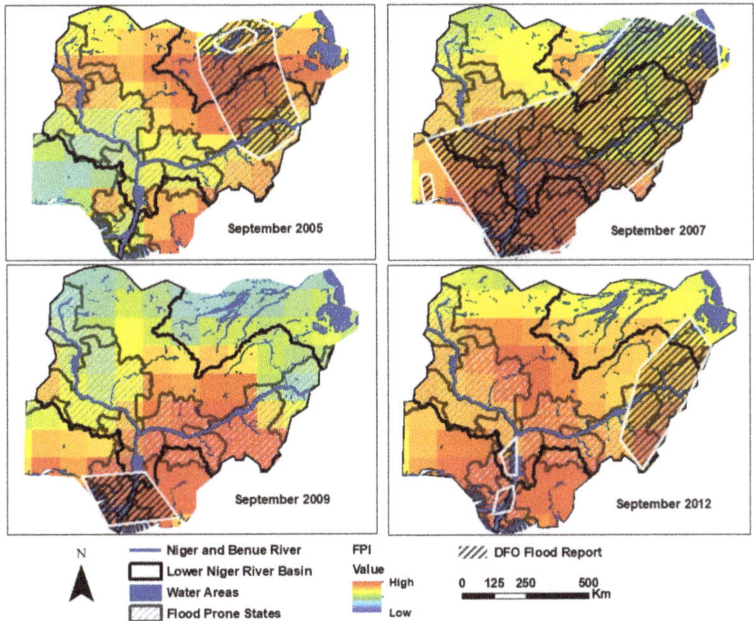

Figure 14. Comparison between GRACE-based *FPI* predicted floods and DFO reported floods.

Table 3. GRACE-based *FPI* values for the flood-prone/worst-hit states in September 2012.

Year	Flood Prone States	FPI
	Adamawa	1
	Anambra	0.5
	Bayelsa	0.4
	Benue	0.8
	Delta	0.5
	Edo	0.4
2012	Kebbi	0.5
	Kogi	0.7
	Kwara	0.5
	Nassarawa	0.4
	Niger	0.8
	Rivers	0.7
	Taraba	0.6

4. Discussion

Reager and Famiglietti [17] demonstrated that GRACE TWSA data can reveal when river basins have been filling with water over several months, when it rains, and the basin becomes full, and floods. The available storage or *Sdef* for LNRB (Figure 10) began to decrease August to October 2012 which represents the peak rainy season in the study area. Further, since soil moisture is critical in the accurate prediction of floods and general runoff [17], the storage deficit serves as an indicator in flood studies. The correlation coefficients of –0.7 for storage deficit and precipitation shows the inverse relationship that exists between the two variables which further supports the conclusion made by Reager and Famiglietti [17] that the storage deficit can be used with traditional methods of precipitation forecasting to determine the likelihood for flooding during the coming weeks. We also analyzed the relationship between storage deficit and *FPI* and found that storage decrease with an increase in the potential for flooding had a correlation coefficient of -0.8.

The *FPI* seems to perform well where flooding is mainly caused by heavy rainfall events [18] which was the case in Nigeria and by extension, LNRB. Heavy rainfall that occurred in August and September 2012 caused the major rivers, especially the Benue River, to overflow its banks, which led to authorities releasing water from the dams located within the basin (Cameroon). For our analysis, the *FPI* captured and predicted the flood events in LNRB in 2012 (Figures 12 and 15) which was also reported by the Nigerian NEMA and DFO (Figure 14) [30,31]. For validation of the GRACE-based *FPI* using the water budget-derived *FPI*, we tested the hypothesis that there is no difference between the indices using the alpha value of 0.05. With a correlation coefficient of 0.9 and P > 0.05 we have no reason to reject the null hypothesis.

The GRACE-based *FPI* and storage deficit, though invaluable, have limitations. The coarse spatial (>200 km^2) and temporal (monthly) resolutions of the GRACE data also makes it limited and unsuitable for forecasting local scale and flash floods [15,18], which may, therefore, make the *FPI* less effective. It has, however, an unequaled capability to monitor available water storage when combined with precipitation forecasting data and could increase warning lead time from one month to two months [17].

However, the relationship between the GRACE and water budget-based *FPI* shows promise when finer spatial and temporal resolutions data are used in deriving water budget TWSC (Equation (1)), thus making it a supplement to the GRACE-based *FPI* and possibly reducing the limitation of the GRACE TWSA.

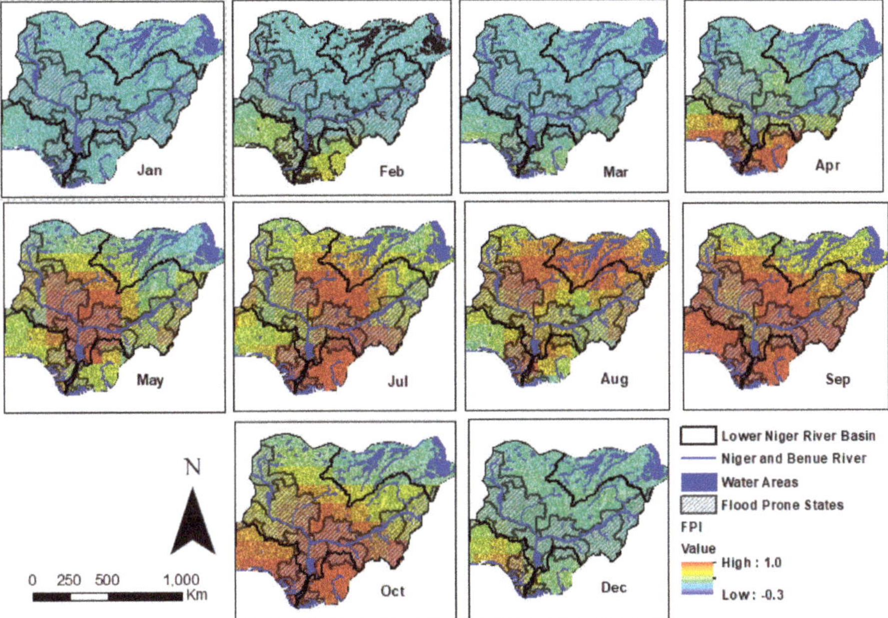

Figure 15. Spatiotemporal distribution of the flood potential index for 2012 in the LNRB. The red and blue areas indicate high and low probability or likelihood of flooding. According to the Nigerian National Emergency Management Agency (NEMA), 30 out of 36 states experienced flooding.

5. Conclusions

We estimated the hydrological conditions of the study area in terms of available storage and precipitation prior to the 2012 catastrophic flood using GRACE TWSA and GPCC data respectively. We also validated the GRACE TWSC using the water budget estimates TWSC, calculated the GRACE-based *FPI* for the basin, quantitatively and qualitatively compared the result to the water budget-based *FPI* and DFO flood report respectively. Based on our findings, we can make the following conclusions.

The GRACE TWSA and the derived *FPI* are both sensitive to precipitation by showing peaks and troughs in their time series which corresponds to wet season (peak) and dry season (trough). Based on the hydrological conditions of the study area in terms of precipitation and antecedent water storage state prior to flooding, the basin had a high amount of rainfall in August 2012 and could not balance the amount of incoming precipitation for September 2012 which then led to flooding.

The GRACE TWSA limitations could be managed assuming the GRACE-based *FPI* is supplemented with the water budget-derived *FPI* and using the water budget TWSC calculated from lower spatial and temporal data.

Therefore, the GRACE-based TWSA, *Sdef* and *FPI* in combination with other precipitation forecasting data and water budget-based TWSC/*FPI* could be utilized for operational flood monitoring in developing countries like Nigeria, where the unavailability of technical manpower, security and the cost of implementing and installing sophisticated flood monitoring/predicting measures could be prohibitive. In terms of cost, most satellite data are relatively cheap and readily available. Additionally, the current issue of security in some of the northern parts of the country prone to flooding might hinder the installation of flood monitoring devices, thus making remote-sensing products a viable option and an invaluable resource in flood studies in regions with little or no flood monitoring data.

Author Contributions: Conceptualization, D.I. and W.Z.; methodology, D.I. and W.Z.; validation, D.I.; formal analysis, D.I.; data curation, D.I.; writing—original draft preparation, D.I.; writing—review and editing, W.Z.; visualization, D.I.; supervision, W.Z.; project administration, W.Z.; funding acquisition, W.Z. and D.I.

Funding: This work was funded by the American Association of University Women (AAUW) scholarship in 2018, and the Chevron international fellowship at the Colorado School of Mines in 2019.

Acknowledgments: We acknowledge the use of data products or imagery from the Land, Atmosphere Near real-time Capability for EOS (LANCE) system operated by NASA's Earth Science Data and Information System (ESDIS) with funding provided by NASA Headquarters. Special thanks to Stephen Semmens, Ashton Krajnovich and Kendall Wnuk for offering their constructive feedback.

Conflicts of Interest: The authors declare no conflict of interest.

References

1. Folorunsho, R.; Awosika, L.F. Flood Mitigation in Lagos, Nigeria through Wise Management of Solid Waste: A Case of Ikoyi and Victoria Islands, Nigeria. Paper Presented at the UNESCO—CSI Workshop Maputo, Maputo, Mozambique, 19–23 November 2001.
2. Olajuyigbe, A.; Rotowa, O.; Durojaye, E. An assessment of flood hazard in Nigeria: The case of mile 12, Lagos. *Mediterr. J. Soc. Sci.* **2012**, *3*, 367–375.
3. Bartholmes, J.C.; Thielen, J.; Ramos, M.H.; Gentilini, S. The European flood alert system EFAS – Part 2: Statistical skill assessment of probabilistic and deterministic operational forecasts. *Hydrol. Earth Syst. Sci.* **2009**, *13*, 141–153. [CrossRef]
4. Scawthorn, C. Modeling flood events in the US. In Proceedings of the Euro Conference on Global Change and Catastrophe Risk Management, HASA, Laxenburg, Austria, 6–9 June 1999.
5. Tariq, M.A.U.R. Risk-Based Planning and Optimization of Flood Management Measures in Developing Countries: Case Pakistan. Ph.D. Thesis, VSSD, Delft University of Technology, Delft, The Netherlands, 2012.
6. Victor, K. Remote Sensing of Floods and Flood-Prone Areas: An Overview. *J. Coast. Res.* **2015**, *31*, 1005–1013.
7. Nasreddine, B.; Feng, Z.; Luca, B.; Yanbo, H.; Yumin, T. Near-Real-Time Flood Forecasting Based on Satellite Precipitation Products. *Remote Sens.* **2019**, *11*, 252. [CrossRef]
8. Sheffield, J.E.F.; Wood, N.; Chaney, K.; Guan, S.; Sadri, X.; Yuan, L.; Olang, A.; Amani, A.; Ali, S.; Demuth, L. A Drought Monitoring and Forecasting System for Sub-Sahara African Water Resources and Food Security. *Bull. Amer. Meteor. Soc.* **2014**, *95*, 861–882. [CrossRef]
9. Zhang, X.; Tang, Q.; Liu, X.; Leng, G.; Li, Z. Soil Moisture Drought Monitoring and Forecasting Using Satellite and Climate Model Data over Southwestern China. *J. Hydrometeor.* **2017**, *18*, 23. [CrossRef]
10. Hossain, F.; Anagnostou, E.N. Assessment of current passive-microwave- and infrared-based satellite rainfall remote sensing for flood prediction. *J. Geophys. Res. Atmos.* **2005**, *110*, D07102. [CrossRef]
11. Brakenridge, R.; Anderson, E. Modis-Based Flood Detection, Mapping and Measurement: The Potential for Operational Hydrological Applications. In *Transboundary Floods: Reducing Risks Through Flood Management*; Marsalek, J., Stancalie, G., Balint, G., Eds.; Nato Science Series IV: Earth and Environmental Sciences; Springer: Dordrecht, The Netherlands, 2006; pp. 1–12.
12. Zhan, X.; Ryu, D.; Crow, W. *Improving Flood Prediction Through the Assimilation of AMSR-E Soil Moisture Retrievals into a Hydrologic Model*; American Geophysical Union, Fall Meeting Moscone Center: San Francisco, CA, USA, 2006; abstract ID: H23E-1545.
13. Adam, D. Gravity measurement: Amazing GRACE. *Nature* **2002**, *416*, 10–11. [CrossRef] [PubMed]
14. Chen, J.L.; Wilson, C.R.; Tapley, B.D.; Ries, J.C. Low degree gravitational changes from GRACE: Validation and interpretation. *Geophys. Res. Lett.* **2004**, *31*, L22607. [CrossRef]
15. Longuevergne, L.; Scanlon, B.R.; Wilson, C.R. GRACE Hydrological estimates for small basins: Evaluating processing approaches on the High Plains Aquifer, USA. *Water Resour. Res.* **2010**, *46*, W11517. [CrossRef]
16. Reager, J.T.; Thomas, B.F.; Famiglietti, J.S. River basin flood potential inferred using GRACE gravity observations at several months lead-time. *Nat. Geosci.* **2014**, *7*, 588–592. [CrossRef]
17. Reager, J.T.; Famiglietti, J.S. Global terrestrial water storage capacity and flood potential using GRACE. *Geophys. Res. Lett.* **2009**, *36*, L23402. [CrossRef]
18. Molodtsova, T.; Molodtsov, S.; Kirilenko, A.; Zhang, X.; VanLooy, J. Evaluating flood potential with GRACE in the United States. *Nat. Hazards Earth Syst. Sci.* **2016**, *16*, 1011–1018. [CrossRef]

19. Sun, Z.; Zhu, X.; Pan, Y.; Zhang, J. Assessing Terrestrial Water Storage and Flood Potential Using GRACE Data in the Yangtze River Basin, China. *Remote Sens.* **2017**, *9*, 1011. [CrossRef]

20. IFRCS—International Federation of Red Cross and Red Crescent Emergency Appeal; Nigeria: Floods. Available online: http://reliefweb.int/sites/reliefweb.int/files/resources/MDRNG01401.pdf (accessed on 10 January 2019).

21. Kwari, J.W.; Paul, M.K.; Shekarau, L.B. The Impacts of Flooding on Socio-Economic Development and Agriculture in Northern Nigeria: A Case Study of 2012 Flooding in Yola and Numan Areas of Adamawa State Nigeria. *Int. J. Sci. Eng. Res.* **2015**, *6*, 1433–1442.

22. Mazawaje, D.F.; Tsenbeya, I.T.; Ismaila, A.B. Analysis of the determinants of floods in Numan Town, Nigeria. *J. Environ. Sci. Eng.* **2014**, *B3*, 264–273. [CrossRef]

23. NASA Earth Observatory—Flooding in Nigeria. Available online: https://earthobservatory.nasa.gov/images/79404/flooding-in-nigeria (accessed on 20 August 2019).

24. Swenson, S.C. GRACE Monthly Land Water Mass Grids NETCDF RELEASE 5.0 Ver. 5.0. PO. DAAC, CA, USA, 2012. Available online: http://dx.doi.org/10.5067/TELND-NC005 (accessed on 20 September 2018).

25. Velicogna, I.; Wahr, J. Measurements of time-variable gravity show mass loss in Antarctica. *Science* **2016**, *311*, 1754–1756. [CrossRef] [PubMed]

26. EartH2Observe project, Plymouth Marine Laboratory Remote Sensing Group, European Union's Seventh Programme for Research, Technological Development and Demonstration under Grant Agreement No. 603608. Available online: https://wci.earth2observe.eu/ (accessed on 10 October 2018).

27. Schneider, U.; Becker, A.; Finger, P.; Meyer-Christoffer, A.; Rudolf, B.; Ziese, M. GPCC Full Data Reanalysis Version 6.0 at 1.0° Monthly Land-Surface Precipitation from Rain-Gauges built on GTS-based and Historic Data 2011. Available online: https://opendata.dwd.de/climate_environment/GPCC/html/fulldata_v6_doi_download.html (accessed on 23 September 2018). [CrossRef]

28. Sakumura, C.; Bettadpur, S.; Bruinsma, S. Ensemble prediction and intercomparison analysis of GRACE time-variable gravity field models. *Geophys. Res. Lett.* **2014**, *41*, 1389–1397. [CrossRef]

29. NASA-JPL. *GRACE D-103133 Gravity Recovery and Climate Experiment Level-3 Data Product User Handbook*; NASA-JPL: Pasadena, CA, USA, 2019.

30. DFO (Dartmouth Flood Observatory), Global Active Archive of Large Flood Events. Available online: http://www.dartmouth.edu/~{}floods/Archives/ (accessed on 14 December 2018).

31. Climate-Data.Org. Climate Adamawa. Available online: https://en.climate-data.org/africa/niger/zinder/adamawa-349376/ (accessed on 22 February 2019).

 *remote sensing*

Article

Latest Geodetic Changes of Austre Lovénbreen and Pedersenbreen, Svalbard

Songtao Ai [1],*, Xi Ding [1], Florian Tolle [2], Zemin Wang [1] and Xi Zhao [1]

[1] Chinese Antarctic Center of Surveying and Mapping, Wuhan University, Wuhan 430079, China; dingxi@chinare.cn (X.D.); zmwang@whu.edu.cn (Z.W.); xi.zhao@whu.edu.cn (X.Z.)

[2] ThéMA, CNRS, Université de Bourgogne Franche-Comté, 25030 Besançon, France; florian.tolle@univ-fcomte.fr

* Correspondence: ast@whu.edu.cn

Received: 24 October 2019; Accepted: 2 December 2019; Published: 4 December 2019

Abstract: Geodetic mass changes in the Svalbard glaciers Austre Lovénbreen and Pedersenbreen were studied via high-precision real-time kinematic (RTK)-global positioning system (GPS) measurements from 2013 to 2015. To evaluate the elevation changes of the two Svalbard glaciers, more than 10,000 GPS records for each glacier surface were collected every year from 2013 to 2015. The results of several widely used interpolation methods (i.e., inverse distance weighting (IDW), ordinary kriging (OK), universal kriging (UK), natural neighbor (NN), spline interpolation, and Topo to Raster (TTR) interpolation) were compared. Considering the smoothness and accuracy of the glacier surface, NN interpolation was selected as the most suitable interpolation method to generate a surface digital elevation model (DEM). In addition, we compared two procedures for calculating elevation changes: using DEMs generated from the direct interpolation of the RTK-GPS points and using the elevation bias of crossover points from the RTK-GPS tracks in different years. Then, the geodetic mass balances were calculated by converting the elevation changes to their water equivalents. Comparing the geodetic mass balances calculated with and without considering snow depth revealed that ignoring the effect of snow depth, which differs greatly over a short time interval, might lead to bias in mass balance investigation. In summary, there was a positive correlation between the geodetic mass balance and the corresponding elevation. The mass loss increased with decreasing elevation, and the mean annual gradients of the geodetic mass balance along the elevation of Austre Lovénbreen and Pedersenbreen in 2013–2015 were approximately 2.60‰ and 2.35‰, respectively. The gradients at the glacier snouts were three times larger than those over the whole glaciers. Additionally, some mass gain occurred in certain high-elevation regions. Compared with a 2019 DEM generated from unmanned aerial vehicle measurement, the glacier snout areas presented an accelerating thinning situation in 2015–2019.

Keywords: mass balance; snow depth; glacier retreat; surface DEM; elevation change

1. Introduction

Glaciers are an important component of the cryosphere, playing an important role in global climate change, and are often considered to be essential climate indicators [1,2]. In light of rapid global climate change, glacier loss is a major contributor to increases in sea level; therefore, glacier mass balance has become an important subject of research [3,4]. There are various methods for estimating glacier mass balance, including direct measurements by stakes and snow pit surveying [5,6], modeling methods based on the high correlation between the mass balance and selected meteorological parameters [7], and geodetic methods involving the comparison of two surfaces at different times [8,9]. To determine the applicable cases for glaciological and geodetic methods, these methods are often compared for validation and calibration [10–13].

Svalbard (74°N–81°N; 10°E–35°E) is covered by a large number of small glaciers and ice caps, which compose 60% of the archipelago [14]. The total glaciated area on Svalbard is 34,560 km^2, which is approximately 6% of the worldwide glacier cover, except for Greenland and Antarctica [15]. Most glaciers in Svalbard are polythermal glaciers, which are sensitive to climate changes; therefore, it is important for scientists to monitor and study the glaciers of Svalbard. Many scientists, especially Norwegian scientists, have performed studies on the glaciers in Svalbard. The glaciers Kongsvegen and Kronebreen have been widely studied for a significant amount of time [16]. Since the Chinese Arctic Yellow River Station was built in 2004, Chinese researchers have focused on Arctic glaciers, carrying out long-term studies on Austre Lovénbreen and Pedersenbreen in Svalbard [17]. Chinese researchers have investigated the volume of Austre Lovénbreen and Pedersenbreen [18] and estimated the mass loss of Pedersenbreen during the periods from 1936 to 1990 and from 1990 to 2009 [9]. The velocities of the two glaciers have also been studied, and the latest research has discussed the fastest ice flow region of Austre Lovénbreen by combining modeling methods with in situ surveying methods [19].

Since the Little Ice Age (LIA), the glaciers in Svalbard have been retreating. Bamber and others suggested that an increased thinning trend occurred in recent years based on aerial surveys performed in 1996 and 2002 [8]. Małecki concluded that mass changes became more negative in central Svalbard glaciers by comparing elevation changes over the periods 1960–1990 and 1990–2009 [20]. Hagen et al. estimated the annual mass balance for the whole of Svalbard to be −0.1 m water equivalent (w.e.) during the period 1979–2000 [21]. Nuth and others estimated that the annual mass balance of the southern and western Spitsbergen glaciers in Svalbard during the period 1936–1990 was −0.30 m w.e, according to geodetic mass balance estimate from aerial photography [22]. Norwegian researchers observed the mass balance of the two glaciers, Austre Broggerbreen and Midtre Lovenbreen, adjacent to Austre Lovénbreen during the period 1966–1988, finding that the ice surface decreased by 8.9 m and 7.5 m, respectively [14]. A French team comprehensively investigated Austre Lovénbreen, and concluded that the annual mass balance of the glacier during the period 1962–2013 was around −0.2 m w.e., and the annual mass balance in 2008–2015 was about −0.4 m w.e. [23]. In our study, we mainly used real-time kinematic global positioning system (RTK-GPS) data to map the surface topography and analyze the interannual geodetic mass balance of Austre Lovénbreen and Pedersenbreen via elevation changes, which is of significance as a reference for traditional glacier mass balance estimates.

2. Study Area

The glaciers Austre Lovénbreen (12.09°E; 78.527°N) and Pedersenbreen (12.175°E; 78.515°N) are located in Svalbard in the Arctic (Figure 1) and are 6.2 km and 10 km away from the Chinese Arctic Yellow River Station, respectively. These two glaciers are recognized as polythermal valley glaciers, lying in a mountainous area, with the highest peak reaching 1017 m. Austre Lovénbreen has an area of 4.5 km^2 with an altitude between 50 m and 550 m [23]. According to previous research, the glacier valley of Pedersenbreen is V-shaped rather than U-shaped [18], with an area of approximately 5.6 km^2 and an altitude between 60 m and 650 m. Pedersenbreen has a narrower snout than Austre Lovénbreen, and both glaciers are relatively flat, and both are covered with small amounts of debris as the elevation increases gradually from the north to the south.

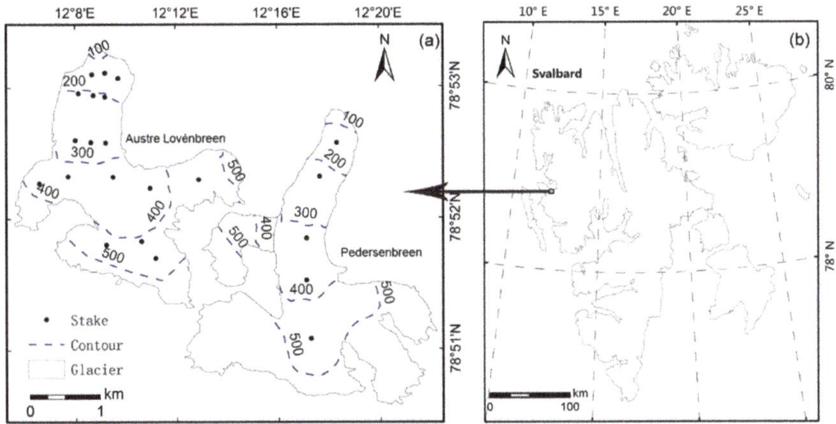

Figure 1. (**a**) Sketch view of Austre Lovénbreen and Pedersenbreen and (**b**) the locations of the glaciers in Svalbard.

The Svalbard area, where the two glaciers are located, has a polar oceanic type climate mainly affected by the North Atlantic current. This climate is characterized by cooler summers and warmer winters than other regions with similar latitudes [24]. According to the weather station in Ny-Ålesund, where the Chinese Arctic Yellow River Station is located, the annual mean temperature during the past 30 years (1981–2010) was −5.2 °C, with the mean temperatures of 3.8 °C and −12 °C of summer and winter, respectively. The annual average precipitation is 427 mm w.e., which is mainly concentrated in winter and autumn dominated by snow [25]. The annual temperature and precipitation during the earlier period (1961–1990) were −6.3 °C and 385 mm, respectively, indicating that an obvious climate change occurred in the different periods [25]. Overall, the entire archipelago has experienced a warming trend of approximately 0.5 °C every decade since 1960 [26].

3. Data and Methods

3.1. Data

The 20-stake observation network on Austre Lovénbreen and five observation stakes along the central line of Pedersenbreen were placed in July 2005 [27]. The movement of these stakes has been annually monitored with high-precision GPS instruments for glacial movement and mass balance studies [17]. In April 2013, April 2014, and May 2015, high-density RTK-GPS points on the surface of Austre Lovénbreen and Pedersenbreen were collected via a snowmobile carrying GPS equipment, and the GPS tracks are shown in Figure 2. The base station is located at the Yellow River Station in Ny-Ålesund. The main purpose of this survey was to map the surface digital elevation model (DEM) and to estimate surface changes. The height data from the GPS measurements were ellipsoidal heights, which needed to be converted into altitudes above sea level. An elevation benchmark near the glaciers (<10 km) was set in Ny-Ålesund. According to a previous study [18], the geoidal height at this point was calculated as 35.158 m, which was used to convert the measured GPS heights to altitudes above sea level. Because we used the relative change of elevations in this study, the tectonic movement at Ny-Ålesund was neglected, since the two glaciers move together with the earth's crust motions.

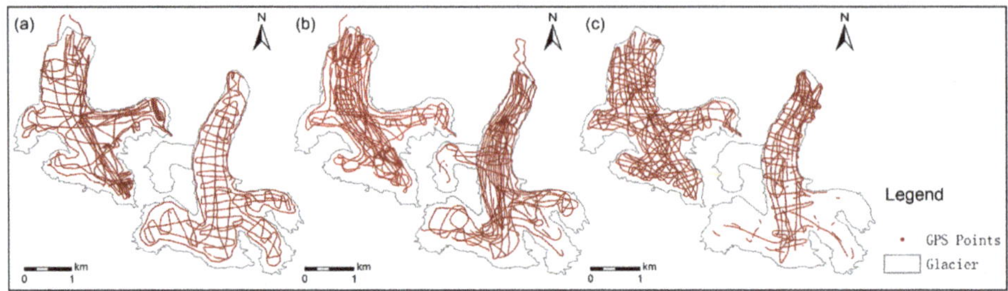

Figure 2. Real-time kinematic global positioning system (RTK-GPS) tracks in different years: (**a**) 2013, (**b**) 2014, and (**c**) 2015.

In order to evaluate the quality of the RTK surveys, all crossover points between different RTK profiles in one survey period were picked out, and the height differences of crossover points are shown in Table 1. In general, these small elevation differences proved the relatively high precision of the field surveys in 2013–2015.

Table 1. Elevation differences of the crossover points between different RTK (real-time kinematic) profiles. The Mean and RMSE (root mean square error) are the primary indicators for data quality evaluation.

	Austre Lovénbreen				Pedersenbreen			
Year	Count	RMSE	Max	Mean	Count	RMSE	Max	Mean
2013	332	0.078 m	0.779 m	−0.020 m	143	0.080 m	0.313 m	−0.007 m
2014	573	0.061 m	0.582 m	0.002 m	527	0.080 m	0.437 m	0.000 m
2015	766	0.097 m	1.147 m	−0.011 m	307	0.077 m	0.367 m	0.004 m

Snow cover measurements were performed on Austre Lovénbreen in April of 2013, 2014, and 2015 by a French team working in the area. A Pico drill was used to measure the depth and density of the snow covering the glacier. In addition, ArcticDEM [28] topographic data and ice-surface DEMs of the glacier snouts created by unmanned aerial vehicle (UAV) photogrammetry in 2019 were employed as supplementary data for the investigation of geodetic changes.

3.2. Comparison of Interpolation Methods

Different DEM resolutions have an important impact on ice-surface-elevation change studies [29]. To reveal the impact of different resolutions on the interpolation results and to select a suitable interpolation resolution, we compared different interpolation resolutions, including 0.2 m, 0.5 m, 1 m, 2 m, and 5 m. According to previous studies, interpolation is an important source of uncertainty in mass change studies using DEMs [30,31]. Different interpolation methods need to be compared to determine the most suitable method for generating the DEM. Several widely used interpolation methods were examined—inverse distance weighting (IDW), ordinary kriging (OK), universal kriging (UK), natural neighbor (NN), spline interpolation, and Topo to Raster (TTR) interpolation. IDW determines z-values using a linearly weighted combination of a set of sample points. The influence of known points on the interpolated values, based on their distance from the output point, can be controlled by defining the power. Kriging is an advanced geostatistical procedure that generates an estimated surface from a scattered set of points via z-values; it requires previous exploratory work on input data to determine the parameters that have an important impact on the interpolation results [32]. Two widely used kriging methods are OK and UK. OK assumes that the variation in the z-values is free of drift. UK is a typical geostatistical method that is based on spatial autocorrelation models. It assumes that the spatial variation in the z-values is determined by the drift and a random error [33] and is suitable for data with an obvious trend. The NN approach interpolates a value by finding the closest subset of input samples to an unknown point and applies weights to them based on their proportionate areas [34]. The surface interpolated by splines passes through the input samples and is smooth everywhere except at the locations of the input samples [35]. TTR is based on thin-plate smoothing splines; it creates smooth, continuous surfaces passing through all the input points and aims to produce a raster of the drainage structure [36]. In this study, we selected a spherical model as the semivariogram model for OK, a linear drift model for UK, and a regularized spline for spline interpolation.

In total, 75% of the RTK-GPS points were randomly selected as training data for interpolating the DEM, and the rest were used as testing data. To evaluate the accuracy of the interpolation results, the errors at the testing points were evaluated by subtracting the interpolated values from the ground records of the vertical coordinates. Root mean square errors (RMSEs) were calculated for every interpolation method to assess their performance. All analyses were performed in ArcGIS 10.3 (Environmental Systems Research Institute, United States).

3.3. Estimation of Geodetic Changes

Based on the interpolation results, the optimal interpolation method was chosen for analyzing the interannual geodetic glacier changes. The surface-elevation changes were analyzed by subtracting one DEM from a later DEM. In addition, the trend of geodetic glacier changes was studied in detail by calculating the elevation differences in the crossover points from the RTK-GPS tracks in different years. The mass change research in our study belonged to the geodetic method and was based on real-time kinematic (RTK)-global positioning system (GPS) data, which are advantageous for long-term studies.

If we ignore the influence of ice flux on the mass balance, elevation changes can be converted into geodetic mass balances independent of dynamics [22]. Glacier mass change needs to be transformed from an ice equivalent into a water equivalent, which means that the density of glacial ice and snow should be estimated. Density assumptions or models for converting the geodetic glacier volume change to mass change have been explored in many studies [37,38].

The net geodetic mass balance of a glacier can be calculated from the area-weighted mass balances of different elevation ranges, similar to the traditional mass balance calculation, as in formula (1):

$$B = \sum_{i=1}^{n} B_i S_i \tag{1}$$

where B is the net geodetic mass balance, and Bi and Si are the geodetic mass balance at different elevation bins and the corresponding percentage of the projected area between two contour lines, respectively. By calculating the average elevation changes obtained from the crossover points in the different elevation bands, the geodetic mass balance can be obtained with a density model of the ice surface.

For an investigation with such a short time interval, a density model is needed to convert the ice-surface changes in a glacier into the water equivalent for a mass balance study, and in situ snow data are important for estimating the geodetic mass balance. Taking snow depth into consideration, the mass balance at a point can be calculated as in formula (2):

$$b = \Delta h \rho_i + (s_2 - s_1)(\rho_s - \rho_i) \tag{2}$$

where b is the annual mass balance at a given point calculated by the geodetic method (represented by the water equivalent); Δh is the elevation change; ρ_i is the ice density (assumed to be 900 kg/m^3); ρ_s is the snow density (assumed to be 400 $\pm$ 100 kg/m^3); and s_1 and s_2 represent the snow depths for the first year and the following year, respectively. The snow depth at a given point can be calculated from the interpolation of the snow depth data obtained from in situ snow depth measurements. Accordingly, the mass balances for an entire glacier can be calculated from interpolation.

4. Results

4.1. Comparisons of Different Interpolation Resolutions and Methods

Taking the RTK-GPS data of Pedersenbreen measured in 2013 as an example, we compared the DEMs generated by IDW with spatial resolutions of 0.2 m, 0.5 m, 1 m, 2 m, and 5 m. More than 2800 testing points were used to examine the vertical accuracy of the DEMs derived from the five interpolation resolutions. All the mean errors in Table 2 are close to zero but do not seem to be related to the resolution. However, the quality of the interpolation result was mainly assessed by RMSE (root mean square error). Table 2 illustrated that the RMSE was smaller with a higher resolution, which means that the elevation extracted near the RTK-GPS tracks was more accurate at a higher resolution. Similar conclusions could also be drawn from other interpolation methods at different resolutions.

Table 2. Errors in the IDW (inverse distance weighting) interpolation results with different resolutions (unit: m).

	0.2 m	0.5 m	1 m	2 m	5 m
Mean	0.004	0.035	0.021	−0.035	−0.023
RMSE	0.359	0.363	0.366	0.393	0.444
Max	4.096	4.158	4.106	3.631	4.104
Min	−2.425	−2.385	−2.582	−2.951	−2.648
Range	6.521	6.543	6.688	6.582	6.752

There were few differences found between the DEMs with resolutions of 0.2 m, 0.5 m, and 1 m based on a comparison of the RMSEs; the differences were at the millimeter level. Considering the limit of the sampling interval of RTK-GPS data, the resolution of the DEM should be appropriate to avoid meaningless interpolation at an over-detailed scale, which would increase the computational burden. Therefore, the final resolution of the DEM in our study was set to 0.5 m. Then, the DEMs generated by different interpolation methods with a resolution of 0.5 m were compared, and the statistics of the errors are shown in Table 3. OK, NN, and spline had similar results; they each had a small RMSE and range, and their mean errors were close to zero, which means that these methods could be considered candidates for generating the DEM. IDW showed the extreme maximum and minimum values and a larger RMSE compared to other methods. The unsatisfied estimation from IDW in this study confirmed the results from other research [39,40]. Although UK had the lowest mean error, the RMSE showed that it was not an optimal method for the surface interpolation of glaciers. TTR had a smaller mean error than NN and spline, but its RMSE was larger than OK, NN, and spline, so we did not consider it when generating the DEMs of the glaciers.

Table 3. Errors in different interpolation methods (unit: m).

	IDW	OK	UK	NN	Spline	TTR
Mean	0.035	0.024	0.011	0.028	0.029	0.027
RMSE	0.362	0.092	0.214	0.095	0.099	0.187
Max	4.158	0.7167	1.345	0.694	0.725	1.567
Min	−2.385	−0.687	−1.418	−0.721	−0.714	−1.690
Range	6.543	1.404	2.763	1.415	1.439	3.257

IDW, OK, UK, NN, and TTR are abbreviations of inverse distance weighting, ordinary kriging, universal kriging, natural neighbor interpolation, and Topo to Raster interpolation, respectively.

We compared the DEMs generated by OK, NN, and spline interpolation (Figure 3), and the hillshade effect was added to the DEMs to evaluate their smoothness. Although OK produced the lowest RMSE, the surface generated by OK was not as smooth as we would expect, whereas the glacier surface generated by NN interpolation was smooth. The surface generated by spline interpolation had some abnormal regions in which the values varied greatly from the other interpolation results. We selected NN interpolation as the most suitable method for analyzing the surface topography in this study. In fact, there was no single ideal interpolation method for all ice terrains; the interpolation method was chosen according to the terrain topography and the type of data analysis needed.

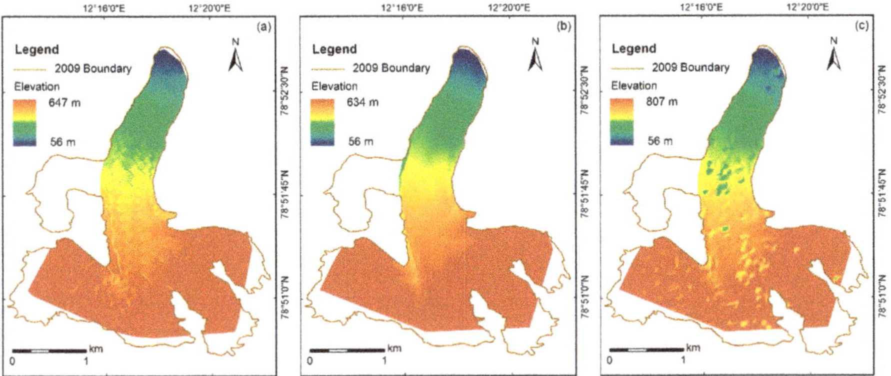

Figure 3. DEMs (digital elevation models) of Pedersenbreen in 2013 generated from different interpolation methods: (**a**) OK (ordinary kriging), (**b**) NN (natural neighbor) interpolation, and (**c**) spline interpolation.

4.2. Glacier Surface Elevation Changes

The DEMs of the two glacier surfaces in 2013–2015 were derived from NN interpolation. According to the 2013 DEM in Figure 3 the surface DEM of Pedersenbreen generated by NN interpolation is smoother than that of the other interpolation methods, and the lowest elevation is approximately 56 m at the northern part of the glacier terminus. The elevation differences between years were acquired by comparing the DEMs, and the elevation differences between 2013 and 2014 are shown in Figure 4. For Austre Lovénbreen, we subtracted the 2013 DEM from the 2014 DEM and found that the range of the surface-elevation differences was extremely large, as it could not be applied to the range of elevation changes of a glacier in a year; similar results were found for Pedersenbreen. The errors of DEM differences were calculated in Table 4; although the mean errors of the two glaciers seem to indicate possible elevation changes in 2013–2014, and the extremely large ranges in Table 4 are not suitable for interpreting the range of elevation changes. The spatial distribution of the elevation differences in Figure 4. is not in accordance with the fact that ice elevations do not change drastically over such a short time interval, assuming the glacier is in a steady state. This discrepancy is ascribed, in part, to the data source of the DEM, as the abnormal regions are mainly distributed at edges with steep topography that do not have GPS tracks in either 2013 or 2014. A similar result was also obtained by comparing the DEMs in other years.

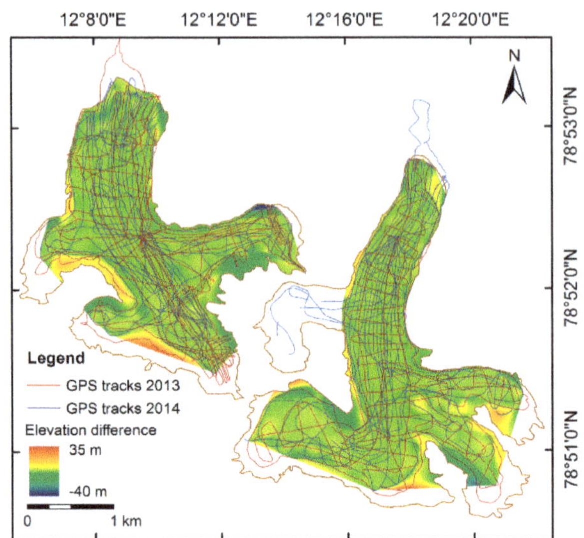

Figure 4. Elevation differences in 2013–2014.

Table 4. DEM (digital elevation model) differences in 2013–2014 (unit: m).

	Austre Lovénbreen	Pedersenbreen
Mean	−0.379	−0.202
STD	4.555	3.654
Min	−41.714	−30.047
Max	30.042	34.976
Range	71.756	65.023

The RTK-GPS points measured in 2013–2015 were not dense enough to cover the entire glacier, resulting in abnormal elevation difference values at certain regions where the RTK-GPS points were sparse. Misleading elevation change results might be obtained by DEM comparisons. Therefore, an alternative method was proposed: calculate the elevation difference in the crossover points from the RTK-GPS tracks in different years, where the glacier surface changes could be partly revealed by those points.

We derived the boundaries of the two glaciers in 2013–2015, as shown in Figure 5. Considering the retreat of the glacier terminus beginning in 2009, and to ensure that we studied the crossover points in glaciated regions, some crossover points located in the glacier terminus were removed according to their boundaries, and the remaining points were used to study the elevation changes.

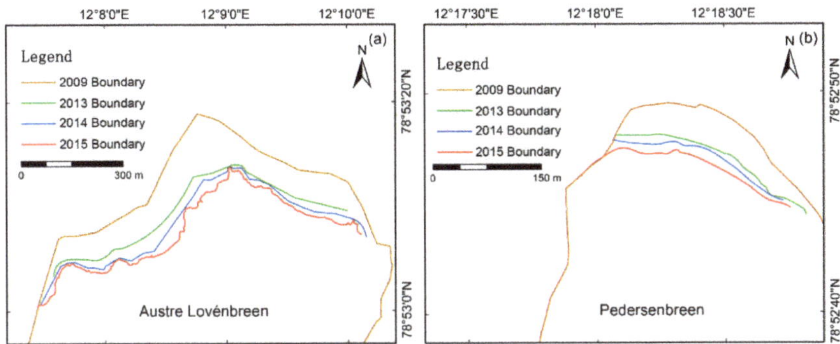

Figure 5. Glacier boundaries of (**a**) Austre Lovénbreen and (**b**) Pedersenbreen.

The elevation differences in the crossover points of the RTK-GPS tracks from different years in the glaciated regions are shown in Figure 6. From 2013 to 2014, the elevation differences ranged from −2.3 to 1.7 m in the study region (Austre Lovénbreen and Pedersenbreen), and the values of most points were negative. From 2014 to 2015, the elevation difference in the two glaciers was −2.0 to 1.9 m, and the values of most points were positive. By observing the distribution of elevation difference at the crossover points, we found that the trend in the elevation difference was more obvious than that calculated by the DEM. The elevation difference gradually increased from north to south, and its value changed from negative to positive. Combined with glacier topography, the altitude had an important impact on elevation changes because elevation changes were negative in most regions with lower altitudes. Higher altitude regions had a less negative elevation change, and, at even higher regions, the elevation increased. In general, the elevations in high-altitude regions were increasing, and the elevations in low-altitude regions were decreasing.

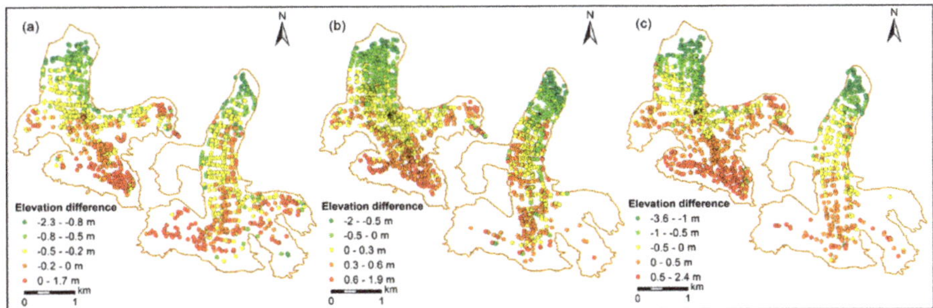

Figure 6. Elevation differences in the crossover points of the RTK-GPS (real-time kinematic global positioning system) tracks from different years: (**a**) 2013–2014, (**b**) 2014–2015, and (**c**) 2013–2015.

We derived a continuous raster of the elevation differences using NN interpolation (Figure 7), and the results illustrated that there was a good relationship between the change in elevation and the corresponding elevation of the two glaciers. In the figure, the color gradient corresponds to the elevation.

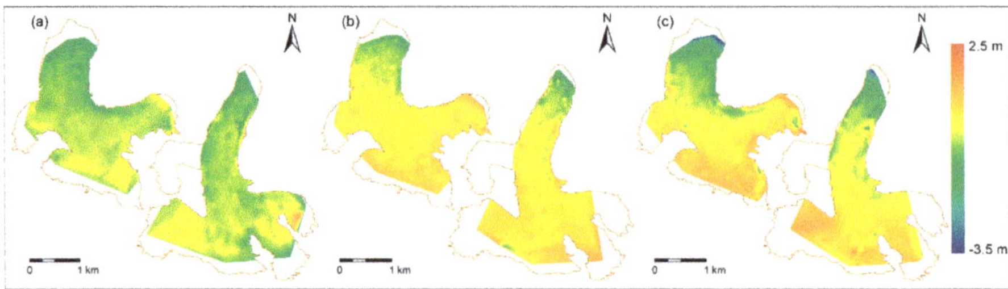

Figure 7. Elevation differences in different years interpolated with the crossover points: (**a**) 2013–2014, (**b**) 2014–2015, and (**c**) 2013–2015.

The regional distribution of the surface-elevation change is obvious. Austre Lovénbreen and Pedersenbreen experienced significant losses at their terminuses. However, there were some obvious regions of accumulation at the lower elevations of Austre Lovénbreen, such as the western margin area and the eastern tributary margins. Pedersenbreen also exhibited an obvious accumulation region at a relatively low elevation (at the eastern margin), which was mainly due to the mass compensation caused by an avalanche around the edge of the glacier in 2015. These regions are marked in Figure 8. There was an ablation region at the east margin of Austre Lovénbreen (marked with the number 4), which was probably caused by an avalanche that occurred in 2013 but not in 2015, which caused the extremely high elevation measured in 2013. Based on the hillshade topography of the Arctic DEM around the glacier (Figure 8), we found that the mass flow path corresponded to the elevation change area at the edges of the glaciers where slopes were steep.

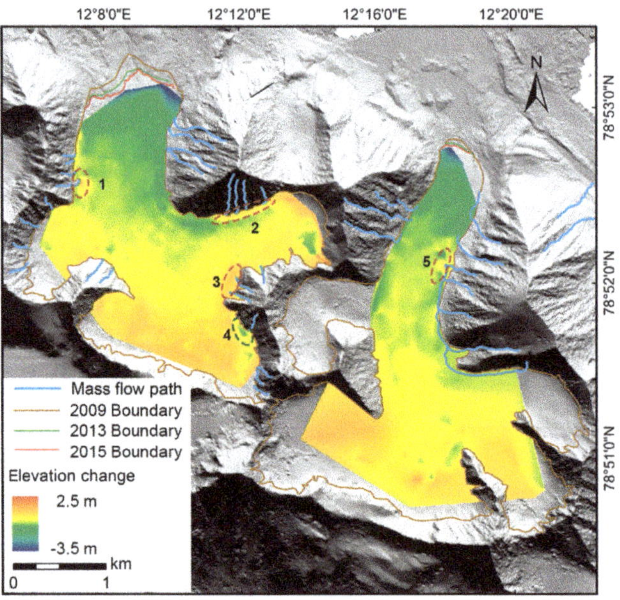

Figure 8. Hillshade topography and elevation changes for Austre Lovénbreen (left) and Pedersenbreen (right) from 2013 to 2015. The dashed ellipse marked regions represent ablation areas (green color, number 4) and accumulation areas (red color, number 1, 2, 3, 5) near the margins of the glaciers.

4.3. Geodetic Glacier Mass Balances

The average annual changes in different elevation bands were obtained from the crossover points, and the results are shown in Table 5. For Austre Lovénbreen, the surface elevation of the entire glacier covered 100–600 m, while Pedersenbreen covered a larger elevation range of about 60–650 m. As shown in the table, there was an obvious relationship between mass change and elevation. The elevation changes in low-altitude areas were negative, indicating that these areas were thinning, and the elevation changes in high-altitude areas were positive, where mass accumulated. In addition, the mass changes varied greatly in different years. In 2013–2014, except for a small amount of accumulation above 500 m, the elevation in most areas of the two glaciers decreased, and the two glaciers seemed to have experienced widespread mass loss. During 2014–2015, the mass accumulation of both glaciers occurred above 300 m, and both glaciers seemed to have a positive mass balance.

Table 5. Elevation changes for Austre Lovénbreen and Pedersenbreen at different elevation intervals (unit: m).

Glacier	Year	<200	200–300	300–400	400–500	≥500
Austre Lovénbreen	2013–2014	−0.919	−0.560	−0.253	−0.097	0.007
	2014–2015	−0.528	−0.084	0.216	0.489	0.562
	2013–2015	−1.594	−0.622	−0.052	0.418	0.520
Pedersenbreen	2013–2014	−0.833	−0.403	−0.311	−0.195	0.027
	2014–2015	−0.489	0.066	0.258	0.367	0.445
	2013–2015	−1.375	−0.336	−0.092	0.202	0.532

The snow depth and density of the two glaciers were not completely recorded in this study, and it was difficult to estimate the precise mass density of the glaciers. Therefore, we tentatively used the elevation changes to indicate geodetic changes over the years. In our study, using the elevation changes at different elevation bins as Bi (the geodetic mass balance at different elevation bins) according to formula (1), together with the data in Table 5, we calculated different years' geodetic changes, which were actually the mean surface elevation changes of the two glaciers, as shown in Table 6.

Table 6. Surface elevation changes in 2013–2015 (unit: m).

Year	Austre Lovénbreen	Pedersenbreen
2013–2014	−0.279	−0.164
2014–2015	0.236	0.309
2013–2015	−0.055	0.172

According to Table 6, Austre Lovénbreen and Pedersenbreen experienced various mass changes in different years. The mass balance was negative in 2013–2014, while it was positive in 2014–2015, but the geodetic balance in 2013–2015 was not completely in accordance with the sum of the mass balances from 2013–2014 and 2014–2015. However, the bias was at the centimeter level, meeting the error level threshold we anticipated. In addition, some differences could be obtained from comparisons; for example, in 2013–2015, Austre Lovénbreen experienced a more serious mass loss than Pedersenbreen, which experienced a mass gain.

To illustrate the relationship between elevation changes and the surface elevation more directly, scatterplots are shown in Figure 9. There appeared to be a significant correlation between the elevation and the elevation change in 2013–2015. In general, both glaciers showed that the correlation was slightly lower in one-year intervals than in a two-year interval. The correlation at the glacier snout differed from the correlation for the entire glacier area; the ice melted more dramatically in the glacier snouts at elevations between 100 m and 200 m, according to the elevation change trend. According to

Figure 9, the mass loss trend of Austre Lovénbreen was more dramatic than that of Pedersenbreen. This finding was consistent with the mass balance results in Table 6.

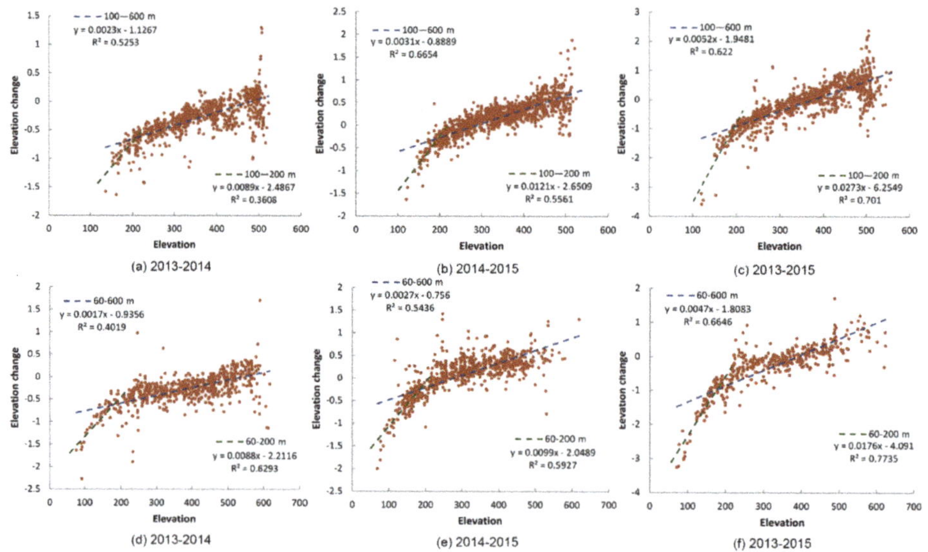

Figure 9. Scatterplot of the ice-surface elevations versus the surface-elevation changes (unit: m) for Austre Lovénbreen (**a–c**) and Pedersenbreen (**d–f**) in different years.

We evaluated the trends of the surface-elevation changes in different years. The corresponding equilibrium line altitudes (ELAs) of Austre Lovénbreen in 2013–2014, 2014–2015, and 2013–2015 were 490 m, 287 m, and 375 m, respectively, and the ELAs of Pedersenbreen in 2013–2014, 2014–2015, and 2013–2015 were 550 m, 285 m, and 385 m, respectively. Shifts in the ELA are often considered to be good indicators for assessing glacial imbalances and estimating mass budget changes [41]. The ELAs in our study differed greatly in different years, and the long-term ELA seemed to be a more reliable indicator. According to the French team, the average ELA of Austre Lovénbreen in 2008–2014 was 431 m [23]. The annual elevation gradient of the geodetic mass balance of Austre Lovénbreen in 2013–2015 was 2.60‰, which means that in the accumulation region, the accumulation increased by 260 mm for every increase of 100 m in the elevation. In contrast, in the ablation area, the ablation increased by 260 mm for every decrease of 100 m in elevation. Accordingly, the annual elevation gradient of the geodetic mass balance of Pedersenbreen was 2.35‰.

Estimating the mass balance of a glacier must consider changes in the entire area; however, field surveys can cover most regions of glaciers, except for high-elevation regions and margins where the slopes are steep. Therefore, in no-data regions, especially at the margin of the glacier snout where the ice melts quickly, it is necessary to make an appropriate estimate of the mass change value. At glacier snouts, the elevation changes need to be corrected. According to the relationship between elevation changes and the elevation (Figure 9), the elevation change from 100 to 200 m for Austre Lovénbreen and the elevation change from 60 to 200 m for Pedersenbreen could be corrected. Some previous studies have made density assumptions of 900 kg/m^3 in the ablation area, and 500 to 600 kg/m^3 in the accumulation area, dominated by firn [15,37,42,43]. If we ignore short-term changes in the vertical firn density, we can assume that the density of the glacier surface is 900 kg/m^3 in the ablation area and 500 kg/m^3 in the accumulation zone when converting elevation changes into mass balances, represented by their water equivalents. Finally, the geodetic mass balances of Austre Lovénbreen and Pedersenbreen in 2013–2014, 2014–2015, and 2013–2015 are shown in Table 7. Because winter snow accumulation in Svalbard does not vary greatly between years, the mass changes of glaciers are mainly dominated by the melting of snow and ice in summer [44]. If we assume that snowfall is invariable, in our study, the geodetic balance in 2013–2014 should be approximately equal to the mass balance in 2013 calculated by the glaciological method, and the geodetic balance in 2014–2015 should be close to the actual mass balance in 2014.

Table 7. Geodetic mass balances in 2013–2015 after correction (unit: m w.e. (water equivalent)).

Year	Austre Lovénbreen	Pedersenbreen
2013–2014	−0.267	−0.165
2014–2015	0.079	0.129
2013–2015	−0.165	0.012

5. Discussion

There are some factors that may influence the accuracy of mass-balance assessments. We let April to April of the next year be a balance year for the mass balance study, while September to September of the next year is usually considered a hydrological year in the Arctic in the classic glaciological method. Different survey time-spans may lead to some discrepancies in comparisons with the glaciological method. Taking Austre Lovénbreen as an example, the net mass balances computed by the French team were −1.111 m w.e. in 2013, 0.010 m w.e. in 2014, and −0.552 m w.e. in 2015 [23]. According to the mass balance obtained by the French team, the geodetic mass balance of Austre Lovénbreen in 2013–2014 in Table 7 should be approximately equal to the sum of the summer mass balance in 2013 and the winter mass balance in 2014. In addition, the geodetic mass balance in 2014–2015 should be close to the sum of the summer mass balance in 2014 and the winter mass balance in 2015. However, our study appeared to underestimate the mass loss in 2013, which might be partly ascribed to ignoring snow depth. On the other hand, it is difficult to calculate the mass balance accurately by a simple linear simulation, as the elevation changes in the lateral zones of the glacier are smaller than the change in the glacier's center [22], and the classic glaciological method may not consider potential subglacier mass changes. Although there may be some discrepancies, the mass balance change trend in our study was consistent with the trend calculated by the French team, which means that the method proposed here to calculate the geodetic mass balance is valid. Similar geodetic methods with high-density GPS data were also applied to other regions. Repeated differential GPS surveys were carried out on Gangju La glacier, Bhutan Himalaya; the annual and decadal geodetic mass balances calculated from the GPS points and snow depth showed consistency with the direct mass balance observed from stakes [6]. Marinsek and Ermolin compared the elevation differences from kinematic GPS surveys on Bahía del Diablo glacier on the Antarctic Peninsula in 2000–2001 and 2010, finding that the geodetic mass balance calculated based on elevation change was close to the glaciological mass balance [45].

Snow depth data of Austre Lovénbreen were obtained from the French team for validation and comparison over the same period (during 2013–2015). The snow depth distributions of Austre Lovénbreen (Figure 10a–c) demonstrated that the snow depth logically increased from north to south, with shallower snow depths at lower elevations and more snow accumulation at higher elevations, representing a positive relationship between snow depth and elevation. According to Figure 10d–f, more snow remained because of glacial accumulation in 2014 than in the other two years combined. In addition, Austre Lovénbreen experienced similar snow accumulation at relatively low elevations in 2013 and 2015, even though 2013 was shallower; in contrast, in the upper cirques of the glacier, there was clearly a deeper snow depth in 2015 than in the two other years. This situation was in accordance with the analyses we performed with RTK-GPS data; hence, we believed that the main difference between 2014 and 2015 was that the snow at relatively high elevations in 2014 did not melt completely during the melting season. In contrast, very little snow survived the summer of 2013, and little or no accumulation was observed. To accurately calculate the mass balance of Austre Lovénbreen, the interpolated snow depths in three years measured by snow cores were removed from the crossover points of the RTK-GPS tracks. Then, ice-surface elevation changes were calculated, as shown in Figure 10d–f. In conjunction with snow depth, the ice changes and snow changes were calculated separately.

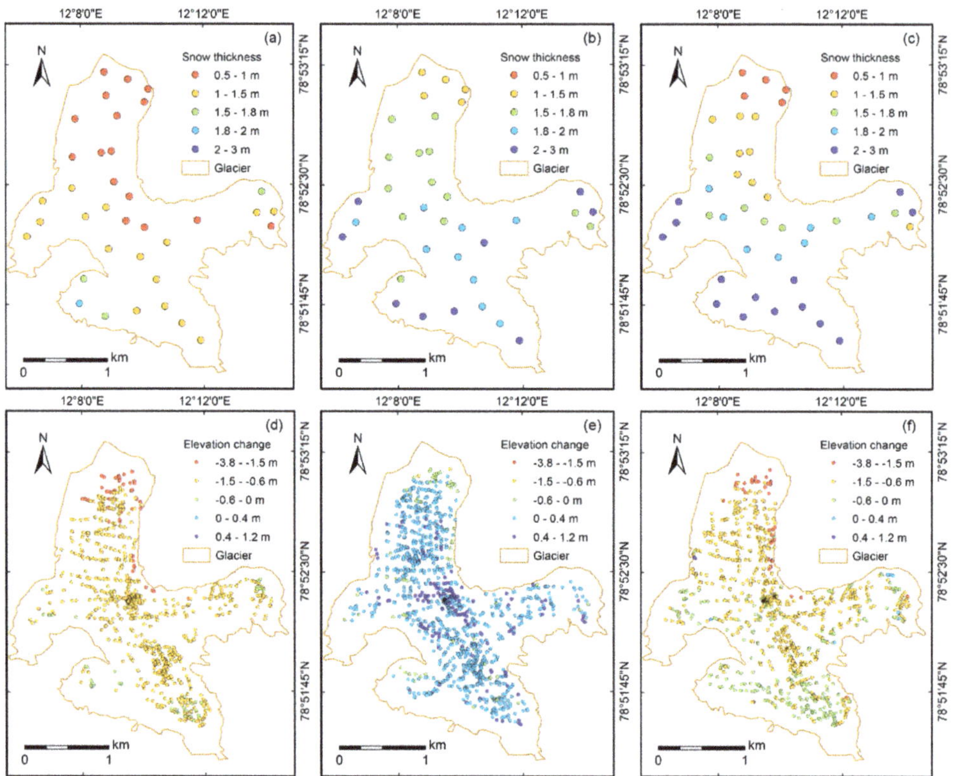

Figure 10. Spatial distributions of snow depth observed in (**a**) 2013, (**b**) 2014, and (**c**) 2015 and ice-surface elevation changes in the crossover points of the RTK-GPS (real-time kinematic global positioning system) tracks after removing the snow depth of Austre Lovénbreen in (**d**) 2013–2014, (**e**) 2014–2015, and (**f**) 2013–2015.

According to formula (2), the mass balances of crossover points in different years could be calculated; in addition, the mass balances for the entire glacier could be calculated from natural neighbor interpolation. As the previous study suggested, the density of the winter snowpack was between 350 and 450 kg m^{-3} for the Austfonna ice cap, Svalbard [46]. With the assumed ice density of 900 kg/m^3 and the assumed snow density of 350 kg/m^3 in this study, the mass changes in the crossover points of the RTK-GPS tracks were estimated with the ice changes and snow changes and then interpolated, as shown in Figure 11. For the region without data, an appropriate approximation of average mass balance was performed by extracting the mass balance values in the surrounding areas covered by the mass balance interpolation results, and the corresponding area percentages of the no-data regions and data covered region were calculated. Then, the mass balances of the entire glacier were estimated from the area-weighted mass balances; the results are shown in Table 8. The mass balance of Austre Lovénbreen in 2013–2014, shown in Table 8, was close to −0.760 m w.e., which was the sum of the summer mass balance in 2013 and the winter mass balance in 2014. The mass balance in 2014–2015 was close to 0.136 m w.e, which was the glaciological mass balance recorded by the French team over the same period. Although some discrepancies still exist in comparison with the actual observation results, the mass balance results calculated when considering snow depth seemed to be more reasonable than those calculated while ignoring snow depth.

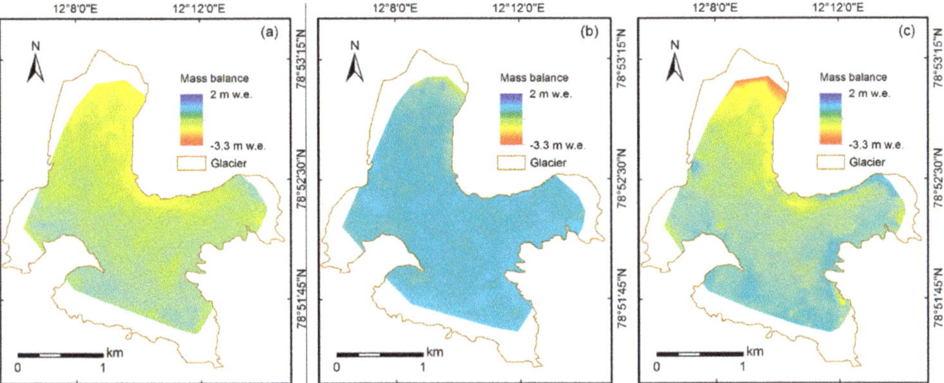

Figure 11. Mass balances calculated considering the snow depth of Austre Lovénbreen in (**a**) 2013–2014, (**b**) 2014–2015, and (**c**) 2013–2015.

Table 8. The geodetic mass balances for Austre Lovénbreen calculated with and without consideration of snow depth and the mass balance calculated by the glaciological method over the same period (unit: m w.e. (water equivalent)).

Year	Ignoring Snow Depth	Considering Snow Depth	Glaciological Mass Balance
2013–2014	−0.267	−0.670	−0.760
2014–2015	0.079	0.184	0.136
2013–2015	−0.165	−0.490	−0.624

There was a large mass loss ascribed to ice melting in 2013. However, the deeper snow in 2014 compensated for the elevation change, leading us to underestimate the ice loss. To explain this phenomenon more directly, we estimated the geodetic mass balance without snow depth data at a point with formula (3):

$$b = \Delta h \rho \tag{3}$$

where b is the mass balance, Δh is the elevation change, and ρ is the estimated average density of the ice and snow mixture. If formula (2) and formula (3) are both true, and the elevation change is not 0 m, then the relationship between the assumed average density and the snow depth can be obtained as formula (4):

$$\rho = \rho_i + (s_2 - s_1)(\rho_s - \rho_i)/\Delta h \tag{4}$$

where ρ_i is the ice density; ρ_s is the snow density; and s_1 and s_2 represent the snow depths for the first year and the following year, respectively. The snow density and ice density are assumed to be invariant. Taking the years of 2013 and 2014 as an example, s_2 is far larger than s_1, and Δh is less than 0 m in most regions; hence, only when the average density ρ is larger than the ice density ρ_i, we can obtain the correct geodetic mass balance. We usually assume the average density to be less than the ice density. However, over a relatively long time interval, the elevation change is considerable, and the snow depth can be neglected in comparison with the ice change; thus, the mass balance can be more accurately calculated with the elevation change data. In fact, over a short time interval, the results of the density assumption are inconsistent with the results obtained, considering limited volumetric changes [37]. This means that, ignoring snow depth, which varied greatly during our research period, might have resulted in a larger bias in the mass balance over a short time interval.

Using a UAV-generated DEM of glacier snout in 2019, we calculated the elevation differences of the glacier snouts between 2013 and 2019 by comparing the DEMs of different years; the results are shown in Figure 12, according to which the two glacier snouts experienced serious mass loss from 2013 to 2019, and the elevation decreased by 17 m in maximum. In order to precisely compare the elevation changes, we only extracted the 2019 surface elevations at crossover points whose positions were derived from RTK-GPS tracks in 2013 and 2015 (Figure 12). At these specific point locations, inside the 2019 DEM coverage, the mean elevation changes in different years were calculated as in Table 9. Although the 2019 DEM covered only the snout area of each glacier, the point elevation changes presented a clear tendency from 2013 to 2019. The mean value of elevation changes in 2015–2019 was two to three times as much as that in 2013–2015, which proved that both glaciers are in an accelerating thinning situation, at least over their snout areas.

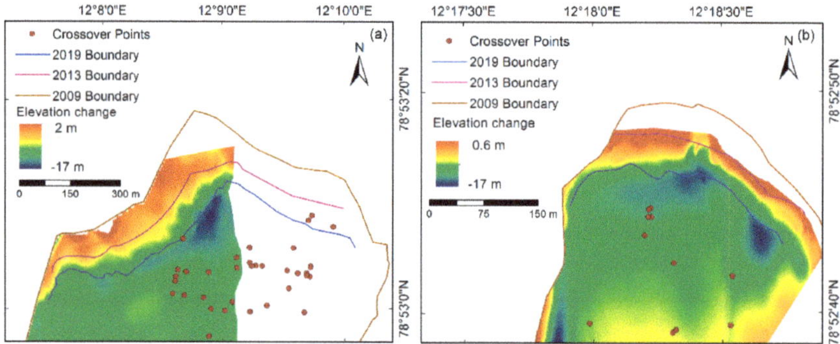

Figure 12. Elevation changes for the glacier snouts of (**a**) Austre Lovénbreen and (**b**) Pedersenbreen from 2013 to 2019. The 2019 boundaries of two glaciers were extracted from UAV (unmanned aerial vehicle) images. The crossover points were extracted from RTK-GPS (real-time kinematic global positioning system) tracks between 2013 and 2015.

Table 9. Mean elevation changes of crossover points at glacier snout areas from 2013 to 2019 (unit: m/a).

Year	Austre Lovénbreen	Pedersenbreen
2013–2015	−0.75	−1.19
2015–2019	−2.23	−2.26
2013–2019	−1.73	−1.90

In this study, we evaluated the smoothness and accuracy of the interpolated glacier surface using different methods, and NN was finally chosen for our RTK-GPS data interpolation. However, the optimum interpolation method depends on the characteristics of the source data, the complexity of the terrain, and the desired properties of the interpolated result. Therefore, we need to evaluate the interpolation methods cautiously in other instances. The terrain itself, the density, and the uncertainty of input data are also important factors in choosing interpolation methods. For example, NN interpolation will ignore details of the terrain, which fits well with glacier surfaces but may not be suitable for complex terrains. Accuracy and smoothness are usually our desired properties. However, the little experience can be obtained from previous studies. For instance, kriging was applied to analyze the mass balance of Storglaciären [47]. Some researchers have used IDW to create a continuous surface of thickness values along the branch lines at the bed of a glacier [48]. In other studies, spline interpolation was chosen to generate the DEM of Alpine glaciers [49]. Bo and others used NN interpolation to build regional DEMs within the Antarctic ice sheet [50]. Mölg concluded that Kriging and Topo to Raster showed robust and reliable results in a mass balance study on the Conejeras glacier, Colombia [51]. Pellitero and others presented a semi-automated method to generate ice thickness from bed topography along a palaeoglacier flowline by applying the standard flow law for ice and generating the 3D surface of the palaeoglacier using multiple interpolation methods, in which IDW and kriging performed well in volume estimation [52]. Kääb chose spline, kriging, and IDW approaches to interpolate surface elevation changes along contour lines on the Svalbard glacial Edgeøya, and the volume change estimations using three interpolation methods were similar [53].

To widely interpret the trends in the geodetic mass balance distribution of Arctic glaciers, we compared the elevation changes and the geodetic mass balance of Austre Lovénbreen and Pedersenbreen. They experienced similar geodetic mass balances: a serious mass loss in 2013–2014 and a slight mass accumulation in 2014–2015. Nevertheless, some differences can be noted, as Austre Lovénbreen displayed stronger thinning than Pedersenbreen. The elevation range distribution of these two glaciers may explain this difference [17] because the area percentage of Pedersenbreen in higher altitude regions is larger than Austre Lovénbreen.

In addition, high-density RTK-GPS measurements require considerable in situ work, considering the complexity of RTK-GPS surveys. Therefore, surveys were only carried out in 2013, 2014, and 2015. Due to this limited time, the mass change results may not be completely consistent with the long-term trends; we found that geodetic balances varied greatly in 2013–2015. For a short-term trend, there seems to be some uncertainty in estimating mass balances by the geodetic method, which requires caution when converting the elevation changes estimated with RTK-GPS data into mass balances. Long-term RTK-GPS data covering the entire glaciers are required for additional comprehensive analyses, which would contribute to future comparative studies with the glaciological method.

6. Conclusions

Based on RTK-GPS data, the surface-elevation changes and the geodetic mass balances of Austre Lovénbreen and Pedersenbreen were preliminarily studied. The following conclusions could be drawn from our analysis.

(1) We compared different interpolation methods, and NN interpolation was more suitable for generating the surface DEMs of glaciers.

(2) The elevation changes across different years were estimated both by the differences in the DEMs and the differences in the RTK-GPS track-derived crossover point elevations; the latter was more suitable for estimating elevation changes linked to the corresponding altitude.

(3) Ignoring the effect of snow depth over a short time interval might lead to a larger bias in the mass balance, especially when the snow of the following year is far deeper than that of the first year during a negative balance period. However, for a study spanning a long time interval (more than 10 years), the snow depth could be neglected in comparison with the ice change. Therefore, caution must be exercised when estimating mass balances with elevation change data from only a few years, and the mass balances estimated with RTK-GPS data might be more accurate for investigations over a longer time interval.

(4) The average ELA and the annual elevation gradient of the geodetic mass balance of Austre Lovénbreen in 2013–2015 were 375 m and 2.60‰, respectively, and those of Pedersenbreen were 385 m and 2.35‰, respectively. The gradients of the two glaciers were three times larger at the glacier snouts, according to the elevation changes at the crossover points from the RTK-GPS tracks. In the study period, Austre Lovénbreen experienced a more serious mass loss than Pedersenbreen, and both glaciers were in accelerating thinning status at snout areas.

Author Contributions: S.A. and Z.W. conceived the study and supervised the experiments. X.D. and F.T. wrote the manuscript. X.D. and X.Z. processed some data and produced some of the figures. S.A., Z.W., and F.T. contributed to the field data collection on the glaciers.

Funding: This research was funded by the National Natural Science Foundation of China (41531069, 41476162).

Acknowledgments: Our study was supported by the National Natural Science Foundation of China (41531069, 41476162). The field data in this article were acquired by researchers from the Chinese Arctic Yellow River Station.

We thank the Chinese Arctic and Antarctic Administration of the State Oceanic Administration for sponsoring the field expeditions around the Chinese Arctic Yellow River Station. Snow cover data were acquired by the French team with the support of the French Polar Institute (IPEV) and the French Agence Nationale de la Recherche (ANR). We are grateful to the team members who supported our in situ observations.

Conflicts of Interest: The authors declare that they have no conflicts of interest to disclose.

References

1. Haeberli, W.; Hoelzle, M.; Paul, F.; Zemp, M. Integrated monitoring of mountain glaciers as key indicators of global climate change: The European Alps. *Ann. Glaciol.* **2007**, *46*, 150–160. [CrossRef]
2. Paul, F.; Bolch, T.; Kaab, A.; Nagler, T.; Nuth, C.; Scharrer, K. The glaciers climate change initiative: Methods for creating glacier area, elevation change and velocity products. *Remote Sens. Environ.* **2015**, *162*, 408–426. [CrossRef]
3. Nuth, C.; Moholdt, G.; Kohler, J.; Hagen, J.O.; Kaab, A. Svalbard glacier elevation changes and contribution to sea level rise. *J. Geophys. Res. Earth Surf.* **2010**, *115*. [CrossRef]
4. Larsen, C.F.; Motyka, R.J.; Arendt, A.; Echelmeyer, K.A.; Geissler, P.E. Glacier changes in southeast Alaska and northwest British Columbia and contribution to sea level rise. *J. Geophys. Res. Earth Surf.* **2007**, *112*. [CrossRef]
5. Braithwaite, R.J. Glacier mass balance: The first 50 years of international monitoring. *Prog. Phys. Geogr.* **2002**, *26*, 76–95. [CrossRef]
6. Tshering, P.; Fujita, K. First in situ record of decadal glacier mass balance (2003–2014) from the Bhutan Himalaya. *Ann. Glaciol.* **2016**, *57*, 289–294. [CrossRef]
7. Woul, M.D.; Hock, R. Static mass-Balance sensitivity of arctic glaciers and ice caps using a degree-Day approach. *Ann. Glaciol.* **2005**, *42*, 217–224. [CrossRef]
8. Bamber, J.L.; Krabill, W.; Raper, V.; Dowdeswell, J.A.; Oerlemans, J. Elevation changes measured on Svalbard glaciers and ice caps from airborne laser data. *Ann. Glaciol.* **2005**, *42*, 202–208. [CrossRef]
9. Ai, S.; Wang, Z.; Tan, Z.; E, D.C.; Yan, M. Mass change study on Arctic glacier Pedersenbreen, during 1936–1990–2009. *Chin. Sci. Bull.* **2013**, *58*, 3148–3154. [CrossRef]
10. Wang, P.; Li, Z.; Li, H.; Wang, W.; Yao, H. Comparison of glaciological and geodetic mass balance at Urumqi glacier No. 1, Tian Shan, Central Asia. *Global Planet. Change* **2014**, *114*, 14–22. [CrossRef]
11. Cox, L.H.; March, R.S. Comparison of geodetic and glaciological mass-Balance techniques, Gulkana Glacier, Alaska, USA. *J. Glaciol.* **2004**, *50*, 363–370. [CrossRef]
12. Paul, F.; Haeberli, W. Spatial variability of glacier elevation changes in the Swiss Alps obtained from two digital elevation models. *Geophys. Res. Lett.* **2008**, *35*. [CrossRef]
13. Fischer, A. Comparison of direct and geodetic mass balances on a multi-annual time scale. *Cryosphere* **2011**, *5*, 107–124. [CrossRef]
14. Hagen, J.O.; Liestol, O. Long-Term glacier mass-balance investigations in Svalbard, 1950–1988. *Ann. Glaciol.* **1990**, *14*, 102–106. [CrossRef]
15. Moholdt, G.; Nuth, C.; Hagen, J.O.; Kohler, J. Recent elevation changes of Svalbard glaciers derived from ICESat laser altimetry. *Remote Sens. Environ.* **2010**, *114*, 2756–2767. [CrossRef]
16. Hagen, J.O.; Eiken, T.; Kohler, J.; Melvold, K. Geometry changes on Svalbard glaciers: Mass-Balance or dynamic response. *Ann. Glaciol.* **2005**, *42*, 255–261. [CrossRef]
17. Xu, M.; Yan, M.; Ren, J.; Ai, S.; Kang, J.; E, D.C. Surface mass balance and ice flow of the glaciers Austre Lovenbreen and Pedersenbreen, Svalbard, Arctic. *Chin. J. Pol. Sci.* **2010**, *21*, 147–159.
18. Ai, S.; Wang, Z.; E, D.; Holmen, K.; Tan, Z.; Zhou, C.; Sun, W. Topography, ice thickness and ice volume of the glacier Pedersenbreen in Svalbard, using GPR and GPS. *Pol. Res.* **2014**, *33*, 18533. [CrossRef]
19. Ai, S.; Ding, X.; An, J.; Lin, G.; Wang, Z.; Yan, M. Discovery of the Fastest Ice Flow along the Central Flow Line of Austre Lovenbreen, a Poly-Thermal Valley Glacier in Svalbard. *Remote Sens.* **2019**, *11*, 1488. [CrossRef]
20. Malecki, J. Elevation and volume changes of seven Dickson Land glaciers, Svalbard, 1960–1990–2009. *Pol. Res.* **2013**, *32*, 18400. [CrossRef]
21. Hagen, J.O.; Melvold, K.; Pinglot, F.; Dowdeswell, J.A. On the net mass balance of the glaciers and ice caps in Svalbard, Norwegian Arctic. *Arct. Antarc. Alpine Res.* **2003**, *35*, 264–270. [CrossRef]

22. Nuth, C.; Kohler, J.; Aas, H.F.; Brandt, O.; Hagen, J.O. Glacier geometry and elevation changes on Svalbard (1936-90): A baseline dataset. *Ann. Glaciol.* **2007**, *46*, 106–116. [CrossRef]

23. Marlin, C.; Tolle, F.; Griselin, M.; Bernard, E.; Saintenoy, A.; Quenet, M.; Friedt, J.M. Change in geometry of a high Arctic glacier from 1948 to 2013 (Austre Lovenbreen, Svalbard). *Geogr. Ann. Ser. A Phys. Geogr.* **2017**, *99*, 115–138. [CrossRef]

24. Moller, M.; Finkelnburg, R.; Braun, M.; Hock, R.; Jonsell, U.; Pohjola, V.A.; Scherer, D.; Schneider, C. Climatic mass balance of the ice cap Vestfonna, Svalbard: A spatially distributed assessment using ERA-Interim and MODIS data. *J. Geophys. Res.* **2011**, *116*. [CrossRef]

25. Forland, E.J.; Benestad, R.; Hanssen-Bauer, I.; Haugen, J.E.; Skaugen, T.E. Temperature and Precipitation Development at Svalbard 1900–2100. *Adv. Meteorol.* **2011**, *2011*, 1–14. [CrossRef]

26. Nowak, A.; Hodson, A. Changes in meltwater chemistry over a 20-Year period following a thermal regime switch from polythermal to cold-Based glaciation at Austre Broggerbreen, Svalbard. *Pol. Res.* **2014**, *33*, 22779. [CrossRef]

27. Ai, S.; E, D.; Yan, M.; Ren, J. Arctic glacier movement monitoring with GPS method on 2005. *Chin. J. Pol. Sci.* **2006**, *17*, 61–68.

28. Porter, C.; Morin, P.; Howat, I.; Noh, M.J.; Bates, B.; Peterman, K.; Keesey, S.; Schlenk, M.; Gardiner, J.; Willis, M.; et al. "ArcticDEM". *Harv. Dataverse* **2018**, *1*. [CrossRef]

29. Gardelle, J.; Berthier, E.; Arnaud, Y. Impact of resolution and radar penetration on glacier elevation changes computed from DEM differencing. *J. Glaciol.* **2012**, *58*, 419–422. [CrossRef]

30. Zemp, M.; Thibert, E.; Huss, M.; Stumm, D.; Rolstad, C.; Nuth, C.; Nussbaumer, S.U.; Moholdt, G.; Mercer, A.; Mayer, C. Reanalysing glacier mass balance measurement series. *Cryosphere* **2013**, *7*, 1227–1245. [CrossRef]

31. Rolstad, C.; Haug, T.; Denby, B. Spatially integrated geodetic glacier mass balance and its uncertainty based on geostatistical analysis: Application to the western Svartisen ice cap, Norway. *J. Glaciol.* **2009**, *55*, 666–680. [CrossRef]

32. Oliver, M.A.; Webster, R. A tutorial guide to geostatistics: Computing and modelling variograms and kriging. *Catena* **2014**, *113*, 56–69. [CrossRef]

33. Armstrong, M. Problems with universal kriging. *Math. Geol.* **1984**, *16*, 101–108. [CrossRef]

34. Guo, Q.; Li, W.; Yu, H.; Alvarez, O. Effects of topographic variability and lidar sampling density on several DEM interpolation methods. *Photogramm. Eng. Remote Sens.* **2010**, *76*, 701–712. [CrossRef]

35. Bojanov, B.D.; Hakopian, H.A.; Sahakian, A.A. *Spline Functions and Multivariate Interpolations*; Kluwer: Norwell, MA, USA, 1993.

36. Hutchinson, M.F. A new procedure for gridding elevation and stream line data with automatic removal of spurious pits. *J. Hydrol.* **1989**, *106*, 211–232. [CrossRef]

37. Huss, M. Density assumptions for converting geodetic glacier volume change to mass change. *Cryosphere* **2013**, *7*, 877–887. [CrossRef]

38. Kaab, A.; Treichler, D.; Nuth, C.; Berthier, E. Brief communication: Contending estimates of 2003–2008 glacier mass balance over the Pamir–Karakoram–Himalaya. *Cryosphere* **2015**, *9*, 557–564. [CrossRef]

39. Erdogan, S. A comparision of interpolation methods for producing digital elevation models at the field scale. *Earth Surf. Process. Landf.* **2009**, *34*, 366–376. [CrossRef]

40. Racoviteanu, A.E.; Manley, W.F.; Arnaud, Y.; Williams, M.W. Evaluating digital elevation models for glaciologic applications: An example from Nevado Coropuna, Peruvian Andes. *Global Planet. Change* **2007**, *59*, 110–125. [CrossRef]

41. Huss, M.; Dhulst, L.; Bauder, A. New long-Term mass-Balance series for the Swiss Alps. *J. Glaciol.* **2016**, *61*, 551–562. [CrossRef]

42. Kaab, A.; Berthier, E.; Nuth, C.; Gardelle, J.; Arnaud, Y. Contrasting patterns of early twenty-First-Century glacier mass change in the Himalayas. *Nature* **2012**, *488*, 495–498. [CrossRef]

43. Gardelle, J.; Berthier, E.; Arnaud, Y. Slight mass gain of karakoram glaciers in the early twenty-First century. *Nat. Geosci.* **2012**, *5*, 322–325. [CrossRef]

44. Hagen, J.O.; Kohler, J.; Melvold, K.; Winther, J.G. Glaciers in Svalbard: Mass balance, runoff and freshwater flux. *Pol. Res.* **2003**, *22*, 145–159. [CrossRef]

45. Marinsek, S.; Ermolin, E. 10 year mass balance by glaciological and geodetic methods of Glaciar Bahia del Diablo, Vega Island, Antarctic Peninsula. *Ann. Glaciol.* **2015**, *56*, 141–146. [CrossRef]

46. Dunse, T.; Schuler, T.; Hagen, J.; Eiken, T.; Brandt, O.; Hogda, K. Recent fluctuations in the extent of the firn area of Austfonna, Svalbard, inferred from GPR. *Ann. Glaciol.* **2009**, *50*, 155–162. [CrossRef]
47. Hock, R.; Jensen, H. Application of kriging interpolation for glacier mass balance computations. *Geogr. Ann. Ser. A Phys. Geogr.* **1999**, *81*, 611–619. [CrossRef]
48. Paul, F.; Linsbauer, A. Modeling of glacier bed topography from glacier outlines, central branch lines, and a DEM. *Int. J. Geogr. Inf. Sci.* **2012**, *26*, 1173–1190. [CrossRef]
49. Mennis, J.L.; Fountain, A.G. A spatio-Temporal GIS database for monitoring alpine glacier change. *Photogramm. Eng. Remote Sens.* **2001**, *67*, 967–975.
50. Bo, S.; Siegert, M.J.; Mudd, S.M.; Sugden, D.; Fujita, S. The Gamburtsev mountains and the origin and early evolution of the Antarctic Ice Sheet. *Nature* **2009**, *459*, 690–693. [CrossRef]
51. Molg, N.; Ceballos, J.L.; Huggel, C.; Micheletti, N.; Rabatel, A.; Zemp, M. Ten years of monthly mass balance of Conejeras glacier, Colombia, and their evaluation using different interpolation methods. *Geogr. Ann. Ser. A Phys. Geogr.* **2017**, *99*, 155–176. [CrossRef]
52. Pellitero, R.; Rea, B.R.; Spagnolo, M.; Bakke, J.; Ivy-Ochs, S.; Frew, C.R.; Hughes, P.; Ribolini, A.; Lukas, S.; Renssen, H. GlaRe, a GIS tool to reconstruct the 3D surface of palaeoglaciers. *Comput. Geosci.* **2016**, *94*, 77–85. [CrossRef]
53. Kaab, A. Glacier Volume Changes Using ASTER Satellite Stereo and ICESat GLAS Laser Altimetry. A Test Study on Edgeoya, Eastern Svalbard. *IEEE Trans. Geosci. Remote Sens.* **2008**, *46*, 2823–2830. [CrossRef]

 *remote sensing*

Article

Sentinel 2 Analysis of Turbidity Patterns in a Coastal Lagoon

María-Teresa Sebastiá-Frasquet [1,*], Jesús A. Aguilar-Maldonado [2], Eduardo Santamaría-Del-Ángel [2] and Javier Estornell [3]

[1] Instituto de Investigación para la Gestión Integrada de Zonas Costeras, Universitat Politècnica de València, C/Paraninfo, 1, 46730 Grau de Gandia, Spain

[2] Facultad de Ciencias Marinas, Universidad Autónoma de Baja California, Ensenada 22860, Mexico; jesusaguilarmaldonado@gmail.com (J.A.A.-M.); santamaria@uabc.edu.mx (E.S.-D.-Á.)

[3] Grupo de Cartografía GeoAmbiental y Teledetección, Universitat Politècnica de València, Camí de Vera s/n, 46022 Valencia, Spain; jaescre@upv.es

* Correspondence: mtsebastia@hma.upv.es

Received: 7 October 2019; Accepted: 4 December 2019; Published: 6 December 2019

Abstract: Coastal lagoons are transitional ecosystems with complex spatial and temporal variability. Remote sensing tools are essential for monitoring and unveiling their variability. Turbidity is a water quality parameter used for studying eutrophication and sediment transport. The objective of this research is to analyze the monthly turbidity pattern in a shallow coastal lagoon along two years with different precipitation regimes. The selected study area is the Albufera de Valencia lagoon (Spain). For this purpose, we used Sentinel 2 images and in situ data from the monitoring program of the Environment General Subdivision of the regional government. We obtained Sentinel 2A and 2B images for years 2017 and 2018 and processed them with SNAP software. The results of the correlation analysis between satellite and in situ data, corroborate that the reflectance of band 5 (705 nm) is suitable for the analysis of turbidity patterns in shallow lagoons (average depth 1 m), such as the Albufera lagoon, even in eutrophic conditions. Turbidity patterns in the Albufera lagoon show a similar trend in wet and dry years, which is mainly linked to the irrigation practice of rice paddies. High turbidity periods are linked to higher water residence time and closed floodgates. However, precipitation and wind also play an important role in the spatial distribution of turbidity. During storm events, phytoplankton and sediments are discharged to the sea, if the floodgates remain open. Fortunately, the rice harvesting season, when the floodgates are open, coincides with the beginning of the rainy period. Nevertheless, this is a lucky coincidence. It is important to develop conscious management of floodgates, because having them closed during rain events can have several negative effects both for the lagoon and for the receiving coastal waters and ecosystem. Non-discharged solids may accumulate in the lagoon worsening the clogging problems, and the beaches next to the receiving coastal waters will not receive an important load of solids to nourish them.

Keywords: Sentinel; Secchi disk; chlorophyll *a*; sediments; phytoplankton

---

1. Introduction

Coastal lagoons are transitional ecosystems between inland and coastal waters. They are shallow water bodies separated from the ocean by a barrier and connected, at least intermittently, to the ocean by one or more restricted inlets [1]. Given these characteristics, they exhibit complex spatial and temporal variability. They are usually part of wetland ecosystems and are among the most endangered ecosystems, especially in coastal areas, due to several anthropogenic threats [2]. These ecosystems are characterized by a high variability due to both natural intrinsic variability and anthropic pressures variability (e.g., man-controlled hydrological cycle, wastewater discharge, etc.). In situ monitoring programs (e.g., Water Framework Directive) have difficulty diagnosing their quality status and the effectiveness of restoration measures. Remote sensing is a complementary tool to the traditional on-site approach that allows constructing a synoptic view that is not possible otherwise. During the last decades, several studies have aimed at monitoring indicator parameters of water quality both in inland and in coastal waters using satellite images. However, the spatial and temporal scale have been constraints for small sized and highly variable ecosystems such as coastal lagoons. High temporal resolution sensors (1–3 days) such as Moderate Resolution Imaging Spectroradiometer (MODIS) or MEdium Resolution Imaging Spectrometer (MERIS) have a limited spatial resolution (250/500 m). Higher spatial resolution sensors such as Landsat Thematic Mapper (TM) (30 m) or SPOT have a low temporal resolution (16 days) not enough for the highly dynamic coastal lagoons [3,4]. The Copernicus Sentinel-2 mission of the European Space Agency (ESA) comprises a constellation of two polar-orbiting satellites, and the first one, Sentinel-2A is operational since June 2015. This mission combines both a high spatial (10–60 m) and a high temporal resolution (5 days) that are necessary to monitor coastal lagoons [4,5].

One of the major environmental problems of coastal lagoons is eutrophication, and one of the most commons parameters used to monitor their ecological status is chlorophyll *a* (Chl*a*) concentration [3,6,7]. Consequently, recent studies have applied the advances in remote sensing to study temporal and spatial evolution of Chl*a* using Sentinel-2 images [8]. A very recent study has applied Sentinel-2 images also to study phycocyanin concentration, which is an indicator of cyanobacterial blooms [7]. Turbidity is also a water quality parameter used as a eutrophication indicator [9]. Turbidity reduces the availability of light underwater, and thus limits light availability for phytoplankton growth and primary productivity [9,10]. Moreover, it is also important for nutrient dynamics, pollutants, and sediment transport [9]. According to the ASTM-International definition, turbidity is an expression of the optical properties of a liquid that causes light rays to be scattered and absorbed rather than transmitted in straight lines through a sample. Turbidity, suspended particulate matter (SPM), and Secchi disk depth are three variables closely related. Frequently, turbidity is used as an estimation of SPM concentration [9,11]. In fact, traditionally, turbidity is estimated visually using a Secchi disk depth or measured directly with nephelometry [10]. The analysis of turbidity is especially important in optically complex waters where phytoplankton and SPM do not covary, and sediment contribution can result in an overestimation of Chl*a* [12,13]. Previous research applied remote sensing to map turbidity in complex coastal waters. The authors of [14] used Landsat 8 and SPOT images in the Mar Menor (Spain); in [4] the authors applied Landsat 5, 7, and 8 in the turbid Gironde and Loire estuaries (France); the authors of [10] used Landsat 8 in Cam Ranh Bay and Thuy Trieu Lagoon (Vietnam), and in [9] the authors applied a multisensory approach in the Danube Delta (Romania). Recently, the trend is to apply Sentinel 2 advantages to monitoring highly variable ecosystems [5,12,13,15].

Determining turbidity in shallow waters requires the use of spectral bands that are sensitive to turbidity and have a limited depth penetration to avoid substantial interference from the bottom [15]. Water absorption increases rapidly from red (645–700 nm) to red edge NIR (700–780 nm) [16]. This absorption limits the light received from the bottom, while it returns light scattered by suspended materials. These bands offer a good balance between turbidity detection and bottom detection [17]. Several studies have already indicated that these spectral bands are appropriate for monitoring turbidity or suspended solids in optically complex regions [15,17,18]. According to [15], the 704 nm wavelength gives the greatest return of light to the sensor at depths between 1 and 2 m. However, at longer wavelengths, sensitivity to suspended material is lost in shallow and very turbid waters [15].

The objective of this research is to analyze the monthly turbidity pattern in a shallow coastal lagoon along two years with different precipitation regime. The selected study area is the *Albufera de Valencia* lagoon (Valencia, Spain). This lagoon faces a eutrophication problem, and it is at risk of disappearing due to the accumulation of sediments. The analysis of turbidity is important to unveil the sediment transport dynamics. For this purpose, we used Sentinel 2 images and in situ data from the monitoring program of the Environment General Subdivision of the regional government, which has been implemented since year 1995. Remote sensing is the only way to obtain a synoptic view of the entire lagoon due to the high spatial complexity and the varying water quality of the more than 60 tributaries.

2. Materials and Methods

2.1. Study Area

The *Albufera de Valencia* lagoon is a shallow turbid coastal lagoon, located in the Mediterranean coast, 10 km south of the city of Valencia (Figure 1) [19,20]. It has an average depth close to 1 m (1–3 m) and covers an area of approximately 24 km^2 [6]. This water body is characterized as hypertrophic, with average annual Chl*a* levels of 167 µg L^{-1} (4–322 µg L^{-1}) and Secchi disk depth of 0.34 m (0.18–1 m) [6].

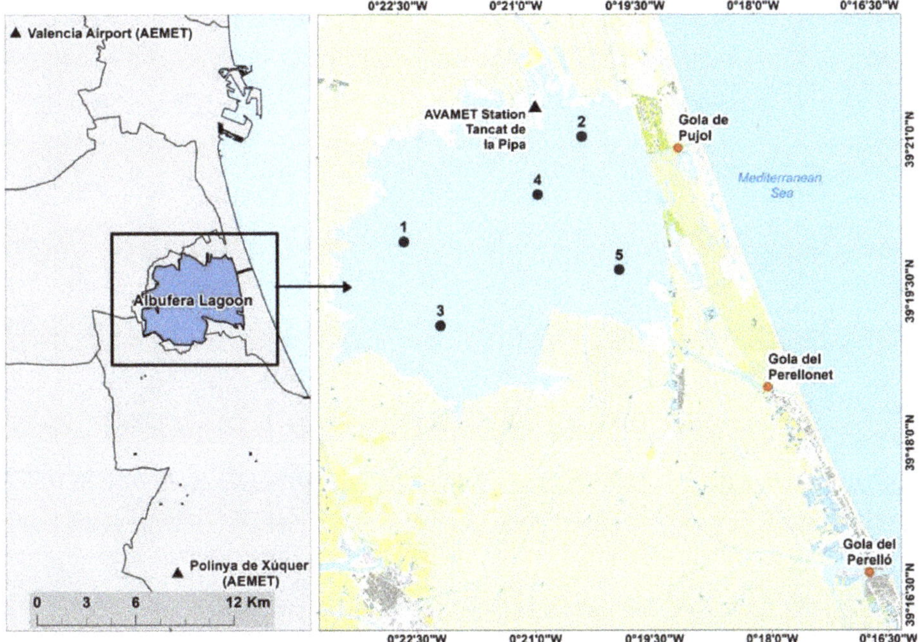

Figure 1. Study area, the *Albufera de Valencia* lagoon and surroundings. Numbered black points are sampling stations from the monitoring program of the Environment General Subdivision of the Valencian government.

It is part of the *Albufera de Valencia* coastal wetland, which is one of the most representative wetlands in the Mediterranean basin, and holds several protection figures at national and international level, such as Spanish Natural Park, Special Protection Areas (SPAs) for birds, Sites of Community Importance (SCIs), and Ramsar Site.

The lagoon is surrounded by an agricultural area with an approximate surface of 223 km^2 primarily used for rice cultivation [6]. The local water council, under the direction of farmers, controls the hydrological cycle in the watershed to meet the needs of rice crop [6,8]. Farming contributes about 60% of the inputs to the Albufera through 63 irrigation channels that carry water from the Turia and Júcar rivers [21,22]. Other sources of water are treated wastewater from the urban and industrial areas nearby, groundwater contributions, direct precipitation on the lagoon, and potential indirect contributions of seawater through sea connections [3].

The lagoon is connected to the Mediterranean Sea through three floodgates, "Golas" in Spanish, from North to South, Gola de Pujol, Gola del Perelló, and Gola del Perellonet (Figure 1). The local water council operates them according to the needs of the rice cycle. They are open from January to March to allow the water level of the lagoon to increase for irrigation. During the rice growing season (April–September) the floodgates remain closed to allow field flooding and with an insignificant flow to the lagoon. Gates open in September to allow rice fields to dry for rice harvest. Finally, gates close again in November to allow flooding of harvested rice fields, which favors the mineralization of nutrients [23].

The eutrophication of the lagoon is an old problem that dates back to the 1960s. Since then, the system shifted from a clear state to a turbid stable state that was consolidated by the almost total disappearance of macrophytes in the early 1970s [24]. The turbid state has prevailed since then, although some studies report short clear water events one or twice a year, with Chl*a* concentrations below 5 mg/m^3 [6]. In addition, sediment deposition threatens the lagoon with clogging showing the importance of studying turbidity patterns.

2.2. Precipitation and Wind Data

The first step was to select one year with total precipitation above the annual average and one year below the annual average, to analyze turbidity patterns in different precipitation regime conditions. The closest stations to the Albufera lagoon with full available data from 1995 to 2018 are the Valencia Airport station (north) and the Polinya del Xúquer station (south), which belong to the State Meteorological Agency (AEMET) (Figure 1). This period was selected because the in situ monitoring data began to be compiled in 1995. Within the Albufera Natural Park, there is a station that belongs to the Valencian Association of Meteorology (AVAMET), called *Tancat de la Pipa* station. There are available data for this station since 2016. We selected the year 2017 as a year below the average precipitation, and 2018 as a year above the average, comparing the data from *Tancat de la Pipa* station with historic records. Wind data were obtained from the *Tancat de la Pipa* station.

2.3. Secchi Disk and Suspended Matter

Secchi disk depth (SDD) (cm) and suspended particulated matter (SPM) (mg/L) were measured monthly from 1995 to 2018 by the monitoring program of the Environment General Subdivision of the Valencian government. There are five sampling stations in the Albufera lagoon, shown as dots in Figure 1. These data are available online: http://www.agroambient.gva.es/es/ (accessed on 6 October 2019).

SDD was measured with a 30 cm diameter black-and-white disk, which was submerged in the water until it was no longer visible to an observer on the surface [25,26]. Secchi disk depth is inversely proportional to the amount of dissolved and/or particulate matter present in the water column; thus, is a turbidity indicator. SPM was determined following the Standard Methods (2005) procedure, 2540D, for surface waters.

SDD and SPM were standardized using the following Equation

$$Z = \frac{x - \bar{x}}{SD},$$ (1)

where x is the month datum of year i, $\bar{x}$ is the month average from 1995 to 2018, and SD is the monthly standard deviation from 1995 to 2018.

The standardized values were classified as follows: (1) values in the interval (−1, 1) indicate normal values; (2) values in the interval (1, 1.6) are above normal conditions, and (3) values (>1.6) are highly anomalous. The limit of the anomalous conditions was based on an Inverse Cumulative Distribution Function (ICDF), in a normal distribution, which defines 1.6 standard deviations as the limit of values without noise with 95% confidence [27,28].

Then, the month average of the standardized values from 1995 to 2018 was calculated to characterize each month. The purpose is to characterize the temporal transparency pattern, which depends on the rice cultivation cycle.

2.4. Satellite Data

We obtained Sentinel 2A and 2B images for the years 2017 and 2018 from the Sentinel Scientific Data Hub available online: https://scihub.copernicus.eu/ (accessed on 6 October 2019) (Table 1). Only cloud-free images were used to observe the spatial variation. The images were subset to the exclusive area of the Albufera lagoon based on a shapefile before further processing.

Table 1. List of Sentinel 2A and 2B images used in this study by date. Only cloud-free images were selected.

Year 2018	Year 2017
11 January	16 January
20 February	5 February
27 March	17 March
26 April	16 April
21 May	16 May
20 June	15 June
10 July	10 July
19 August	4 August
13 September	13 September
3 October	13 October
27 November	22 November
22 December	17 December

Software SNAP version 5 (Brockmann Consult) was used for image processing. All images were downloaded in L1C product in order to use the same atmospheric correction for all of them, by means of the Sen2Cor processor. This processor provides good results in eutrophic waters [8,20,29].

Following [5] results, we used band 5 (705 nm) to estimate turbidity with 20 m of spatial resolution. The reflectance values of band 5 (705 nm) were spatially standardized following Equation (1), where x is the month datum of sampling station i pixel, $\bar{x}$ is the month average of all Albufera lagoon pixels, and SD is the monthly standard deviation of all Albufera lagoon pixels. The spatially standardized results were transformed into raster format for mapping. The purpose was to characterize the spatial turbidity pattern under different precipitation regime.

Chl*a* concentration was estimated from L1C products with the "Case 2 Regional Coast Colour" (C2RCC) processor of the SNAP software. Chl*a* concentration was mapped to better understand the contribution of phytoplankton to turbidity patterns in the Albufera lagoon.

The Spearman correlation test was used to test the statistical significance of the correlation between remote sensing and in situ data. We contrasted the 2017 and 2018 standardized reflectance values (band 5, 705 nm) with the monthly standardized data of SDD for the complete study period (1998 to 2018) for each sampling station. The remote sensing data of each pixel containing a sampling station was extracted to compare with the historical in situ data.

3. Results

In Figure 2, monthly precipitation in 2017 and 2018 is compared for the following three meteorological stations: *Polinyà del Xúquer* (south of study area), Valencia Airport (north of study area), and *Tancat de la Pipa* (study area) (Figure 1). The three stations show the same precipitation trend and similar values, except in the autumn of 2018 where *Tancat de la Pipa* experienced more rain. Then, the average monthly precipitation from 1995 to 2018 was built with the average of the nearest stations with available data, Valencia Airport and *Polinya del Xúquer*. In this Mediterranean-type climate region, the main rainy period is autumn and the average annual precipitation is 487.7 mm. In Figure 3, the average monthly precipitation is represented (bars) against the monthly precipitation of years 2017 (orange line) and 2018 (black line). The last data was obtained from *Tancat de la Pipa* station. In this station, the total precipitation for 2017 was 307.0 mm being approximately 180 mm lower than average annual precipitation. The total precipitation for 2018 was 709.8 mm, which was more than 200 mm above the average annual precipitation. During the autumn months, September to November, accumulated precipitation was only 45.8 mm in 2017, while in 2018 it was 561.0 mm exceeding the annual average. October 2018 recorded the maximum precipitation with 287.6 mm, with 232.2 mm measured in a single day (18 October 2018). In this area, prevailing wind direction shows a marked seasonal variability. During the warm months the winds of the East and Southeast (winds to the west) prevail, while during the rest of the year the winds of the West prevail (winds to the east), especially from the Southwest (Northwest only in October).

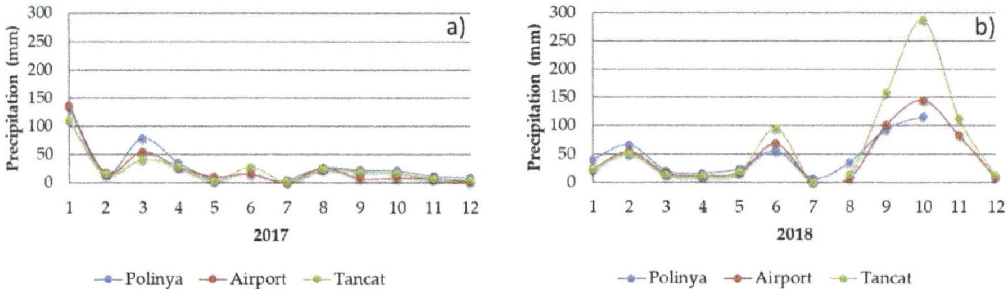

Figure 2. Monthly precipitation (**a**) 2017 and (**b**) 2018, comparison of the three meteorological stations: *Polinyà del Xúquer* (Polinya), Valencia Airport (Airport), and *Tancat de la Pipa* (Tancat).

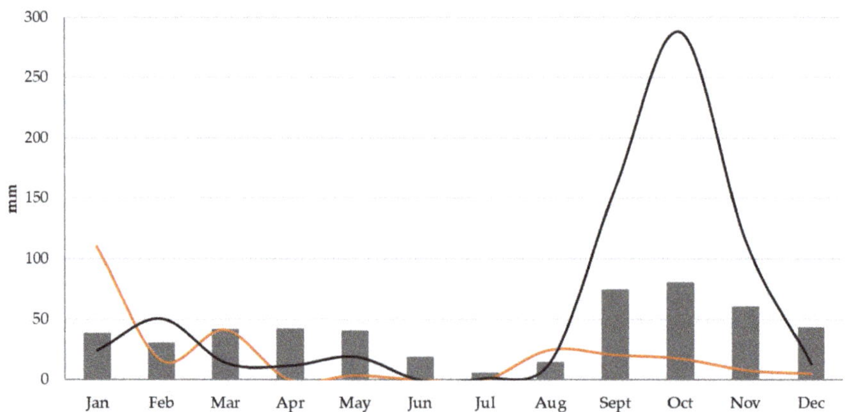

Figure 3. Average monthly precipitation (1995 to 2018) calculated from *Polinyà del Xúquer* and Valencia Airport stations (grey bars). Monthly precipitation 2017 (orange line) and 2018 (black line) at *Tancat de la Pipa* station.

Table 2 summarizes data from the in situ monitoring program of the Environment General Subdivision of the Valencian government, from January 2017 to December 2018. There is approximately one measure of each variable (SPM, SDD, and Chl*a*) per month. However, some data is missing; for instance, December 2018 only has SDD data. The highest SPM values were observed in May and June (June 2018 no data available), with values even higher than 100 mg/L, and SDD of about 17 cm in all the sampling stations. The highest Chl*a* values were observed in October 2017 (average about 150 mg m^{-3}), and in October and November 2018 (average about 120 and 150 mg m^{-3} respectively). To analyze if there is a monthly pattern associated to the irrigation cycle in the Albufera lagoon, we studied the in situ data of the entire period from 1995 to 2018. In order to detect anomalies above or below the Albufera lagoon baseline, we calculated the standardized monthly averages of SDD (blue bars) and SPM (brown bars) (Figure 4), in the five in situ sampling stations. In this figure, values above zero standard deviations show higher values than the average, and values below zero are lower than the average. SDD and SPM are inversely correlated variables [9,11], so months with high SDD have low SPM. In general, from April to October SPM values are above the average, and the maximum values are observed in May–June and October. However, sampling station 1 shows a different pattern, with SPM values from March to August above the average, and the maximum values in April and August. This can be explained due to East winds during warm months that may have a resuspension and accumulation effect in this shallow area.

Table 2. Data from the monitoring program of the Environment General Subdivision of the Valencian government. Suspended particulate matter (SPM), Secchi disk depth (SDD), and chlorophyll *a* (Chl*a*). nd = no data (missing data).

Data	SPM (mg/L)					SDD (cm)					Chla (mg m^{-3})				
Station	1	2	3	4	5	1	2	3	4	5	1	2	3	4	5
18 January 2017	28	34	40	60	60	35	30	33	25	30	10	3.7	6.9	6.4	3.9
9 February 2017	nd	<1	<1	nd	<1	nd	30	25	nd	25	nd	48.8	39.7	nd	<0.1
14 February 2017	39	41	36	55	81	35	nd	nd	nd	nd	20.1	14.1	15.6	11.1	18.3
20 March 2017	30	23	13	20	28	40	40	45	50	40	43.3	50.9	13.2	2.5	<0.1
3 April 2017	42	38	37	40	40	35	38	40	38	38	30.5	35	8.9	13.8	2.8
9 May 2017	98	nd	90	100	82	nd	nd	nd	nd	nd	46.6	nd	33.1	32.5	22.4
12 June 2017	67	140	97	60	105	18	20	14	17	20	33.9	27.6	88.9	74.7	57.7
11 July 2017	80	48	86	92	86	30	30	25	30	23	42.6	18.2	39.9	51	0.7
7 August 2017	25	nd	35	35	nd	30	30	30	30	30	46.7	nd	49.1	49	nd
11 September 2017	35	35	26	38	43	30	30	30	30	25	<0.1	37.5	<0.1	48.2	<0.1
17 October 2017	nd	76	nd	67	72	30	20	nd	20	25	65.5	157.4	nd	197.9	172
20 November 2017	58	78	84	31	74	30	37.5	30	30	37.5	89	60.6	70.9	69.2	72.6
21 December 2017	nd	41	43	nd	48	32	30	30	nd	30	61.2	73.5	75	nd	160
23 January 2018	52	52	50	nd	60	28	30	35	nd	25	84.3	71.4	82	nd	83.6
19 February 2018	<1	52	82	29	85	23	35	35	50	35	57.1	38.3	67.3	24.3	46.3
1 March 2018	73	64	83	88	103	20	31	30	30	29	163.1	91.1	90.1	98.6	<0.1
17 April 2018	66	116	88	92	90	23	25	23	25	25	<0.1	125.6	67	81.4	45.6
15 May 2018	108	130	130	130	130	30	17	17	20	17	30.6	127.1	167.8	167.2	229.7
13 June 2018	<1	nd	nd	nd	nd	40	25	25	25	25	22.5	nd	nd	nd	nd
11 July 2018	28	22	18	34	40	30	35	35	30	30	101.1	42.9	26.3	41.3	36.2
20 August 2018	22	25	30	35	21	25	35	42	35	37	108.4	26.2	23.5	25.7	23.1
18 September 2018	nd	8	nd	9	25	25	35	nd	35	35	nd	61.6	nd	101.1	38.4
17 October 2018	46	38	53	51	50	nd	30	30	30	30	nd	130.1	92.3	137	117.3
12 November 2018	28	20	20	18	15	nd	30	30	30	30	nd	157.8	143.3	172.1	130.5
11 December 2018	nd	nd	nd	nd	nd	nd	35	40	nd	50	nd	nd	nd	nd	nd
17 December 2018	nd	nd	nd	nd	nd	nd	25	25	25	25	nd	nd	nd	nd	nd
Average	49	49	52	52	58	28	29	29	29	29	55.7	63.7	57.3	67.1	66.6
SD	32	40	38	36	38	14	11	14	15	11	41.9	50	45.8	59	63.1

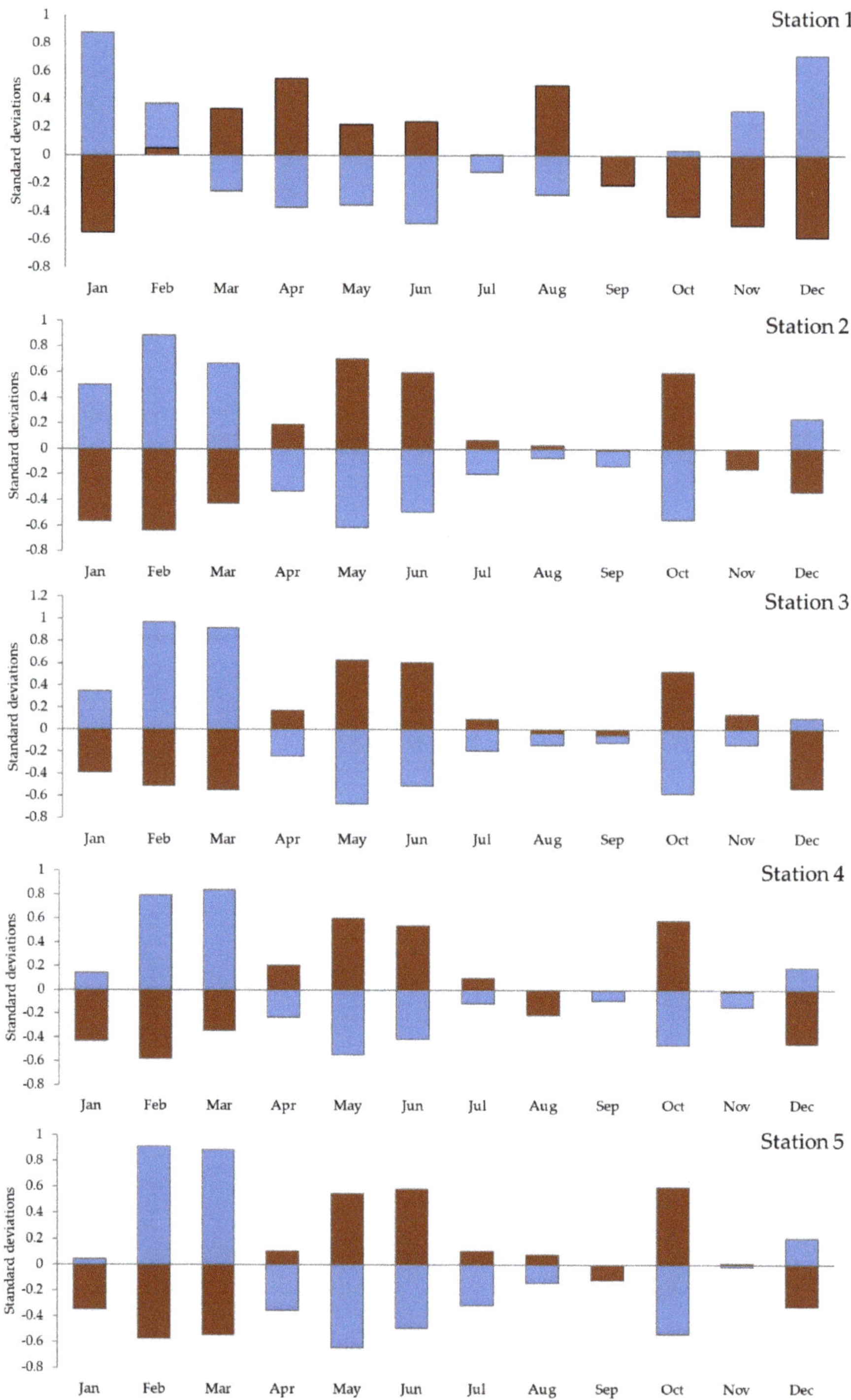

Figure 4. Standardized monthly averages of Secchi disk depth (blue bars) and suspended particulated matter (brown bars) for the period 1995 to 2018, in the five sampling stations of the Albufera lagoon.

Monthly turbidity is mapped in Figure 5 (year 2018) and Figure 6 (year 2017) to better analyze the spatial pattern. Turbidity is represented as standardized reflectances of band 5 (705 nm) from Sentinel 2A and 2B. This reflectance represents turbidity as follows: values in the interval (−1, 1) indicate average values (blue color); values in the interval (1, 1.6) are above average conditions (yellow color), and values (>1.6) are highly anomalous (red color). Applying the spatially standardized anomalies approach is important to be able to detect deviations from the baseline.

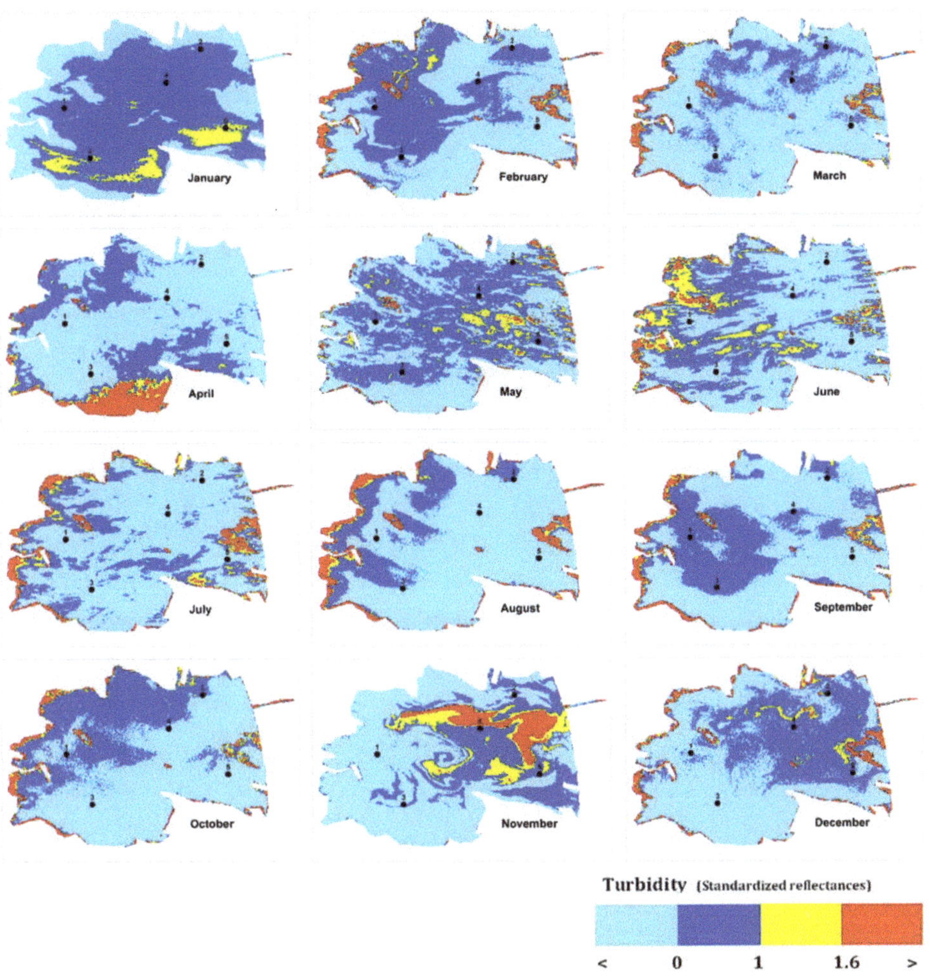

Figure 5. Monthly standardized reflectances band 5 (705 nm) from Sentinel 2A and 2B, year 2018, in the Albufera lagoon. Turbidity is represented as follows: values in the interval (−1, 1) indicate average values; values in the interval (1, 1.6) are above average conditions, and values (>1.6) are highly anomalous.

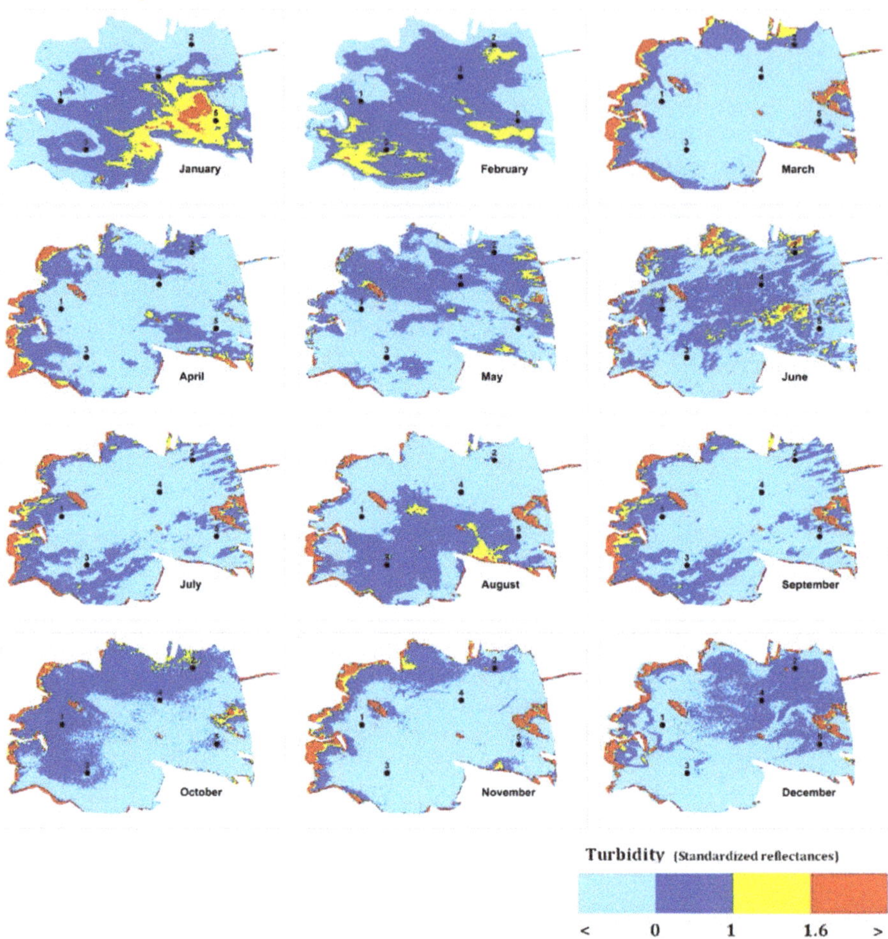

Figure 6. Monthly standardized reflectances band 5 (705 nm) from Sentinel 2A and 2B, year 2017, in the Albufera lagoon. Turbidity is represented as follows: values in the interval (−1, 1) indicate average values; values in the interval (1, 2) are above average conditions, and values (>1.6) are highly anomalous.

The spatial distribution of turbidity is quite heterogeneous. Despite the five in situ sampling stations are located all around the lagoon, the high spatial variability is much better captured with remote sensing. The correlation between remote sensing and in situ data was analyzed with the Spearman correlation test. We contrasted the 2017 and 2018 standardized reflectance values (band 5, 705 nm) with the monthly standardized data of SDD for the complete study period (1998 to 2018) for each sampling station (Table 3). We wanted to test if turbidity patterns mapped with remote sensing in the studied years followed the monthly historical pattern. According to p-values, the correlation was statistically significant (p-value < 0.05) for all sampling stations except sampling station 2, 2018.

It is important to remember that high turbidity values can be due to inorganic particulated matter (sediments) but also to high phytoplankton values [12,13]. Monthly Chl$a$ concentration in the Albufera lagoon is mapped in Figure 7 (year 2018) and Figure 8 (year 2017). Chl$a$ is used a phytoplankton biomass indicator. In general, the highest Chl$a$ values do not coincide with the highest turbidity values, which indicated the major importance of inorganic particles during high turbidity events. For instance, April 2018 is characterized by high Chl$a$ values while turbidity is under the average (<0) in nearly all of the lagoon. Phytoplankton biomass behavior showed differences between a wet year (2018) and a dry year (2017). In 2018, the highest phytoplankton biomass (Chl*a*) was observed in April and affected nearly the entire lagoon. In 2017, the highest biomass from May to July also affected nearly the entire lagoon. Both years had a second Chl*a* maximum in October.

Table 3. Correlation between the monthly standardized data of Secchi disk depth and standardized band 5 (705 nm) of Sentinel (for each year $n = 12$).

Sampling Stations	2017		2018	
	Spearman Correlation	*p*-Value	Spearman Correlation	*p*-Value
1	0.613	0.034	0.614	0.034
2	0.860	0.000	0.557	0.060
3	0.887	0.000	0.665	0.018
4	0.897	0.000	0.658	0.020
5	0.622	0.031	0.594	0.042

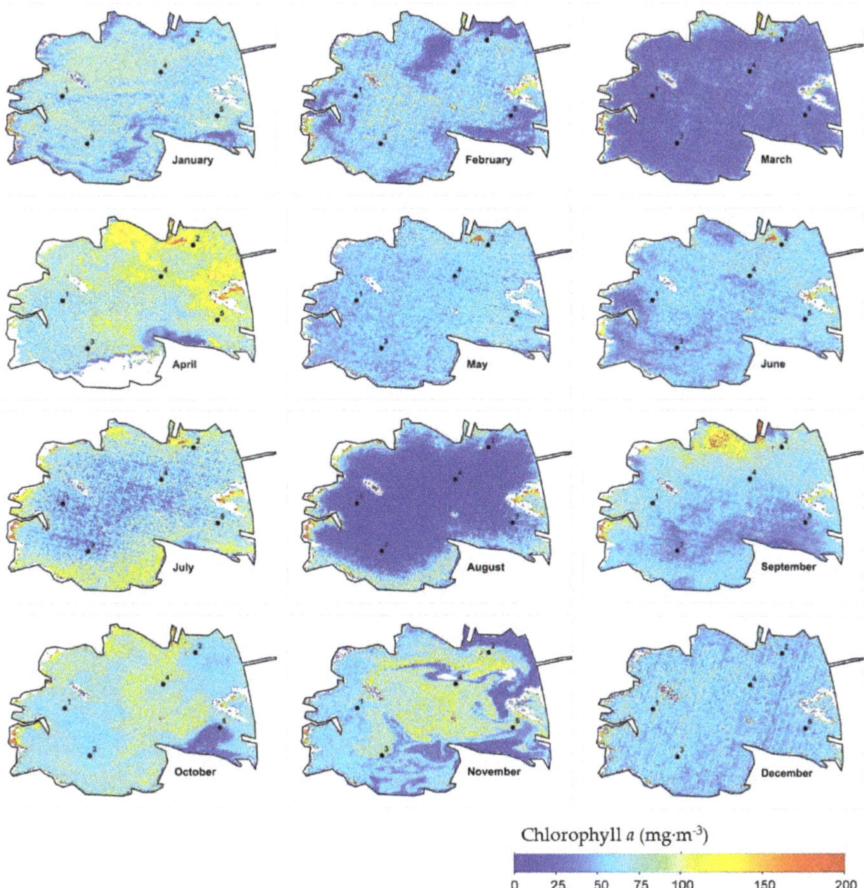

Figure 7. Monthly chlorophyll *a* concentration in the Albufera lagoon 2018.

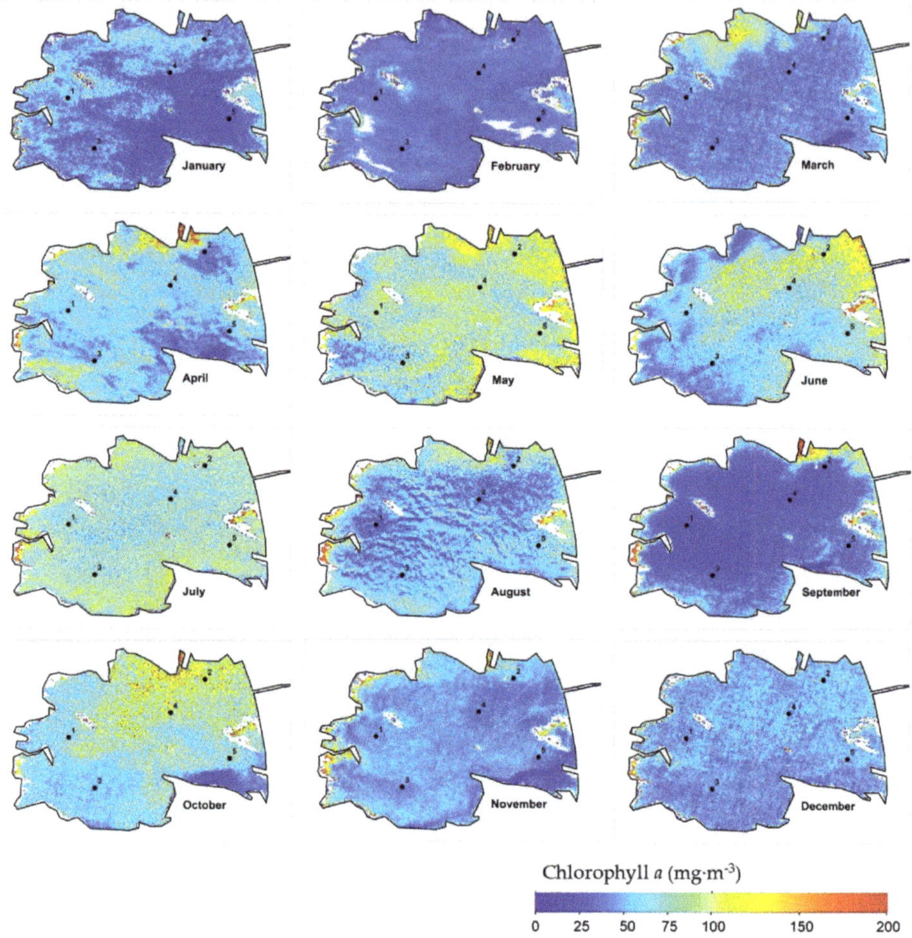

Figure 8. Monthly chlorophyll *a* concentration in the Albufera lagoon 2017.

To better analyze temporal variability and the effect of extreme meteorological events, we mapped turbidity and Chl*a* before and after the most important storm of the study period (Figure 9). This storm was on October 18 and total precipitation was 232.2 mm. Before the precipitation, Chl*a* levels were above 75 mg m^{-3} in nearly the entire lagoon. After the precipitation, a generalized decrease was observed.

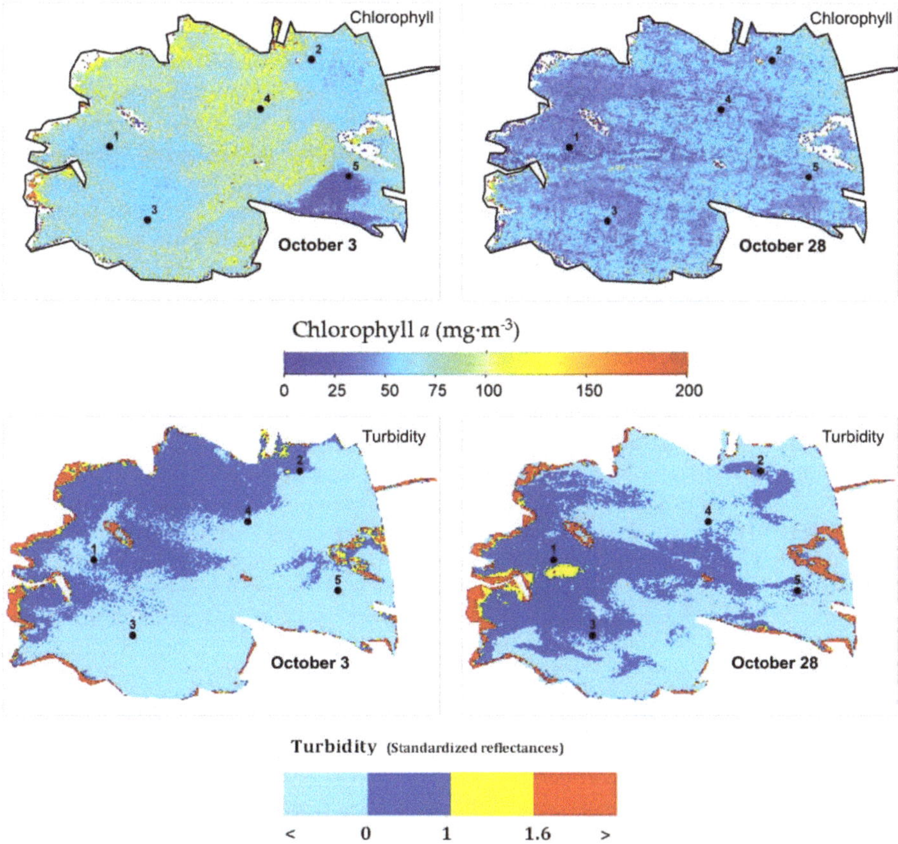

Figure 9. Chlorophyll *a* concentration and turbidity before and after a storm in the Albufera lagoon. The storm was on October 18 and total precipitation was 232.2 mm.

4. Discussion

In our study, we applied the standardized anomalies approach to the analysis of spatial and temporal patterns. According to the anomalies theory, the baseline is interpreted as the boundary on which if a value is above it is described as a positive anomaly (or increase), while if a value is below it indicates a negative anomaly (or decrement) [27,28]. The baseline was calculated from the period 1995 to 2018, the available historical data that defines the recent average behavior. Thanks to that analysis, in Figure 4, we can clearly distinguish the seasonal pattern. The temporal pattern in the Albufera lagoon is highly dependent on the rice cycle regulation of water inflows. SPM is higher from April to October in all sampling stations (except sampling station 1 from March to August), and thus the SDD is lower from April to October (Figure 4). The rice growing season is approximately from March-April to September. This period is characterized by high residence time of water in the lagoon since floodgates are closed and freshwater inputs are minimum [3]. In September, floodgates are opened to dry the fields for harvesting. The rainy season starts in September in this Mediterranean area when the floodgates are open; this favors water renewal. During the study period, from 1995 to 2018, the lowest water transparency is in May–June and October in sampling stations 2 to 5. Sampling station 1 exhibits slightly different behavior. A lower water transparency is maintained from March to September and transparency only shows a recovery during November to January. This station is located in the western area of the lagoon, which is the shallowest part (<0.9 m).

In general, the turbidity temporal and spatial pattern is similar in a wet year (2018, Figure 5) than in a dry year (2017, Figure 6). Thanks to the spatially standardized anomalies approach, it is important easy to detect deviations from the baseline. The highest turbidity values were observed on the west shore of the lagoon during most of the year. This agrees with the lagoon hydrological sectors proposed by [30]. According to them, the Northwest and West sectors have the lowest water circulation, while the Northeast and Southeast areas have the highest due to the proximity of the gates. The highest Chl*a* values are also observed very close to the western shore, as observed also by [3], but also the northern shore reaches very high values.

The spatial distribution of turbidity observed in Figures 5 and 6 is closely related to meteorological events. From September to November 2018, several heavy rain events carried more sediments to the lagoon through surface runoff. In [22] the authors explained that heavy storms were one of the main factors explaining the variation in the limnology of the Albufera lagoon. Storms may last only a few hours in this Mediterranean area, and a single storm could double the annual mean rainfall (e.g., October 2018 precipitation was higher than 2017 annual precipitation). During these storms, the potential for soil infiltration is low, so runoff is very important. We observed high turbidity both in the western sector of the lagoon and near the outflowing channels (eastern sector). In these areas, the phytoplankton and sediments can be transported to the sea because the floodgates (Golas) are open. These high turbidity values are mapped in yellow color (values above the average) and in red (highly anomalous values). From July to September, during the rice growing season, when freshwater inflows to the lagoon are greatly reduced, the most important variable is east wind. The wind dominant direction from sea to land causes the accumulation of suspended material in the western area of the lagoon. In April a false anomaly is observed, which was due to cloud presence. The study images were selected taking into account the lowest cloud coverage to avoid these interferences, but no better image was available in April 2018.

In recent decades, a clear water phase (CWP) has been observed yearly, but it does not show a regular pattern, either temporally or spatially in the lagoon [30,31]. During this phase cyanobacteria plankton is substituted by other microalgae, especially diatoms, which are consumed by filter-feeders such *Daphnia magna* [31]. The authors of [8] studied with Landsat images a CWP event that happened in March 2000. They observed that the re-eutrophication process started from the northwest shoreline, which is the area with lowest circulation [30]. A CWP was reported in January 2017 [7]. As shown in Figure 5, we observed an area of high transparency next to the west shoreline and a highly turbid area in the southeast part of the lagoon. In this month, an important rain event was the most possible cause of sediment transport towards the floodgates. In [7] the authors found two annual minima of cyanobacteria (March and September), which is the dominant phytoplankton in this hypereutrophic lagoon. These minima coincide with the maximum area of transparency in Figure 5, and with low Chl*a* values in Figures 7 and 8. However, in 2017 the lowest Chl*a* values were detected in February. The authors of [7] observed one cyanobacteria maximum in May. Then, they describe a sharp decline in primary production that contrasts with other authors such as [3], who found that Chl*a* concentrations increase from May to August 2006 due to the low water circulation. We can appreciate in Figures 4 and 5 an increase in turbidity from March to May, and a decrease in turbidity from May to August, which is more marked in 2018 (Figure 5). In our results, the Chl*a* pattern is different each studied year, in 2017 high Chl*a* levels are constant from May to July, while in 2018 there is an important decrease after April. The main difference between both years was an important precipitation event of 82.6 mm on 3 June 2018. This shows the importance of meteorological events on the lagoon dynamics. To analyze this further, Figure 9 shows Chl*a* concentration before and after the most important precipitation in October 2018. The decrease in Chl*a* after the storm and the water quality improvement can be explained by rapid flushing. If we compare Figures 5 and 6 with Figures 7 and 8, the highly anomalous values of turbidity cannot be attributed to Chl*a*, which suggests the importance of inorganic particulated matter, and indicates sediment transport.

The analysis of turbidity gives information about organic and inorganic suspended materials, that is, about phytoplankton and inorganic particles. Previous remote sensing research [8,32] focuses mainly on Chl*a* study, which is an indicator of phytoplankton biomass. Our study of turbidity patterns provides important supplementary information to those previous studies. The authors of [30] demonstrated that flushing pulses are key to improve water quality and to remediate eutrophication. In our study, we demonstrated that during important rain events the turbidity pattern shows higher values towards the floodgates "Golas". Then, it is important that during rain events the connection between the lagoon and the sea remains open to allow sediment discharge and prevent clogging of the lagoon. Dredging the lagoon to remove the sediments has been considered by the managers for several years to solve both eutrophication and clogging problems [33]. However, dredging is a desperate measure, very costly, and with environmental consequences. An improved water management, with increased flushing pulses frequency would be a good management measure that could help in alleviating not only eutrophication problems but also lagoon clogging. For that reason, it is essential to maintain the freshwater inflow to this lower part of the Júcar and Turia rivers. In recent years, three constructed wetlands have been developed in the Albufera lagoon (*Tancat de la Pipa*, *Tancat de Mília*, and *Estany de la Plana*), but their functioning is not maximizing the removal of phytoplankton, phosphorus, and nitrogen [6,34]. A better understanding of turbidity patterns can provide relevant information to choose the most suitable location for future restoration measures.

5. Conclusions

In our study, we applied the standardized anomalies approach to the analysis of spatial and temporal patterns of turbidity. This methodology allows comparing variables measured with different units, such as SPM and SDD in this study, and detecting deviations from a baseline. Thanks to this approach we can define the seasonal pattern of turbidity, which is not possible by the analysis of an isolated year or a reduced number of study years. In addition, we can define the areas with the highest values above the spatial baseline, which means we can identify the lagoon areas with the most anomalous values.

Turbidity patterns in the Albufera lagoon show a similar trend in wet and dry years, which is mainly linked to the irrigation practice of rice paddies. High turbidity periods are linked to higher water residence time and closed floodgates. However, precipitation and wind also play an important role in the spatial distribution of turbidity. During storm events, phytoplankton and sediments are discharged to the sea, if the floodgates remain open. Fortunately, the rice harvesting season, when the floodgates are open, coincides with the beginning of the rainy period. Nevertheless, this is a lucky coincidence. It is important to develop a conscious management of floodgates, because having them closed during rain events can have several negative effects both for the lagoon and for the receiving coastal waters and ecosystem. Non-discharged solids may accumulate in the lagoon worsening the clogging problems, and the beaches next to the receiving coastal waters will not receive an important load of solids to nourish them.

Author Contributions: Conceptualization, methodology, investigation, data curation and writing-original draft preparation M.-T.S.-F and J.A.A.-M.; software and formal analysis J.A.A.-M.; resources and funding acquisition M.-T.S.-F.; writing review and editing M.-T.S.-F. and J.E. Visualization and supervision M.-T.S.-F., J.E. and E.S.-D.-Á.

Funding: María-Teresa Sebastiá-Frasquet was a beneficiary of the CAS18/00107 post-doctoral research grant, supported by the Spanish Ministry of Education Culture and Sports during her stay at the Universidad Autónoma de Baja California (Mexico); image processing was developed partially during her stay. J.A.A.-M. was a beneficiary of the doctorate scholarship with the announcement number 291025, supported by the Council of Science and Technology of Mexico (CONACYT by its acronym in Spanish).

Acknowledgments: The authors want to thank María Sahuquillo, from the Environment General Subdivision of the Valencian government, and Paloma Mateache, Natural Park director for their insightful knowledge of the lagoon dynamics and help in interpreting the results. The authors also want to thank the anonymous reviewers who helped to improve the original manuscript.

Conflicts of Interest: The authors declare no conflict of interest. The funders had no role in the design of the study; in the collection, analyses, or interpretation of data; in the writing of the manuscript, or in the decision to publish the results.

References

1. Riera, R.; Tuset, V.M.; Betancur-R, R.; Lombarte, A.; Marcos, C.; Pérez-Ruzafa, A. Modelling alpha-diversities of coastal lagoon fish assemblages from the Mediterranean Sea. *Prog. Oceanogr.* **2018**, *165*, 100–109. [CrossRef]
2. Sebastiá-Frasquet, M.-T.; Altur, V.; Sanchis, J.-A. Wetland Planning: Current Problems and Environmental Management Proposals at Supra-Municipal Scale (Spanish Mediterranean Coast). *Water* **2014**, *6*, 620–641. [CrossRef]
3. Doña, C.; Chang, N.B.; Caselles, V.; Sánchez, J.M.; Camacho, A.; Delegido, J.; Vannah, B.W. Integrated satellite data fusion and mining for monitoring lake water quality status of the Albufera de Valencia in Spain. *J. Environ. Manag.* **2015**, *151*, 416–426. [CrossRef]
4. Gernez, P.; Lafon, V.; Lerouxel, A.; Curti, C.; Lubac, B.; Cerisier, S.; Barillé, L. Toward Sentinel-2 High Resolution Remote Sensing of Suspended Particulate Matter in Very Turbid Waters: SPOT4 (Take5) Experiment in the Loire and Gironde Estuaries. *Remote Sens.* **2015**, *7*, 9507–9528. [CrossRef]
5. Caballero, I.; Navarro, G.; Ruiz, J. Multi-platform assessment of turbidity plumes during dredging operations in a major estuarine system. *Int. J. Appl. Earth Obs. Geoinf.* **2018**, *68*, 31–41. [CrossRef]
6. Onandia, G.; Gudimov, A.; Miracle, M.R.; Arhonditsis, G. Towards the development of a biogeochemical model for addressing the eutrophication problems in the shallow hypertrophic lagoon of Albufera de Valencia, Spain. *Ecol. Inform.* **2015**, *26*, 70–89. [CrossRef]
7. Sòria-Perpinyà, X.; Vicente, E.; Urrego, P.; Pereira-Sandoval, M.; Ruíz-Verdú, A.; Delegido, J.; Soria, J.M.; Moreno, J. Remote sensing of cyanobacterial blooms in a hypertrophic lagoon (Albufera of València, Eastern Iberian Peninsula) using multitemporal Sentinel-2 images. *Sci. Total Environ.* **2020**, *698*, 134305. [CrossRef] [PubMed]
8. Sòria-Perpinyà, X.; Urrego, P.; Pereira-Sandoval, M.; Ruiz-Verdú, A.; Peña, R.; Soria, J.M.; Delegido, J.; Vicente, E.; Moreno, J. Monitoring the ecological state of a hypertrophic lake (Albufera of València, Spain) using multitemporal Sentinel-2 images. *Limnetica* **2019**, *38*, 457–469.
9. Güttler, F.N.; Niculescu, S.; Gohin, F. Turbidity retrieval and monitoring of Danube Delta waters using multi-sensor optical remote sensing data: An integrated view from the delta plain lakes to the western–northwestern Black Sea coastal zone. *Remote Sens. Environ.* **2013**, *132*, 86–101. [CrossRef]
10. Quang, N.H.; Sasaki, J.; Higa, H.; Huan, N.H. Spatiotemporal Variation of Turbidity Based on Landsat 8 OLI in Cam Ranh Bay and Thuy Trieu Lagoon, Vietnam. *Water* **2017**, *9*, 570. [CrossRef]
11. Kari, E.; Kratzer, S.; Beltrán-Abaunza, J.M.; Harvey, E.T.; Vaičiute, D. Retrieval of suspended particulate matter from turbidity-model development, validation, and application to MERIS data over the Baltic Sea. *Int. J. Remote Sens.* **2017**, *38*, 1983–2003. [CrossRef]
12. Kuhn, C.; de Matos Valerio, A.; Ward, N.; Loken, L.; Oliveira Sawakuchi, H.; Kampel, M.; Richey, J.; Stadler, P.; Crawford, J.; Striegl, R.; et al. Performance of Landsat-8 and Sentinel-2 surface reflectance products for river remote sensing retrievals of chlorophyll-a and turbidity. *Remote Sens. Environ.* **2019**, *224*, 104–118. [CrossRef]
13. Liu, H.; Li, Q.; Shi, T.; Hu, S.; Wu, G.; Zhou, Q. Application of Sentinel 2 MSI Images to Retrieve Suspended Particulate Matter Concentrations in Poyang Lake. *Remote Sens.* **2017**, *9*, 761. [CrossRef]
14. Erena, M.; Domínguez, J.A.; Aguado-Giménez, F.; Soria, J.; García-Galiano, S. Monitoring Coastal Lagoon Water Quality through Remote Sensing: The Mar Menor as a Case Study. *Water* **2019**, *11*, 1468. [CrossRef]
15. Caballero, I.; Stumpf, R.P.; Meredith, A. Preliminary Assessment of Turbidity and Chlorophyll Impact on Bathymetry Derived from Sentinel-2A and Sentinel-3A Satellites in South Florida. *Remote Sens.* **2019**, *11*, 645. [CrossRef]
16. Vanhellemont, Q.; Ruddick, K. Acolite for Sentinel-2: Aquatic applications of MSI imagery. In Proceedings of the 2016 ESA Living Planet Symposium, Prague, Czech Republic, 9–13 May 2016.
17. IOCCG. *Remote Sensing of Ocean Colour in Coastal, and Other Optically-Complex, Waters*; Reports of the International Ocean-Colour Coordinating Group, No. 3; Sathyendranath, S., Ed.; IOCCG: Dartmouth, NS, Canada, 2000.

18. Toming, K.; Kutser, T.; Laas, A.; Sepp, M.; Paavel, B.; Nõges, T. First experiences in mapping lake water quality parameters with Sentinel-2 MSI imagery. *Remote Sens.* **2016**, *8*, 640. [CrossRef]

19. Soria, J.M.; Miracle, M.R.; Vicente, E. Aporte de nutrientes y eutrofización de la Albufera de Valencia. *Limnetica* **1987**, *3*, 227–242.

20. Soria, X.; Delegido, J.; Urrego, E.P.; Pereira-Sandoval, M.; Vicente, E.; Ruíz-Verdú, A.; Moreno, J. Validación de algoritmos para la estimación de la clorofila-a con Sentinel-2 en la Albufera de València. In Proceedings of the XVII Congreso de la Asociación Española de Teledetección, Murcia, Spain, 3–7 October 2017; pp. 289–292.

21. Usaquén Perilla, O.L.; García Gómez, A.; Álvarez Díaz, C.; Revilla Cortezón, J.A. Methodology to assess sustainable management of water resources in coastal lagoons with agricultural uses: An application to the Albufera lagoon of Valencia (Eastern Spain). *Ecol. Indic.* **2012**, *13*, 129–143. [CrossRef]

22. Soria, J.M.; Vicente, E.; Miracle, M.R. The influence of flash floods on the limnology of the Albufera of Valencia lagoon (Spain). *Int. Ver. Theor. Angew. Limnol. Verh.* **2000**, *27*, 2232–2235. [CrossRef]

23. Romo, S.; García-Murcia, A.; Villena, M.J.; Sanchez, V.; Ballester, A. Phytoplankton trends in the lake of Albufera de Valencia and implications for its ecology, management, and recovery. *Limnetica* **2008**, *27*, 11–28.

24. Vicente, E.; Miracle, M.R. The coastal lagoon Albufera de Valencia: An ecosystem under stress. *Limnetica* **1992**, *8*, 87–100.

25. Wang, S.; Lee, Z.; Shang, S.; Li, J.; Zhang, B.; Lin, G. Deriving inherent optical properties from classical water color measurements: Forel-Ule index and Secchi disk depth. *Opt. Express* **2019**, *27*, 7642–7655. [CrossRef] [PubMed]

26. Wernand, M.R. On the history of the Secchi disc. *J. Eur. Opt. Soc.-Rapid Publ.* **2010**. [CrossRef]

27. Santamaría-del-Ángel, E.; Sebastia-Frasquet, M.T.; Gonzalez-Silvera, A.; Aguilar-Maldonado, J.; Mercado-Santana, A.; Herrera-Carmona, J. Uso Potencial de las Anomalías Estandarizadas en la Interpretación de Fenómenos Oceanográficos Globales a Escalas Locales. In *Costas y Mares Mexicanos: Construyendo la Línea Base para su Futuro Sostenible, Oceanografía Fisicoquímica*; Rivera-Arriaga, E., Sánchéz-Gil, P., Gutiérrez, J., Eds.; EPOMEX: Campeche, Mexico; Universidad Autónoma de Colima: Colima, Mexico, 2019.

28. Aguilar-Maldonado, J.A.; Santamaría-del-Ángel, E.; Gonzalez-Silvera, A.; Sebastiá-Frasquet, M.T. Detection of Phytoplankton Temporal Anomalies Based on Satellite Inherent Optical Properties: A Tool for Monitoring Phytoplankton Blooms. *Sensors* **2019**, *19*, 3339. [CrossRef] [PubMed]

29. Pereira-Sandoval, M.; Ruescas, A.; Urrego, P.; Ruiz-Verdú, A.; Delegido, J.; Tenjo, C.; Soria-Perpinyà, X.; Vicente, E.; Soria, J.; Moreno, J. Evaluation of Atmospheric Correction Algorithms over Spanish Inland Waters for Sentinel-2 Multi Spectral Imagery Data. *Remote Sens.* **2019**, *11*, 1469. [CrossRef]

30. Soria, J.M. Past, present and future of la Albufera of Valencia Natural Park. *Limnetica* **2006**, *25*, 135–142.

31. Miracle, M.R.; Sahuquillo, M. Changes of life-history traits and size in Daphnia magna during a lear-water phase in a hypertrophic lagoon (Albufera of Valencia, Spain). *Int. Ver. Theor. Angew. Limnol. Verh.* **2002**, *28*, 1203–1208.

32. Sòria-Perpinyà, X.; Miracle, M.R.; Soria, J.; Delegido, J.; Vicente, E. Remote sensing application for the study of rapid flushing to remediate eutrophication in shallow lagoons (Albufera of Valencia). *Hydrobiologia* **2019**, *829*, 125–132. [CrossRef]

33. Rodrigo, M.A.; Alonso-Guillén, J.A. Assessing the potential of Albufera de València Lagoon sediments for the restoration of charophyte meadows. *Ecol. Eng.* **2013**, *60*, 445–452. [CrossRef]

34. Martín, M.; Oliver, N.; Hernández-Crespo, C.; Gargallo, S.; Regidor, M. The use of free water surface constructed wetland to treat the eutrophicated waters of lake L'Albufera de Valencia (Spain). *Ecol. Eng.* **2013**, *50*, 52–61. [CrossRef]

 *remote sensing* 

Article

Operational Monitoring and Damage Assessment of Riverine Flood-2014 in the Lower Chenab Plain, Punjab, Pakistan, Using Remote Sensing and GIS Techniques

Asif Sajjad [1], Jianzhong Lu [1,*], Xiaoling Chen [1], Chikondi Chisenga [1,2], Nayyer Saleem [1] and Hammad Hassan [1]

[1] State Key Laboratory of Information Engineering in Surveying, Mapping and Remote Sensing, Wuhan University, Wuhan 430079, China; asifsajjad@whu.edu.cn (A.S.); xiaoling_chen@whu.edu.cn (X.C.); cchisenga@must.ac.mw (C.C.); saleemnayyer@whu.edu.cn (N.S.); hhrizvi@whu.edu.cn (H.H.)
[2] Department of Earth Sciences, Ndata School of Climate and Earth Sciences, Malawi University of Science and Technology, Limbe P.O. Box 5196, Malawi
* Correspondence: lujzhong@whu.edu.cn; Tel.: +86-27-6877-8755; Fax: +86-27-6877-8229

Received: 2 January 2020; Accepted: 20 February 2020; Published: 21 February 2020

Abstract: In flood-prone areas, the delineation of the spatial pattern of historical flood extents, damage assessment, and flood durations allow planners to anticipate potential threats from floods and to formulate strategies to mitigate or abate these events. The Chenab plain in the Punjab region of Pakistan is particularly prone to flooding but is understudied. It experienced its worst riverine flood in recorded history in September 2014. The present study applies Remote Sensing (RS) and Geographical Information System (GIS) techniques to estimate the riverine flood extent and duration and assess the resulting damage using Landsat-8 data. The Landsat-8 images were acquired for the pre-flooding, co-flooding, and post-flooding periods for the comprehensive analysis and delineation of flood extent, damage assessment, and duration. We used supervised classification to determine land use/cover changes, and the satellite-derived modified normalized difference water index (MNDWI) to detect flooded areas and duration. The analysis permitted us to calculate flood inundation, damages to built-up areas, and agriculture, as well as the flood duration and recession. The results also reveal that the floodwaters remained in the study area for almost two months, which further affected cultivation and increased the financial cost. Our study provides an empirical basis for flood response assessment and rehabilitation efforts in future events. Thus, the integrated RS and GIS techniques with supporting datasets make substantial contributions to flood monitoring and damage assessment in Pakistan.

Keywords: floods; Landsat-8; remote sensing; GIS; disaster mapping; damage assessment; Lower Chenab Plain

---

1. Introduction

Flood disasters are among the most frequent and destructive of all-natural disasters, posing a potential threat to life and property. Every year, human lives, agricultural activities, and infrastructures are seriously affected by shattering flood disasters around the globe [1–3]. In the past three decades, flood disasters affected nearly 2.8 billion people and resulted in over 200,000 causalities with substantial damages to property and economy [4]. Floods account for ~47% of all weather-related disasters that occur across the world [5]. Climate change and rapid population increases in floodplains enforce and boost the frequency and magnitude of riverine flood damages [6,7]. As flood disasters increase in magnitude, the Asian region continues to face a large number of flood hazards and associated losses in lives and all kinds of infrastructure and economic progress [4,7]. In recent years, the south Asia region has been experiencing intense flooding with increased frequency, especially in Pakistan [8–10].

In Pakistan, flood events are recognized as a major natural hazard that historically originates from the Indus and Chenab rivers [11–13]. Since the creation of Pakistan, 1955, 1959, 1973, 1976, 1988, 1992, 1995, 1996, 1997, 2006, 2010, and 2014 are recorded as the years for destructive floods that resulted in adverse impacts on human lives, property, and the country's economy [9,14,15]. Floods in Pakistan mostly occur in the monsoon months of July and September, due to heavy rains and the melting of snow upstream in the Himalaya region. This results in riverine floods that produce tremendous detrimental impacts on human lives, agriculture, and infrastructure [16–18]. For instance, the 2010 riverine flood caused a large inundation that covered an area of 70,238 km^2, with 884,715 affected houses [8,12,18]. Likewise, the 2011 flood affected 5.88 million people and damaged standing crops and infrastructure covering an area of 16,440 km^2 and 1882 km^2, respectively [19]. In 2014, heavy rains coupled with the melting of glaciers in the upstream part of Chenab caused flash flooding in mountainous areas and riverine floods in the upper and lower Chenab plain [20–22]. It was considered one of the worst riverine flood catastrophes in terms of damages to standing crops, housing, and infrastructure along the floodplain areas of the Chenab river [13,16,22,23]. This indicates that flooding is a serious problem in Pakistan that requires significant efforts to reduce its effects, mainly through effective post-disaster monitoring and management, especially in the Chenab plain [22–24]. As such, we provide a scientific basis that provides a rapid flood 2014 mapping and monitoring using Landsat-8 data, with a focus on the lower Chenab plain. We first provide the limitation of previous studies on the riverine flood of 2014 that focused less on post-flood mapping, monitoring, and damage assessment. These previous studies mainly focused on the upper Chenab plain [13,16,21,22] without focusing on the lower Chenab plain. This is despite the fact that the lower Chenab plain is a fertile flood-prone plain and considered an economically underdeveloped region in the Punjab province of Pakistan. Despite being an understudied area, negligence adds to the problems with a lack of localized flood management strategies. Therefore, accurate post-flood mapping, monitoring, and evaluating damages are the utmost requirements for rapid flood risk assessment in lower Chenab plain. Secondly, to a basis for rapid flood risk mapping, we make use of multispectral remote sensing open data in conjunction with Geographical Information System (GIS) techniques for monitoring and evaluating damage assessment in the severely flood-affected areas of the lower Chenab plain, Punjab, Pakistan.

The use of remote sensing and GIS techniques is chosen as it contributes to an exploration of flood causes and additionally provides accurate mapping of flood extents that enables detailed investigations of flood instances [18,19,25]. Moreover, these techniques can be applied in damage assessment to standing crops and infrastructure. Effective flood modeling and mapping are important for flood assessment, loss estimation, and sustainable land use planning along flood plains to mitigate flood risk effectively [25,26]. Multi-temporal remote sensing images provide a wide source of low-cost information with a reliable accuracy that can be beneficially utilized for flood mapping [27,28]. Furthermore, multispectral remote sensing-derived indexes and GIS-based classifications can be exploited to detect flooded areas [18,29,30]. a wide range of open-source remote sensing data has permitted valuable and accurate historical data, which is essential for a comprehensive study on flood disaster mapping and damage assessment [31]. For example, moderate resolution Landsat data have provided free, up-to-date satellite images across the globe since 1972, which can be used for flood disaster monitoring and damage assessment [32,33]. High-resolution Google Earth (GE) images provide historical and recent data that can also be used for flood monitoring and temporal flood mapping [34]. GE images can also be used as input datasets to digitize land uses, particularly useful in small areas [5], and for the validation of supervised classified images [35]. Thus, the current study uses remote sensing and GIS techniques to detect flood inundation and duration, and evaluate flood damages on different land uses, such as agricultural land and built-up areas. We used the identified relationship between land uses and flood instances to estimate the overall damage, flood extent, and propose emergency flood management for future floods on the lower Chenab plain.

2. Materials and Methods

2.1. Study Area

The study area is the lower Chenab plain, which is located in the downstream part of river Chenab, central Pakistan. It is located between 70°41′13.2″E and 29°6′0″N, and 71°37′58.8″E and 30°31′33.6″N, within the three main districts of Punjab province, namely the Multan, Muzaffar Garh, and Bahawalpur districts (Figure 1). It is one of the fertile plains of Punjab province, where floods occur almost every year. The Chenab river originates from the Indian state of Himachal Pradesh. Then, it flows through Indian-occupied Kashmir and enters at Marala Headwork into the province of Punjab, Pakistan. The southern part of the study area includes the Panjnad Barrage where the Sutlej River joins the Chenab, as shown in Figure 1. The total length of the River Chenab is about 974 km, of which 729 km fall within Pakistan. Of the 729 km, we selected approximately 215 km as our study area. We then divided the 215 km into three equal sections, each approximately 71 km long (Figure 1), to enable the visualization and detailed investigation of the results. We also created a 6-km-wide buffer around the Chenab River that covers a distance of about 215 km from north to south as the final extent of our study area. The mean annual Chenab discharge is 1.52 million m^3 [36,37]. The highest temperature documented in the study area is 50 °C in the month of June with the lowest recorded temperature of 2 °C in the month of January [20]. The average yearly rainfall is 157 mm, with July receiving a maximum of 45 mm and October receiving a minimum of only 2mm [20]. The main source of livelihood in the study area is agriculture with main crops being sugarcane, cotton, maize, rice, and fodder.

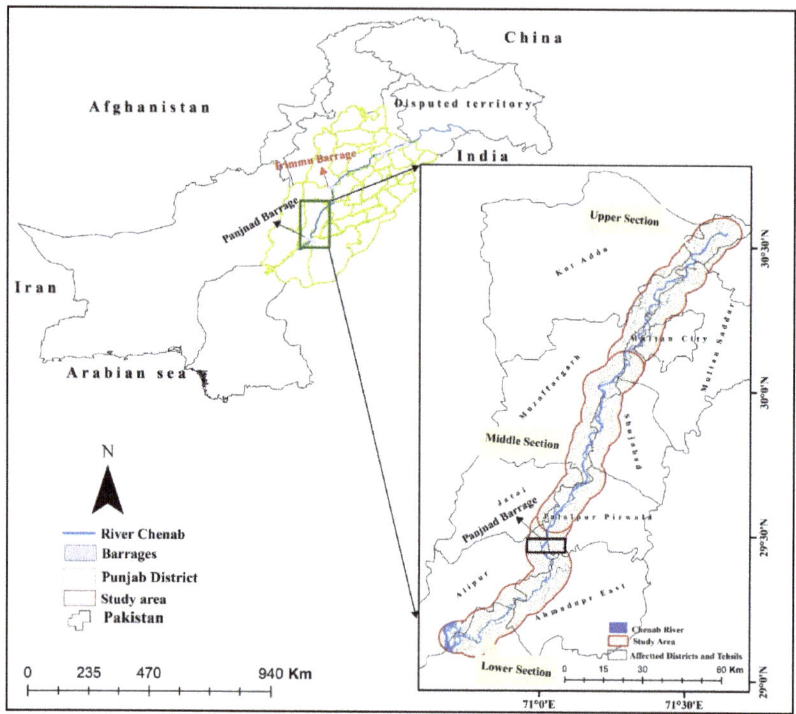

Figure 1. Location of the lower Chenab plain in Pakistan, and the zoomed area represents the river Chenab system, and also shown is the 6-km buffer zone and the extent of our study area.

2.2. Materials

The main data source of this study is Landsat 8 imagery (Table 1), acquired from the United States Geological Survey (USGS). Landsat images have a temporal resolution of ~16 days that has provided free up-to-date images across the globe since 1972 [22,33,38,39]. However, the study area provides a unique opportunity with a high temporal resolution of ~8 days since the area lies within two adjacent Landsat paths of 150 and 151 (Figure 2; Table 1), which enables us to make an in-depth analysis of the flood duration and inundation. In total, we used 9 temporal images acquired between 7 August and 27 November 2014 (Table 1). We used 8 temporal images for flood monitoring. Furthermore, we used 3 temporal images for flood damage assessment using land use/cover classification during flood instances in lower Chenab plain (Table 1).

We also used GPS and the Google Earth (GE) platform to collect spatial training datasets, which were used to support land use/cover classification. The GPS data were collected after the flood waters had ceased, as it was challenging to do field survey during flooding time, due to flood inundation. These points were also used to validate classification and inundation results. Due to its high resolution, the GE platform provides a better visualization of the real-world scenario of land use/cover change in the study area [34]. Furthermore, GE images of three flood instances (and with similar acquisition dates as the Landsat-8 images) were taken for cross-comparison with the Landsat-based land use/cover classification. The Pre-flood image was acquired on 4 August, 2014, the Co-flood image was taken on 13 September, 2014, and the closest available Post-flood image was acquired on 11 November, 2014. To further understand post-flood situation in the study area, a comprehensive field survey was conducted to identify damages and affected areas and observe the Post-flood effects and rehabilitation process.

Table 1. Specifications of used Landsat-8 data.

Path	Row	Date of Acquisition	Flood Instances	Use	Resolution (m)
151	39 and 40	7 August 2014	Pre-flood	Land-use classification	30
150	39 and 40	17 September 2014	Co-flood	Land-use classification and Flood monitoring	30
151	39 and 40	24 September 2014	Co-flood	Flood monitoring	30
150	39 and 40	3 October 2014	Post-flood	Flood monitoring	30
151	39 and 40	10 October 2014	Post-flood	Flood monitoring	30
150	39 and 40	19 October 2014	Post-flood	Flood monitoring	30
151	39 and 40	26 October 2014	Post-flood	Flood monitoring	30
151	39 and 40	11 November 2014	Post-flood	Flood monitoring	30
151	39 and 40	27 November 2014	Post-flood	Land-use classification and Flood monitoring	30

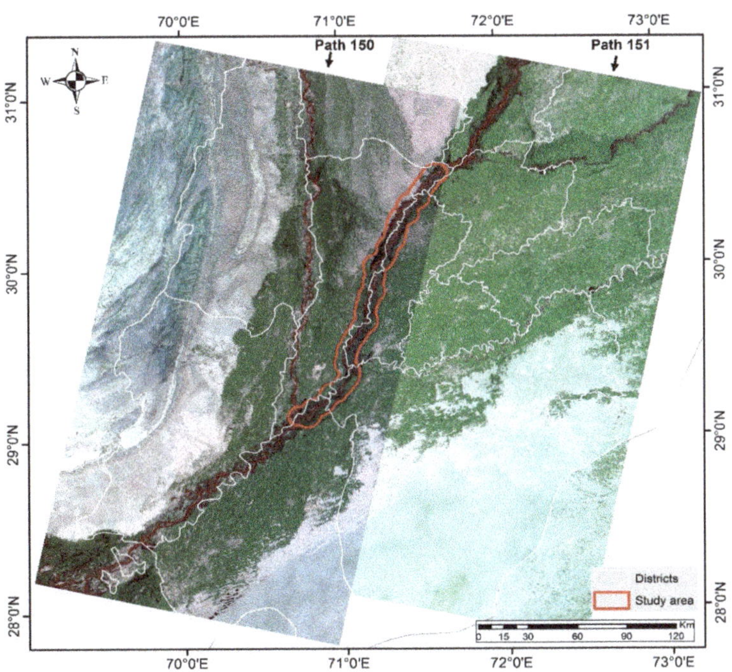

Figure 2. Landsat satellite images: Path 150 and 151 show the study area and districts.

2.3. Methods

The overview of the used methodology is presented in Figure 3. Firstly, we applied radiometric correction, layer stacking and mosaicking and resampling to the Landsat-8 images before use. Then, we converted the digital number values on the Landsat images to reflectance using ENVI's Radiometric Calibration Tool [18]. Finally, we used Arc GISs' spatial analysis tools for Layer stacking and mosaicking, and resampling. The pre-processed Landsat-8 pre-flood, co-flood, and post-flood images were subjected to supervised classification for land use/cover mapping in ArcGIS 10.5, as shown in Figure 3. Landsat data are most common and is often used for different land use/cover mapping and water extraction mapping [13,40]. The Supervised classification method is used to extract information from Landsat data [29,41]. In addition, a maximum likelihood (ML) approach was applied for land use/land cover classification of flood instances. The ML approach is widely used and is easy to apply for land use/cover classification. The ML approach has also been applied to land use/cover mapping of pre-flooding, co-flooding, and post-flooding [11,18,22,39–42]. We identified water, agriculture land, and vegetation, built up, sand, barren land, and deposited material as the six land-use classes in the study region. To facilitate the ML approach for classification, a total of 360 spatial training samples were collected from a GPS field survey and comparatively high-resolution GE images, and were further assigned to the pixel values of the most probable land use/cover class. All spatial training datasets were compiled and prepared in Microsoft Excel and were imported into the Arc GIS 10.5 environment. The classified images were converted into a shapefile format and used as input for the GIS-based spatial overlay analysis. This allows us to spatially compare and intersect all classified land uses to provide a clear picture of flood inundated and damaged land uses in lower Chenab.

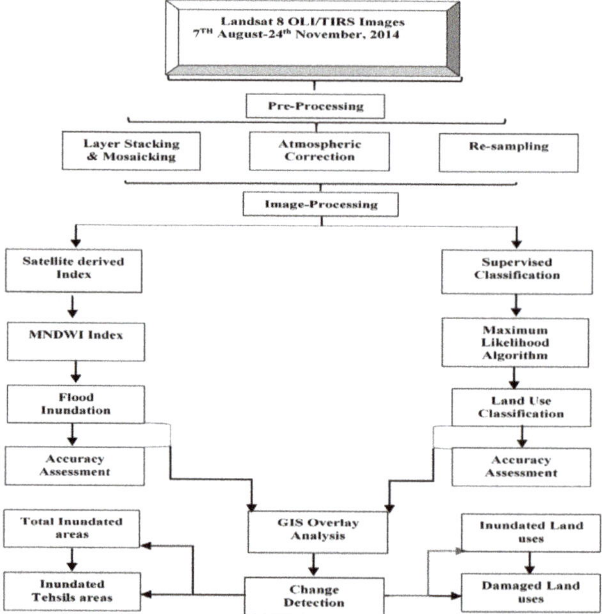

Figure 3. Methodological framework for the flood inundation mapping and damage assessment using multi-temporal Landsat-8 images.

For damage assessment, we defined a damaged area as an area that experienced a change from 'agricultural' and 'built up' in pre-flood to 'deposited material' in post-flood. a pre-flood image was used as a reference and a change detection technique was used to evaluate damages to agriculture and built-up areas in the study area.

2.3.1. MNDWI Index

The Modified Normalized Difference in Water Index (MNDWI) was used to delineate the spatial pattern of flood-2014 inundation along the lower Chenab plain. The MNDWI index is widely used for the rapid delineation of floodwater required for flood monitoring and assessment, and has been compared to other indices, i.e., the NDWI and Water Ratio Index (WRI) [43,44]. The Normalized Difference Water Index (NDWI) presented by McFeeters [45], is very efficient at delineating water information but has difficulties in case a built-up area exists in water environment. Therefore, Xu [46] presented an effective MNDWI index, which is much better at distinguishing between water and built-up areas. The MNDWI is calculated using green (Band 3) and shortwave infrared (Band 6) wavelengths to delineate water, as given by Equation (1):

$$MNDWI = \frac{(\text{Band3}) - (Band\ 6)}{(Band3) + (Band6)} \tag{1}$$

where GREEN (Band 3) = Green wavelength (0.53–0.59mm) and SWIR (Band 6) = Short-wave infrared Wavelength (1.57–1.65 mm)

This index is utilized for the removal of built-up area noise, and it uses Band 3 wavelength to maximize water reflectance. The resulting value ranges from −1 to +1. The low water reflectance and high reflectance of built-up in Band 6 result in positive values of water and negative values of built up in the MNDWI image. The limitation of the MNDWI index is that it does not efficiently distinguish between hill shadow and water body [47,48]. From literature, the MNDWI index has been used for the extraction of water on a flat plain with scattered built-up areas using Landsat 8 OLI and SAR data, achieving high accuracy and better performance. Thus, it was also strongly recommended, in comparison with other indexes, to be applied on Landsat images [49–52].

2.3.2. Classification and Inundation Validation

The classified images and index-derived inundation were validated using GPS and Google earth (GE) points. For this, the random sample points' tool within the Arc GIS spatial analyst was used to extract random points on the classified land uses of inundated and non-inundated images. These points were then converted into a kml format and overlaid on GE. The values of these points were evaluated using visual interpretation and expert knowledge. The accuracy of classified images was assessed using 557 Arc GIS random points and 150 GPS points. We have used 30 Arc GIS random points of class, namely: water, agriculture, built-up, barren, sand, and deposited material, to each flood instance image. In addition to this, we have further used 25 GPS points of above-mentioned classes to only post-flood instance for the validation process. The accuracy of inundation maps was assessed using 400 ArcGIS random points and 200 GPS points. Furthermore, for each inundation map, we have used 50 ArcGIS random points and 25 GPS points for the validation of water and non-water classes. Lastly, the results from the comparison of random samples and GPS with GE images were used to create the confusion matrix. The confusion matrix was used to evaluate the accuracy of the ML land use/cover classification and MNDWI inundation. We also used the Kappa Coefficient (KC) as an indicator to validate the qualitative agreement, either positive or negative, between classified samples and ground-truth points. It is normally calculated from a statistical assessment to evaluate the proportional improvement by the classifier over ground-truth samples to land-use classes.

3. Results

3.1. Accuracy Assessment

The overall inundation accuracy obtained from all images is above 90%, as shown in Table 2. The highest overall accuracy of 92% is obtained from the peak-flood image (17 September) and the post-flood image (27 November), while the least overall accuracy of 88% is acquired from the post-flood (11 November) image. Similarly, the obtained average user and producer accuracy of both classes is nearly 90%. The highest user accuracy of the water class is attained from the peak-flood image (17 September), which is 96%, and the least is 85%, which is obtained from the post-flood (11 November) image. On the other hand, the highest user accuracy of the non-water class is achieved from the post-flood (27 November) image, which is 93%, and the least is 86%, which is from the peak-flood (17 September) image. Furthermore, the highest attained producer accuracy of the water class is 94% from the post-flood (26 October) image, and the least is 87% from the peak-flood (17 September) image. The highest obtained non-water class producer accuracy is 96% from the peak-flood (17 Sepember) image and the least is 86%, which is obtained from the post-flood (11 November) image. The highest

KC accuracy of 85% is achieved from the post-flood (27 November) image and the least KC accuracy is obtained from the post-flood (11 November) image. The average KC accuracy of all the images is above 80% (Table 2).

The highest overall supervised classification accuracy obtained is 92% from the post-flood image. Whereas, the lowest accuracy of 85% was obtained from the co-flood image, as shown in Table 3. The KC accuracy shows that the highest accuracy of 90% is obtained from the post-flood (27 November) image and that the lowest of 81% is obtained from the co-flood (17 September) image. Similarly, the user and producer accuracies were also calculated, which are listed in Table 3. The highest user accuracy was attained by water, barren land, and sand classes of 92%, 96%, and 99%, respectively. Likewise, the highest producer accuracy was obtained by the water, built-up area, and deposited material classes at 99%, 98%, and 95%, respectively. The overall land-use classification accuracy suggests that classified images are reliable for further analysis.

Table 2. Accuracy assessment of the modified normalized difference water index (MNDWI)-derived flood 2014-inundation.

Classes	17 September 2014				24 September 2014				3 October 2014				10 October 2014			
	PA	UA	OA	KC	PA	UA	OA	KC	PA	UA	OA	KC	PA	UA	OA	KC
Water	0.87	0.96	0.92	0.84	0.91	0.93	0.91	0.83	0.88	0.92	0.90	0.81	0.91	0.90	0.90	0.81
Non-Water	0.96	0.86			0.93	0.91			0.91	0.88			0.90	0.92		

Classes	19 October 2014				26 October 2014				11 November 2014				27 November 2014			
	PA	UA	OA	KC	PA	UA	OA	KC	PA	UA	OA	KC	PA	UA	OA	KC
Water	0.91	0.91	0.90	0.80	0.94	0.86	0.90	0.81	0.91	0.85	0.88	0.76	0.93	0.91	0.92	0.85
Non-water	0.90	0.92			0.87	0.90			0.86	0.92			0.92	0.93		

PA = Producer Accuracy; UA = User Accuracy; OA = Overall Accuracy; KC = Kappa Coefficient

Table 3. Accuracy assessment of supervised classified land uses.

Classes	Pre-Flood Classified Image (7 August 2014)				Co-Flood Classified Image (17 September 2014)				Post-Flood Classified Image (17 September 2014)			
	PA	UA	OA	KC	PA	UA	OA	KC	PA	UA	OA	KC
water	0.95	0.92	0.86	0.83	0.92	0.92	0.85	0.81	0.99	0.88	0.92	0.90
built up	0.95	0.80			0.83	0.76			0.98	0.84		
vegetation	0.70	0.88			0.87	0.80			0.92	0.92		
barren	0.85	0.92			0.78	0.88			0.88	0.96		
sand	0.80	0.80			0.85	0.88			0.83	0.99		
Deposited material	0.95	0.86			-	-	-	-	0.92	0.92		

PA = Producer Accuracy; UA = User Accuracy; OA = Overall Accuracy; KC= Kappa Co-efficient.

3.2. Flood Mapping and Monitoring

The temporal flood 2014 extent maps were prepared and used to determine the most inundated areas in the study area, as a tool for flood monitoring. Figure 4a,b shows the flood inundation with a flood peak on 17 September, which remained stable until the 24 September, and then the inundation gradually decreased till post-flood 27 November. Figure 4a and Table 4 show that northwestern Muzaffargarh and northeastern Multan Saddar part was the most inundated and affected region along with the southeastern ShuJabad and Jalalpur Pirwala. The flooded area receded in three phases: In the first phase from 24 September to 10 October, the southern part gradually receded from west to east. The central part began to recede from 10–26 October 2014 in the second phase, and in the third phase, almost all the flood water receded until 27 November 2014. Hence, the flood duration result shows that Muzaffargarh and Multan Saddar remained inundated for almost two months and reported the most affected areas by flood-2014 in lower Chenab plain.

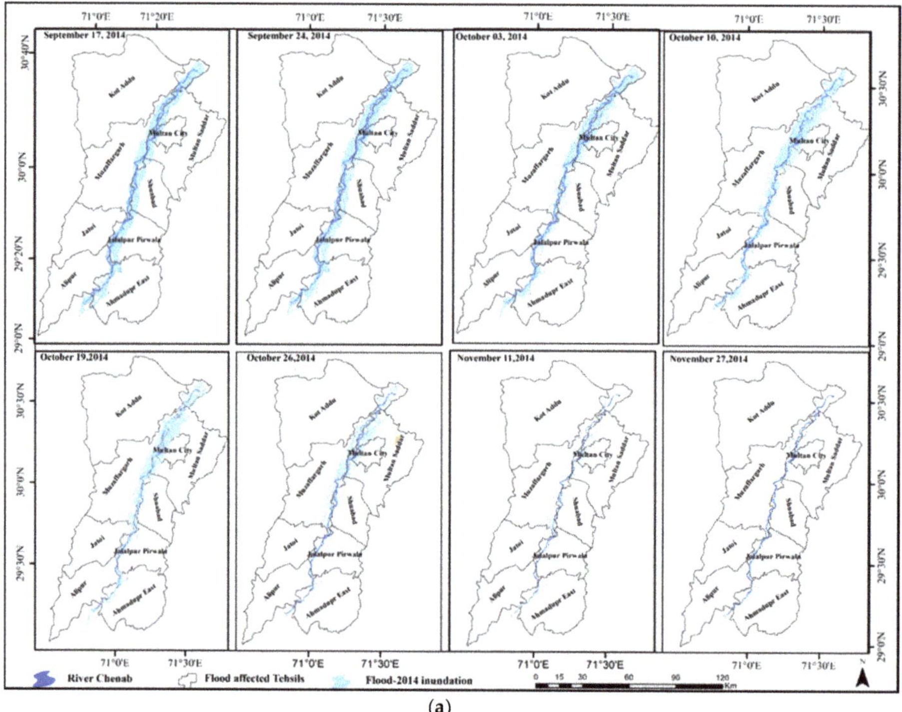

(a)

Figure 4. *Cont.*

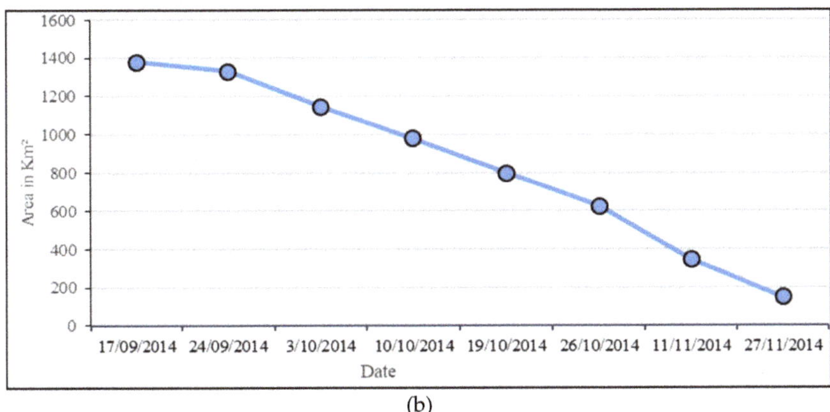

(b)

Figure 4. (**a**) Comprehensive flood inundation and recession in the lower Chenab plain. (**b**) Flood recession in the lower Chenab plain from 17 September (peak-flood) to 27 November 2014 (post-flood).

Table 4. Inundated areas of the lower Chenab plain in Km^2.

Districts	Tehsils	17 September 2014	24 September 2014	3 October 2014	10 October 2014	19 October 2014	26 October 2014	11 November 2014	27 November 2014
Muzaffargarh	Alipur	58.87	57.02	50.86	34.31	26.24	15.66	12.34	11.14
	Muzaffargarh	530.53	507.99	485.48	409.12	363.56	308.75	166.25	68.93
	Kot Addu	2.41	2.77	2.45	2.82	2.13	1.55	0.71	0.42
	Jatoi	11.39	15.37	13.29	19.48	15.23	11.25	8.21	1.38
Multan	Jalalpur Pirwala	212.51	191.89	141.34	97.86	56.34	40.16	21.45	16.01
	ShuJabad	169.51	159.95	96.19	88.48	64.73	40.36	25.13	10.39
	Multan Saddar	209.21	223.26	212.56	205.03	188.45	147.9	74.32	22.80
	Multan City	37.43	42.58	41.95	39.72	39.67	35.78	17.65	3.7
Bahawalpur	Ahmadupr East	148.57	129.13	103.80	83.6	44.05	24.94	19.34	15.79
Total Area		1380.43	1330.96	1147.92	980.54	800.4	626.35	345.4	146.56

The peak-flood (17 September) extent is compared in Figure 5a,b. In Figure 5a, the MNDWI index shows the accumulated flood extent that covers an area of 1380km^2. Whereas, in Figure 5b, the classified image shows that about 1330.03 km^2 of the area was inundated, which represents a 2% deviation from the MNDWI results. The result shows that both the MNDWI index and supervised classification produce almost similar inundation areas in lower Chenab plain. Furthermore, these inundation results, when incorporated with the GE images acquired at the same time as the Landsat-8 images, also showed the severe spatial pattern of inundation, and they also validate our estimated flood inundation extent, confirming that flood waters remained for almost two months and receded very slowly (Section 3.3).

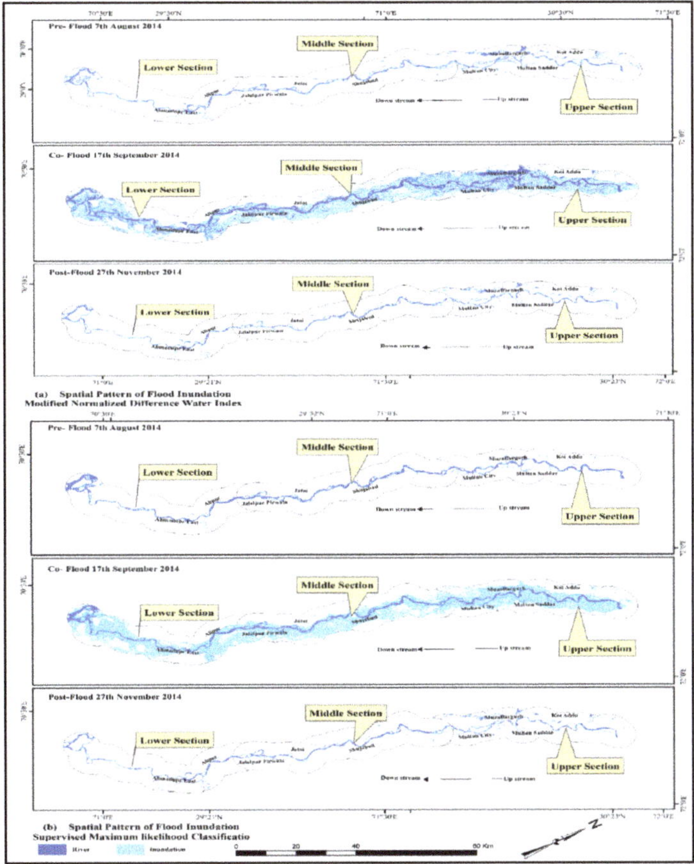

Figure 5. Spatio-temporal flood inundation (**a**) using the MNDWI index (**b**) using Supervised Classification.

3.3. Land Use and Land Cover Changes

Figures 6–8 show the results of the supervised ML classification for land use/land cover mapping of Pre-flood (07 August), Co-flood (17 September), and Post-flood (27 November) images in the lower Chenab plain. We classified water, built-up, agriculture land and vegetation, barren land, sand, and deposited material on the pre- and post-flood images. However, deposited material was not identified in co-flood image due to intensive flood inundation (Figure 7). The results show significant changes in all classified land uses in relation to flood instances. In pre-flood, only 8% was covered by the water body, 17.39% of built up, 55.37% of vegetation/agriculture land, 4.84% of barren land,

8.32% of sand, and 5.92% of deposited material within a total area of 2536.11 km^2 (Figure 6). The water body showed only 8% in the pre-flood situation. However, after the flood occurrence in the month of September, 2014, an abrupt change appeared in the water body that represents a massive increase to 50% of the area, as shown in Figure 7. The massive water had not only inundated but also severely affected all other land uses. The built-up area considerably decreased from 17.39% to 11.85% in the study area. Likewise, we noticed a massive increase in water that affected the agriculture/vegetation covered area, which decreased from the initial total area of 2536.11 km^2, from 52.37% to 32.40%. The noticeable decrease of sand was from 8.3% to 1.1%. The slight decrease is noticed in barren land from 4.8% to 4%. Thus, the massive increase in water caused a large part of the agricultural/vegetation area to be flooded and destroyed most of the standing crops. In addition, the built-up area was also severely affected and flooded.

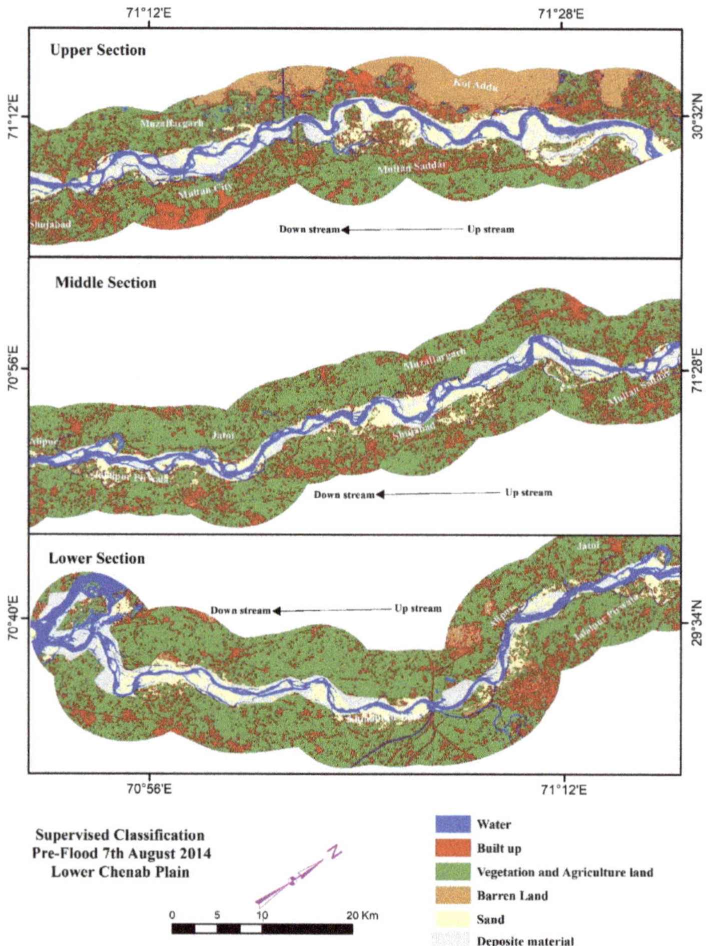

Figure 6. Land use/Land cover map of the pre-flood image developed using Landsat-8 images and supervised classification.

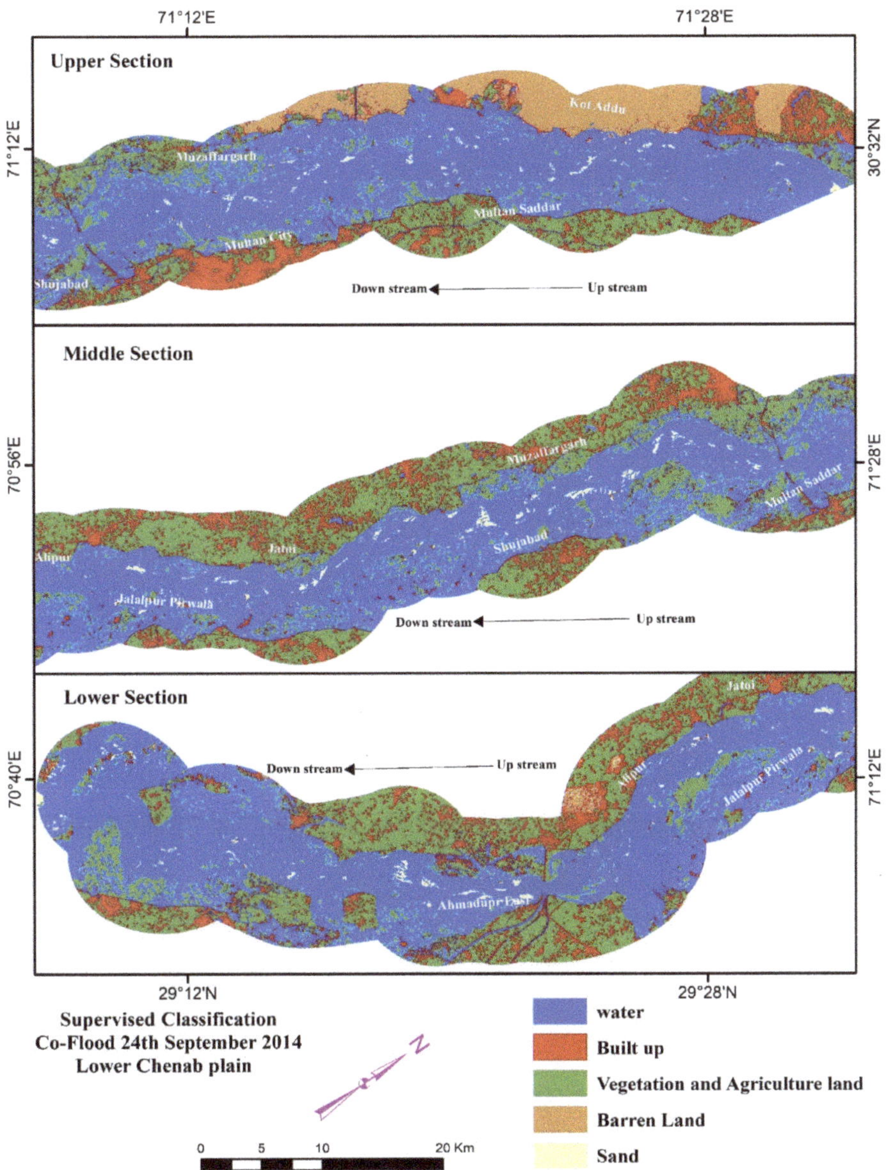

Figure 7. Land use/Land cover map of the co-flood image, 17 September, 2014, developed using Landsat-8 images and supervised classification.

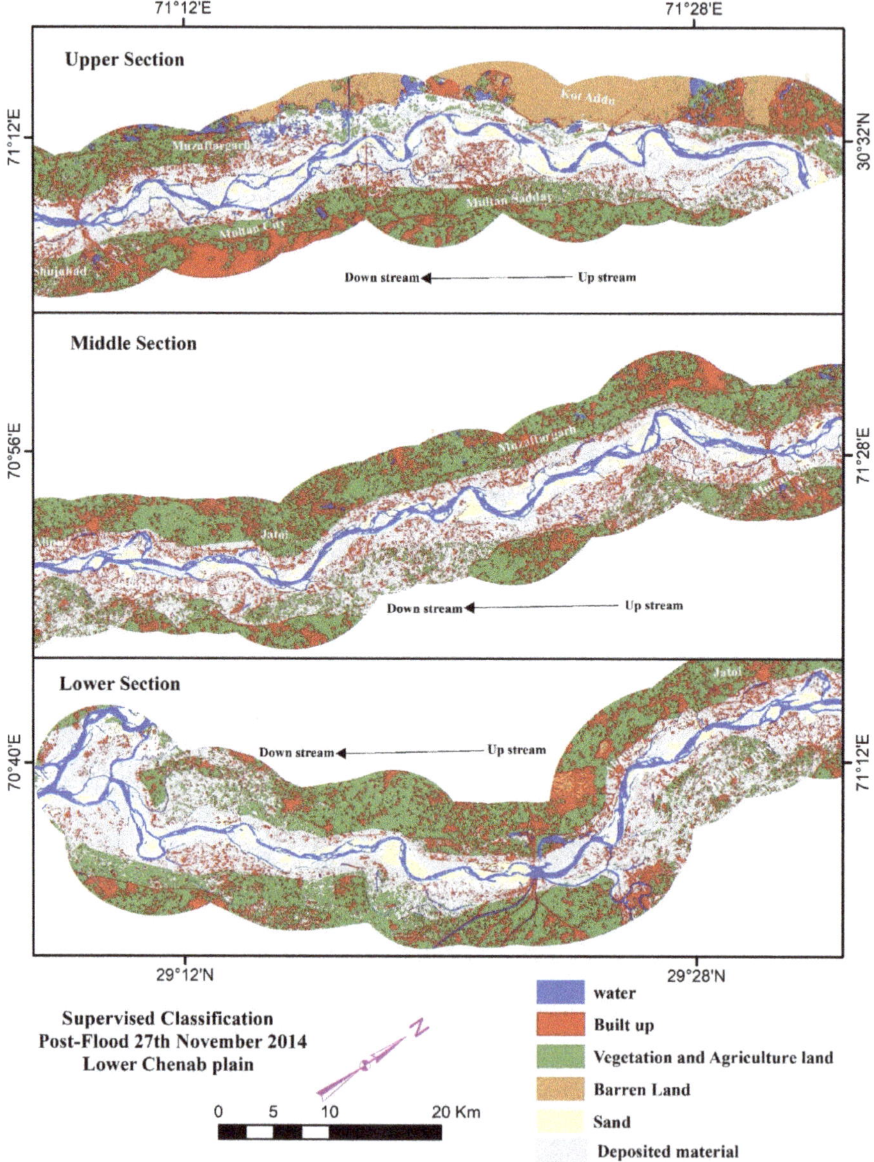

Figure 8. Land use/Land cover map of the post-flood image developed using Landsat-8 images and supervised classification.

In the post-flood period (Figure 8), water receded to its original pre-flood stage and decreased to only 6% from 50% in the co-flood image. Similarly, built up also regained its original pre-flood stage and increased to 17.79%. After the flood occurred, water receded back but large amounts of sediment and other materials remained, representing a dramatic increase from 5.92% to 30%. Similarly, an increase in the vegetation and agriculture area can be noticed from 32.40% to 37.50%. The barren land area did not change much as it remained constant in each instance. Sand slightly increased to 4.4% from 1.17% of total 2536.11 km^2 area, as shown in Figures 8 and 9. The total area in km^2 of each class is statistically represented in Figure 9. Hence, all classified images show a similar pattern of land use/cover classes compared to high resolution GE images (Figure 10).

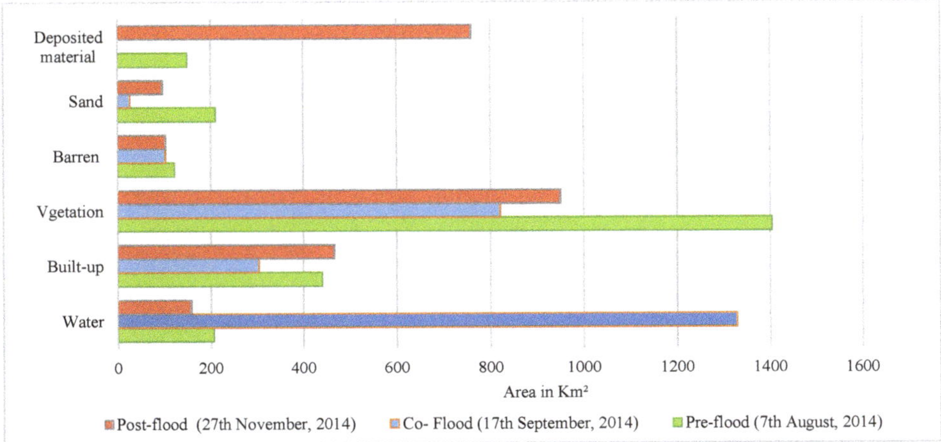

Figure 9. Change detection of land use/land cover in the lower Chenab plain using Landsat-8 flood instances images.

The classification result and GE images show the same pattern of land use changes in the study area. The agriculture and built-up areas along the river are clearly visible in the pre-flood image (Figure 10A). Conversely, the co-flood image (Figure 10B) depicts the abnormal change in water area that has increased drastically and caused an inundation to all the surrounding agricultural and built-up areas. As shown in Figure 10C the post-flood image shows the huge amount of deposited material along the river and its surrounding areas that dramatically increased and destroyed standing crops, which caused late sowing of agricultural crops, and further leads to low productivity in the study area.

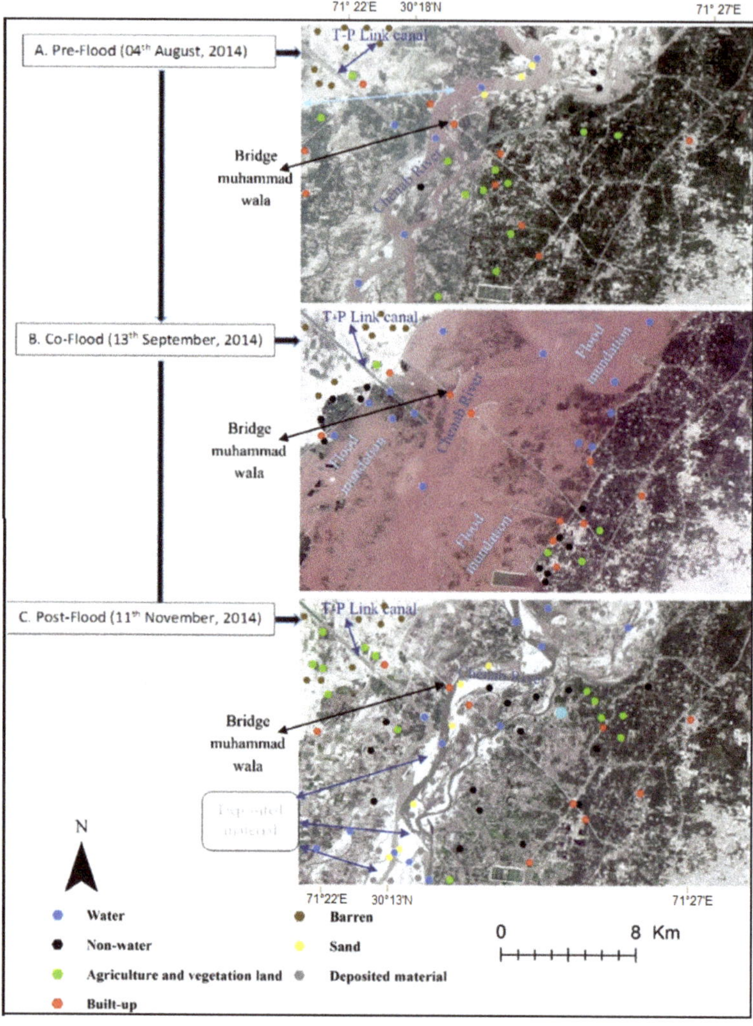

Figure 10. Google Earth 7.3.2.5776. (A. 4 August, 2014, B. 13 September, 2014, C. 11 November, 2014). Section 1, the lower Chenab plain. 30.290058° N,71.378478° E, Eye alt 11.42 mi. Also shown are validation points.

3.4. Damage Assessment

The results revealed that agriculture/vegetation areas were the most severely inundated, ~495 km^2, and while inundation in the built-up area covered 229 km^2, from the total inundated area of 1330 km^2, as shown in Figure 11a,b. The damage result further showed that a total of 361.43km^2 of agricultural land and 187.36km^2 of built-up areas were damaged due to deposited material. This result shows that the Multan district is the most affected/damaged district from flood-2014, as also shown in Figure 12a. Further details on the damaged agriculture land and built-up areas of the involved districts are listed in Figure 12b.

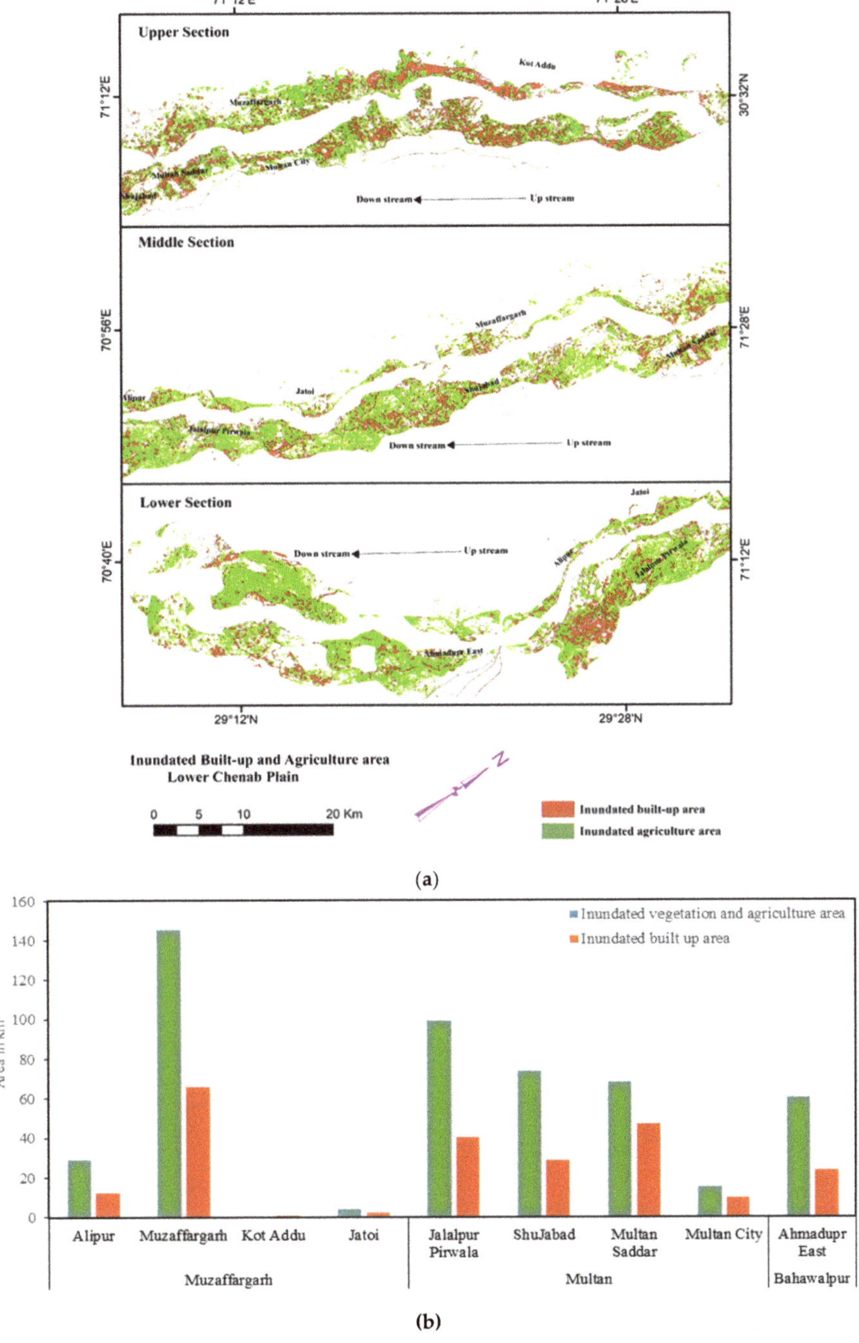

(**a**)

(**b**)

Figure 11. (**a**) Inundated agriculture and vegetation areas and built-up areas in the lower Chenab plain. (**b**) a graph of inundated agriculture and vegetation area, and built-up areas in the lower Chenab plain.

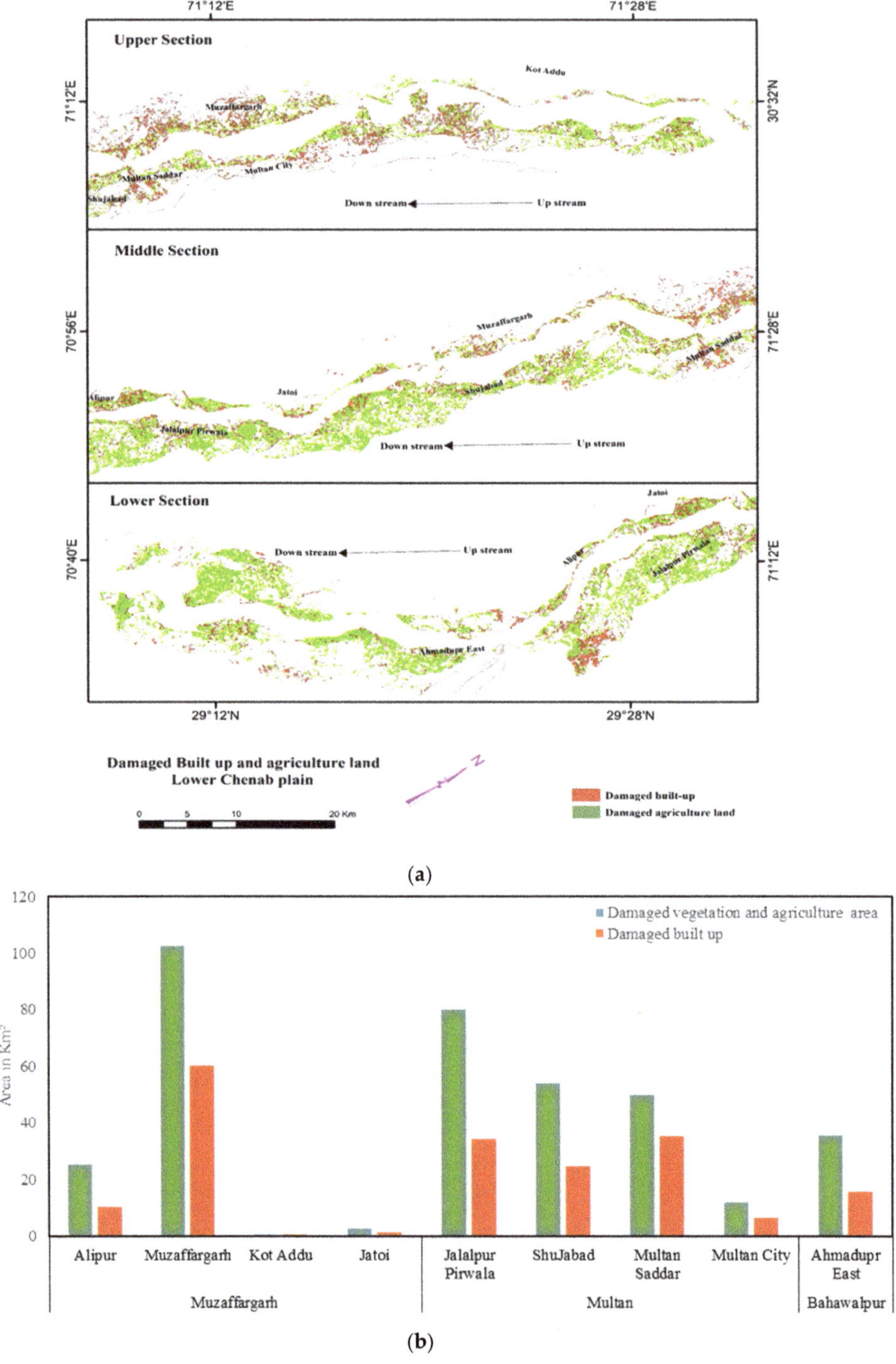

(a)

(b)

Figure 12. (**a**) Damaged agriculture and vegetation areas and built-up areas in the lower Chenab plain. (**b**) a graph of damaged agriculture and vegetation areas and built-up areas in the lower Chenab plain.

4. Discussion

Pakistan is a flood-prone country with historical records of various magnitudes of flood events [37,53,54]. In the past decade, flood disasters in Pakistan have surpassed all other disasters in terms of the frequency of occurrences, and also killed over 5700 people coupled with severe damages to the country's economy [5,12,15,17]. In the year of 2014, the flood started from late monsoon rainy season when the river Chenab inflow significantly increased from its upstream tributaries and resulted in high discharge, which exceeded the limits of the river flowing capacity and subsequently caused a huge inundation in the study area, as shown in Figures 7 and 10b.

Our results show that open satellite data coupled with an ML-supervised classification approach and the MNDWI allow the delineation of flood inundated and damaged areas with high overall accuracy. To obtain high accuracy levels, appropriate satellite data is critical. Firstly, temporal relationship between flood occurrence and satellite characteristics is an important parameter in flood modeling. For instance, a low-resolution (~250 m), multispectral Moderate Resolution Imaging Spectroradiometer (MODIS) satellite with its daily revisit time has already been used to obtain a co-flood image for certain flood events in Pakistan, but with questionable accuracy [19,55]. However, due to absence of any co-flood open access high resolution, e.g., Sentinel-1, for the study area, we opted to use moderate spatial resolution Landsat data (~30 m) for flood monitoring and assessment [9,18,22]. Despite a relatively better resolution, Landsat data are limited in such a way that they lack the timely acquisition of geospatial data, which sometimes reduces their suitability for flood monitoring and inundation mapping [26,39,56]. In the study area however, two adjacent Landsat satellite paths (150 and 151) made it possible to acquire high temporal resolution Landsat data (~7 days), against a single temporal resolution of 15 days for an area [9,18,57]. As such, the high temporal resolution of Landsat data enables a detailed investigation of operational flood mapping and monitoring in a study area. Secondly, existences of cloud cover can restrain the availability of flood instances images. Certainly, SAR (Sentinel 1,2,3) and RADAR satellites can easily penetrate clouds and acquire images in comparison to optical satellites (Landsat, MODIS) [26,58]. In our cases, the free SAR data were not available and the study area was entirely cloud free during the 2014 flood, since the causative meteorological events happened in upper Chenab plain. These events generated the flood peak that happened almost 10 days prior to the flooding in lower Chenab plain. Thirdly, consideration of land use/cover change of the flooded area is also an important factor in choosing the satellite data. The inundated agriculture and built-up area can easily be detected by most Radar SAT and optical satellites [59]. With multispectral Landsat images, we can recognize flooded areas even after several days of flood occurrence [41,42,56]. Nevertheless, we used Landsat-8 data as it is also possible to detect sediment material over agriculture fields and built-up areas with reliable accuracy [26,59]. Despite that, high resolution SAR data and a field survey are necessary for accurate and reliable flood mapping and damage assessment [59,60]. Finally, we have to consider the accuracy of the applied methods to the satellite data in order to ascertain the reliability of the flood monitoring results. This study produces an overall accuracy for the MNDWI index of almost 90% (Table 2), while classified images had an average overall accuracy of about 88% (Table 3). The water class attained the highest overall accuracy in this study, which has the ability to obtain a highest possible accuracy of 100% [49]. Despite that sand also shows reliable accuracy, in some occasions, wet sand was confused with water and vegetation/agriculture land. This is also noted in some instances where agriculture and built-up areas are also confused in transition areas due to mixed pixels.

During the flood-2014 event, the peak-flood water arrived on 17 September, 2014, and then remained constant for a while. The water slightly receded at a rate of 7 km^2/day, until 24 September, while a recharge of flood water continues from upstream areas. Here, the flood water started to decrease at a rate of 23km^2/day until 26 October, exposing a once flooded area of 636.35 km^2. On 27 November, the river regained its pre-flood stage and flowed normally. However, inundation remained for almost two months in the study area, as shown in Figure 4. As shown in Figures 11 and 12, the flood-2014 caused huge inundation and damages to agriculture land and built up in the study area.

The field survey reveals that most of the local people live in mud houses, which are highly vulnerable to flood disaster and have no flood resilience capacity [25]. As a result, already vulnerable mud houses were badly destroyed by the flood, besides infrastructure being destroyed by the gushing flood-2014. Moreover, the survey reveals that the month of November is considered agriculture land preparation and sowing season for new Rabi season crops and simultaneously harvesting season of certain kharif crops, such as sugarcane and rice in the study area. As a result, vegetation/agriculture land slightly increased to 37% in the post-flooding instance. Likewise, the built-up area also increased to 17.7%, which regained its pre-flood situation, probably due to the receded flood water and reconstruction. Furthermore, the survey shows that the Chenab river carried a huge amount of sediment material from its upstream mountainous areas, and it was ultimately deposited along the lower Chenab plain in 2014. However, deposited material immensely increased to 30% in the post-flooding instance (Figure 10). In our case, the field survey provided real-time reliable information, which is immensely required for comprehensive flood disaster assessment.

The moderate 30m resolution Landsat data that we used indicated that tracking inundated and damaged agriculture areas and built-up areas is reliable and acceptable. This mostly applies to damage assessment, which could be used for post-flood rehabilitation and a relief operation. However, for mapping the of detailed built up and agriculture area with a high overall accuracy, we must utilize high-resolution satellite data (RADARSAT, SAR), aerial photography, and extensive field survey, which are very often time and cost intensive and the required high-resolution satellite images are not open access. In this work, we did a field survey in order to evaluate real time damages, geo-location information, and the validation of Remote Sensing (RS) results. Thus, we correlated our survey findings for the validation of RS results and formulated an emergency damage assessment. Our damage analysis aims to provide a rapid and low-cost assessment of damaged agricultural and built-up areas but does not provide direct information on monetary losses and indirect damages, which may occur after flooding.

Landsat-8 data are affected by clouds, spectral sensitivity, and moderate spatial resolution and can only detect surface reflectance changes. Conversely, Radar satellite data can easily penetrate clouds and will perform properly in all weather conditions and detect changes in vegetation structure and moisture [59,60]. Radar data is more reliable when extracting flood inundation and damages in agricultural and built-up areas [25]. This well-known advantage allows Radar data to find the inundated and damaged areas in the post-flood stage in order to carry out an emergency flood damage assessment [58–60]. However, Radar data is costly and is not applicable for historical data analysis. However, despite such limitations, the integration of field and ancillary data allowed us to extract flood inundation and damage assessment with reliable accuracy. We noticed a number of factors that affected the tracking of damage assessment in this study. The lower Chenab floodplain contains numerous mangoes and other Bela forests. The co-flood Landsat image reveals that flood inundation beneath thin mangoes and other tree canopy covers was easily detected but thick or dense canopy covers' pixels were not identified due to the moderate resolution. This resulted in addition to the overall omission error and also caused the least identification of inundated areas in change detection analysis during flood instances. Secondly, a misclassification error was found in the classified images as the open land classified as the deposited material class in the post-flood image due to the moderate resolution and spectral sensitivity. Then, this error was removed and corrected through field-survey findings. Finally, the moderate 30-m spatial resolution of the Landsat images contributed to mix-up and misclassification in built-up areas where floodwaters pass through housing structures, roads, and other features. This increases the brightness of otherwise dark pixels (water), resulting in an increase in omission errors, which is a limitation for built-up damage assessments.

5. Conclusions

In this study, we presented a low cost and user-friendly flood monitoring and damage assessment with the integration of open access optical remote sensing and appropriate processing methods jointly exploited with field data. In particular, we used Landsat-8 data and processed them with open-source, GIS-based, supervised classification for damage assessment and a satellite-derived MNDWI index for flood inundation and monitoring. The classified images and inundation mapping produced an excellent overall accuracy of about 90%, which is validated and shows reliable results. we conclude that these methods have been proven to be useful for estimating and understanding a future flood phenomenon with its diverse impacts. These methods do not require more time and provide near real-time information using the user input with indigenous knowledge and expertise. Furthermore, the classified result reveal that the agriculture sector has been the most affected land use/cover in the study area. In a large context, people are mostly engaged with agricultural activities, and this pattern of flood impacts are a major concern as it directly influences the livelihood of the floodplain community. The index result shows that about 75% of the area experienced severe flooding, which lies mostly in the southern and central part of study area, and further revealed that flood inundation remained for almost two months. Finally, despite the fact that flood disasters are recurrent phenomena, our study proves that the used datasets and methods can be useful for emergency, real-time, automated flood monitoring and damage assessment in order to formulate emergency flood disaster management, particularly for relief and response operations. In contrast, traditional flood monitoring and damage assessment with paid on-demand data provides comparatively accurate results, but it is a more resource- and time-consuming process.

Despite that our study is local to Pakistan, a few points from our results can be applied in other areas and thereby contribute to science and global perspectives on flooding. In this paper, we have shown that the existing supervised classification methods, combined with the MNDWI and ground-based validation point, can solve a problem of flooding and provide flood management solutions to local authorities. The combination of supervised classification and the MNDWI has contributed to the accurate mapping of flood extents using Landsat imagery. Despite that these two methods are mostly applied separately, the combined application in flood mapping can help to produce reliable results, with complimentary properties. Furthermore, in the unique agricultural environment along the river, the approach of combining two classification methods can provide insights into flood instances and inundation in both build up and agricultural areas, enabling a rapid check on damages to both built up and agricultural crops and thereby providing an estimation of loss that can support ground-based research. We have also shown that despite a low temporal resolution of Landsat images, the exploitation of two adjacent Landsat paths can provide a high temporal resolution, which can be applied not only to flood management and monitoring but also other fields that require high temporal resolution data, especially where another dataset is not available. This exploitation increased the temporal resolution by two-fold, from ~15 days to ~ 7 days, which can also be factored into operational flood monitoring for emergency responses, such as early response and relief operations. Thus, our study has given another perspective to flood monitoring and management using the available and free satellite imagery.

Author Contributions: Conceptualized overall research design, A.S., J.L.; performed image analysis, prepared the flood inundation maps, land use/land cover maps, flood damage assessment, and change assessment, A.S.; validation, A.S., J.L., C.C.; investigation, X.C., C.C., N.S., H.H.; writing—original draft preparation, A.S.; writing—review and editing, A.S., J.L., X.C., C.C.; project administration, J.L., X.C.; funding acquisition, J.L., X.C. All authors have read and agreed to the published version of the manuscript.

Funding: This work was funded by the National Key Research and Development Program (2018YFC1506506, 2017YFB0504103), the Frontier Project of Applied Foundation of Wuhan (2019020701011502), the Natural Science Foundation of Hubei Province (2019CFB736), the fundamental Research Funds for the Central Universities (2042018kf0220), and the LIESMARS Special Research Funding.

Acknowledgments: The authors wish to thank the USGS (http://earthexplorer.usgs.gov/) and Google Earth (https://earth.google.com) for providing the needed data for this study. The authors would also like to thank those who reviewed the article incomprehensibly and provided valuable suggestions to improve the manuscript.

Conflicts of Interest: The authors declare no conflict of interest.

References

1. Ward, P.J.; Jongman, B.; Weiland, F.S.; Bouwman, A.; van Beek, R.; Bierkens, M.F.; Ligtvoet, W.; Winsemius, H.C. Assessing flood risk at the global scale: Model setup, results, and sensitivity. *Environ. Res. Lett.* **2013**, *8*, 044019. [CrossRef]
2. Schumann, G.; Bates, P.D.; Apel, H.; Aronica, G.T. Global Flood Hazard Mapping, Modeling, and Forecasting: Challenges and Perspectives. *Glob. Flood Hazard Appl. Model. Mapp. Forecast.* **2018**, 239–244.
3. Iqbal, M.S.; Dahri, Z.H.; Querner, E.P. The impact of climate change on flood frequency and intensity in the Kabul River basin. *Geosciences* **2018**, *8*, 114. [CrossRef]
4. Rahman, A.; Khan, A.N. Analysis of flood causes and associated socio-economic damages in the Hindu Kush region. *Nat. Hazards.* **2011**, *59*, 1239–1260. [CrossRef]
5. Halgamuge, M.N.; Nirmalathas, A. Analysis of large flood events: Based on flood data during 1985–2016 in Australia and India. *Int. J. Disaster Risk Reduct.* **2017**, *24*, 1–11. [CrossRef]
6. Milly, P.C.D.; Wetherald, R.T.; Dunne, K.A.; Delworth, T.L. Increasing risk of great floods in a changing climate. *Nature* **2002**, *415*, 514–517. [CrossRef] [PubMed]
7. Islam, A.S.; Bala, S.K.; Haque, M. Flood inundation map of Bangladesh using MODIS time-series images. *J. Flood Risk Manag.* **2010**, *3*, 210–222. [CrossRef]
8. Syvitski, J.P.; Brakenridge, G.R. Causation and avoidance of catastrophic flooding along the Indus River, Pakistan. *GSA Today* **2013**, *23*, 4–10. [CrossRef]
9. Gaurav, K.; Sindha, R.; Panda, P.K. The Indus flood of 2010 in Pakistan: a perspective analysis using remote sensing data. *Nat. Hazards.* **2011**, *59*, 1815–1826. [CrossRef]
10. Mahmood, S.; Rahman, A.; Sajjad, A. Assessment of 2010 flood disaster causes and damages in district Muzaffargarh, Central Indus Basin, Pakistan. *Environ. Earth Sci.* **2019**, *78*, 63. [CrossRef]
11. Sajjad, A.; Lu, J.; Chen, X.; Chisenga, C.; Mahmood, S. The riverine flood catastrophe in August 2010 in South Punjab, Pakistan: Potential causes, extent and damage assessment. *Appl. Ecol. Environ. Res.* **2019**, *17*, 14121–14142.
12. Federal Flood Commission Islamabad (FFCI). *Annual flood Report*; Ministry of Water and Power, Government of Pakistan: Islamabad, Pakistan, 2015.
13. Atif, I.; Mahboob, M.; Waheed, A. Spatio-Temporal Mapping and Multi-Sector Damage Assessment of 2014 Flood in Pakistan using Remote Sensing and GIS. *Indian J. Sci. Technol.* **2015**, *8*, 35. [CrossRef]
14. Ali, S.; Li, D.; Congbin, F. Twenty first century climatic and hydrological changes over upper Indus Basin of Himalayan region of Pakistan. *Environ. Res. Lett.* **2015**, *10*, 14007. [CrossRef]
15. Rahman, A.; Khan, A.N. Analysis of 2010-flood causes, nature and magnitude in the Khyber Pakhtunkhwa, Pakistan. *Nat. Hazards.* **2013**, *66*, 887–904. [CrossRef]
16. Mahmood, S.; Rani, R. Extent of 2014 flood damages in Chenab Basin Upper Indus Plain. In *Natural Hazards-Risk Assessment and Vulnerability Reduction*; Intech Open: London, UK, 2018. [CrossRef]
17. Hashmi, H.N.; Siddiqui, Q.T.M.; Ghuman, A.R.; Kamal, M.A.; Mughal, H. a critical analysis of 2010 floods in Pakistan. *Afr. J. Agric. Res.* **2012**, *7*, 1054–1067.
18. Khalid, B.; Cholaw, B.; Alvim, D.S.; Javeed, S.; Khan, J.A.; Javed, M.A.; Khan, A.H. Riverine flood assessment in Jhang district in connection with ENSO and summer monsoon rainfall over Upper Indus Basin for 2010. *Nat. Hazards* **2018**, *92*, 971–993. [CrossRef]
19. Haq, M.; Akhtar, M.; Muhammad, S.; Paras, S.; Rahmatullah, J. Techniques of Remote Sensing and GIS for flood monitoring and damage assessment: a case study of Sindh province, Pakistan. *Egypt. J. Remote Sens. Space Sci.* **2012**, *15*, 135–141. [CrossRef]
20. Pakistan Meteorological Department (PMD). *Annual Report*; Regional Meteorological Observatory: Lahore, Pakistan, 2015.
21. Ashraf, M.; Shakir, A.S. Prediction of river bank erosion and protection works in a reach of Chenab River, Pakistan. *Arab. J. Geosci.* **2018**, *11*, 145. [CrossRef]
22. Chohan, K.; Ahmad, S.R.; Islam, Z.; Adrees, M. Riverine flood damage assessment of cultivated lands along Chenab River using GIS and remotely sensed data: a case study of district Hafizabad, Punjab, Pakistan. *J. Geogr. Inf. Syst.* **2015**, *7*, 506–526. [CrossRef]
23. National Disaster Management Authority((NDMA). *Annual Flood Report*; Regional office: Islamabad, Pakistan, 2015.
24. Punjab Provincial Disaster Management Authority (PPDMA). *Annual Flood Report*; Regional Office: Lahore, Pakistan, 2015.

25. Moel, H.D.; Alphen, J.V.; Aerts, J. Flood maps in Europe–methods, availability and use. *Nat. Hazards Earth Syst. Sci.* **2009**, *9*, 289–301. [CrossRef]

26. Musa, Z.N.; Popescu, I.; Mynett, A. a review of applications of satellite SAR, optical, altimetry and DEM data for surface water modelling, mapping and parameter estimation. *Hydrol. Earth Syst. Sci.* **2015**, *19*, 3755–3769. [CrossRef]

27. Zhang, P.; Lu, J.; Feng, L.; Chen, X.; Zhang, L.; Xiao, X.; Liu, H. Hydrodynamic and Inundation Modeling of China's Largest Freshwater Lake Aided by Remote Sensing Data. *Remote Sens.* **2015**, *7*, 4858–4879. [CrossRef]

28. Bhatt, C.M.; Rao, G.S.; Farooq, M.; Manjusree, P.; Shukla, A.; Sharma, S.V.S.P.; Kulkarni, S.S.; Begum, A.; Bhanumurthy, V.; Diwakar, P.G.; et al. Satellite-based assessment of the catastrophic Jhelum floods of September 2014, Jammu & Kashmir, India. Geomatics. *Nat. Hazards Risk* **2016**, *8*, 309–327.

29. Chignell, S.M.; Anderson, R.S.; Evangelista, P.H.; Laituri, M.J.; Merritt, D.M. Multi-temporal independent component analysis and Landsat 8 for delineating maximum extent of the 2013 Colorado front range flood. *Remote Sens.* **2015**, *7*, 9822–9843. [CrossRef]

30. Lu, J.; Li, H.; Chen, X.; Liang, D. Numerical study of remote sensed dredging impacts on the suspended sediment transport in China's Largest Freshwater Lake. *Water* **2019**, *11*, 2449. [CrossRef]

31. Giustarini, L.; Chini, M.; Hostache, R.; Pappenberger, F.; Matgen, P. Flood Hazard Mapping Combining Hydrodynamic Modeling and Multi Annual Remote Sensing data. *Remote Sens.* **2015**, *7*, 14200–14226. [CrossRef]

32. Wulder, M.A.; Masek, J.G.; Cohen, W.B.; Loveland, T.R.; Woodcock, C.E. Opening the archive: How free data has enabled the science and monitoring promise of Landsat. *Remote Sens. Environ.* **2012**, *12*, 2–10. [CrossRef]

33. Sanyal, J.; Lu, X.X. Application of Remote Sensing in Flood Management with Special Reference to Monsoon Asia: a review. *Nat. Hazards* **2004**, *33*, 283–301. [CrossRef]

34. Hu, Q.; Wu, W.; Xia, T.; Yu, Q.; Yang, P.; Li, Z.; Song, Q. Exploring the use of Google Earth imagery and object-based methods in land use/cover mapping. *Remote Sens.* **2013**, *5*, 6026–6042. [CrossRef]

35. Twumasi, N.Y.D.; Shao, Z.; Altan, O. Mapping Built-Up Areas Using Two Band Ratio Onlandsat Imagery of Accra In Ghana From 1980 To 2017. *Appl. Ecol. Environ. Res.* **2019**, *17*, 13147–13168.

36. *Government of Punjab (GoP) Punjab Development Statistics*; Bureau of Statistics Government of Punjab: Lahore, Pakistan, 2015.

37. Siddiqui, M.; Haider, S.; Gabriel, H.F.; Shahzad, A. Rainfall–runoff, flood inundation and sensitivity analysis of the 2014 Pakistan flood in the Jhelum and Chenab river basin. *Hydrol. Sci. J.* **2018**, *63*, 13–14. [CrossRef]

38. Hansen, M.C.; Loveland, T.R. a review of large area monitoring of land cover change using Landsat data. *Remote Sens. Environ.* **2012**, *122*, 66–74. [CrossRef]

39. Munasinghe, D.; Cohen, S.; Huang, Y.F.; Tsang, Y.P.; Zhang, J.; Fang, Z.F. Intercomparison of Satellite Remote Sensing-Based Flood Inundation Mapping Techniques. *J. Am. Water Resour. Assoc.* **2018**, *54*, 834–846. [CrossRef]

40. Alphan, H.; Doygun, H.; Unlukaplan, Y.I. Post-classification comparison of land cover using multitemporal Landsat and ASTER imagery: The case of Kahramanmaras, Turkey. *Environ. Monit. Assess.* **2009**, *151*, 327–336. [CrossRef]

41. Rokni, K.; Ahmad, A.; Selamat, A.; Hazini, S. Water feature extraction and change detection using multitemporal Landsat imagery. *Remote Sens.* **2014**, *6*, 4173–4189. [CrossRef]

42. AlFaisal, A.; Kafy, A.A.; Roy, S. (Integration of Remote Sensing and GIS Techniques for Flood Monitoring and Damage Assessment: a Case Study of Naogaon District, Bangladesh. *J. Remote Sens. GIS* **2018**, *7*, 236.

43. Joyce, K.E.; Belliss, S.E.; Samsonov, S.V.; McNeill, S.J.; Glassey, P.J. a review of the status of satellite remote sensing and image processing techniques for mapping natural hazards and disasters. *Prog. Phys. Geogr.* **2009**, *33*, 183–207. [CrossRef]

44. Revilla-Romero, B.; Hirpa, F.A.; Pozo, J.T.; Salamon, P.; Brakenridge, R.; Pappenberger, F.; De Groeve, T. On the use of global flood forecasts and satellite-derived inundation maps for flood monitoring in data-sparse regions. *Remote Sens.* **2015**, *7*, 15702–15728. [CrossRef]

45. McFeeters, S.K. The use of the normalized difference water index (NDWI) in the delineation of open water features. *Int. J. Remote Sens.* **1996**, *17*, 1425–1432. [CrossRef]

46. Xu, H. Modification of normalized difference water index (NDWI) to enhance open water features in remotely sensed imagery. *Int. J. Remote Sens.* **2006**, *27*, 3025–3033. [CrossRef]

47. Fisher, A.; Flood, N.; Danaher, T. Comparing Landsat water index methods for automated water classification in eastern Australia. *Remote Sens. Environ.* **2016**, *175*, 167–182. [CrossRef]

48. Ho, L.T.K.; Umitsu, M.; Yamaguchi, Y. Flood hazard mapping by satellite images and SRTM DEM in the Vu Gia-Thu Bon alluvial plain, central Vietnam. *Arch. Photogramm Remote Sens.* **2010**, *38*, 275–279.

49. Elhag, M. Consideration of Landsat-8 spectral band combination in typical Mediterranean forest classification in Halkidiki, Greece. *Open Geosci.* **2017**, *9*, 468–479. [CrossRef]

50. Acharya, T.D.; Subedi, A.; Lee, D.H. Evaluation of Water Indices for Surface Water Extraction in a Landsat 8 Scene of Nepal. *Sensors.* **2018**, *18*, 2580. [CrossRef] [PubMed]

51. Zhou, S.L.; Zhang, W.C. Flood monitoring and damage assessment in Thailand using multi-temporal HJ-1A/1B and MODIS images. *IOP Conf. Ser. Earth Environ. Sci.* **2017**, *57*. [CrossRef]

52. Notti, D.; Giordan, D.; Caló, F.; Pepe, A.; Zucca, F.; Pedro Galve, J. Potential and Limitations of Open Satellite Data for Flood Mapping. *Remote Sens.* **2018**, *10*, 1673. [CrossRef]

53. Mahmood, S.; Khan, A.H.; Mayo, S.M. Exploring underlying causes and assessing damages of 2010 flash flood in the upper zone of Panjkora River. *Nat. Hazards* **2016**, *83*, 1213–1227. [CrossRef]

54. Mahmood, S.; Khan, A.H.; Ullah, S. Assessment of 2010 flash flood causes and associated damages in Dir Valley, Khyber Pakh-tunkhwa Pakistan. *Int. J. Disaster Risk Reduct.* **2016**, *16*, 215–223. [CrossRef]

55. Memon, A.A.; Muhammad, S.; Rahman, S.; Haq, M. Flood monitoring and damage assessment using water indices: a case study of Pakistan flood-2012. *Egypt. J. Remote Sens Space Sci.* **2015**, *18*, 99–106. [CrossRef]

56. Shuhua, Q.I.; Brown, D.G.; Tian, Q.; Jiang, L.; Zhao, T.; Bergen, K.M. Inundation Extent and Flood Frequency Mapping Using LANDSAT Imagery and Digital Elevation Models. *GIScience Remote Sens.* **2009**, *46*, 101–127.

57. Frazier, P.S.; Page, K.J. Water body detection and delineation with Landsat TM data. *Photogramm. Eng. Remote Sens.* **2000**, *66*, 1467.

58. Uddin, k.; Matin, M.A.; Meyer, F.J. Operational Flood Mapping Using Multi-Temporal Sentinel-1 SAR Images: a Case Study from Bangladesh. *Remote Sens.* **2019**, *11*, 1581. [CrossRef]

59. Pradhan, B.; Pirasteh, S.; Shafie, M. Maximum flood prone area mapping using RADARSAT images and GIS: Kelantan river basin. *Int. J. Geoinformatics* **2009**, *5*, 11–23.

60. Giordan, D.; Notti, D.; Villa, A.; Zucca, F.; Caló, F.; Pepe, A.; Dutto, F.; Pari, P.; Baldo, M.; Allasia, P. Low cost, multiscale and multi-sensor application for flooded area mapping. *Nat. Hazards Earth Syst. Sci.* **2018**, *18*, 1493–1516. [CrossRef]

 *remote sensing*

Article

The Status of Earth Observation Techniques in Monitoring High Mountain Environments at the Example of Pasterze Glacier, Austria: Data, Methods, Accuracies, Processes, and Scales

Michael Avian [1,*], Christian Bauer [2], Matthias Schlögl [1,3], Barbara Widhalm [1], Karl-Heinz Gutjahr [4], Michael Paster [5], Christoph Hauer [5], Melina Frießenbichler [1], Anton Neureiter [6], Gernot Weyss [6], Peter Flödl [5], Gernot Seier [2] and Wolfgang Sulzer [2]

[1] Department of Earth Observation, Zentralanstalt für Meteorologie und Geodynamik, 1190 Vienna, Austria; matthias.schloegl@zamg.ac.at (M.S.); barbara.widhalm@zamg.ac.at (B.W.); melina.friessenbichler@zamg.ac.at (M.F.)

[2] Institute of Geography and Regional Sciences, University of Graz, 8010 Graz, Austria; christian.bauer@uni-graz.at (C.B.); gernot.seier@uni-graz.at (G.S.); wolfgang.sulzer@uni-graz.at (W.S.)

[3] Institute of Mountain Risk Engineering, University of Natural Resources and Life Sciences, 1190 Vienna, Austria

[4] Research Group Remote Sensing and Geoinformation, JOANNEUM RESEARCH, 8010 Graz, Austria; karlheinz.gutjahr@joanneum.at

[5] Christian Doppler Laboratory for Sediment Research and Management, Institute of Hydraulic Engineering and River Research, University of Natural Resources and Life Sciences, 1190 Vienna, Austria; michael.paster@boku.ac.at (M.P.); christoph.hauer@boku.ac.at (C.H.); peter.floedl@boku.ac.at (P.F.);

[6] Department of Climate Research, Zentralanstalt für Meteorologie und Geodynamik, 1190 Vienna, Austria; anton.neureiter@zamg.ac.at (A.N.); gernot.weyss@zamg.ac.at (G.W.)

* Correspondence: michael.avian@zamg.ac.at; Tel.: +43-1-36026-2513

Received: 28 February 2020; Accepted: 13 April 2020; Published: 15 April 2020

Abstract: Earth observation offers a variety of techniques for monitoring and characterizing geomorphic processes in high mountain environments. Terrestrial laserscanning and unmanned aerial vehicles provide very high resolution data with high accuracy. Automatic cameras have become a valuable source of information—mostly in a qualitative manner—in recent years. The availability of satellite data with very high revisiting time has gained momentum through the European Space Agency's Sentinel missions, offering new application potential for Earth observation. This paper reviews the status of recent techniques such as terrestrial laserscanning, remote sensed imagery, and synthetic aperture radar in monitoring high mountain environments with a particular focus on the impact of new platforms such as Sentinel-1 and -2 as well as unmanned aerial vehicles. The study area comprises the high mountain glacial environment at the Pasterze Glacier, Austria. The area is characterized by a highly dynamic geomorphological evolution and by being subject to intensive scientific research as well as long-term monitoring. We primarily evaluate landform classification and process characterization for: (i) the proglacial lake; (ii) icebergs; (iii) the glacier river; (iv) valley-bottom processes; (v) slope processes; and (vi) rock wall processes. We focus on assessing the potential of every single method both in spatial and temporal resolution in characterizing different geomorphic processes. Examples of the individual techniques are evaluated qualitatively and quantitatively in the context of: (i) morphometric analysis; (ii) applicability in high alpine regions; and (iii) comparability of the methods among themselves. The final frame of this article includes considerations on scale dependent process detectability and characterization potentials of these Earth observation methods, along with strengths and limitations in applying these methods in high alpine regions.

Keywords: laserscanning; UAV—structure from Motion; multi-spectral satellite data; synthetic Aperture Radar; glacier lake evolution; glacier river; slope processes; rock fall; cryosphere

1. Introduction

Glaciers and their changes are well recognized as crucial indicators for climate change [1–4]. Glacier retreat has many, potentially severe, impacts on human life, as exemplified by its influence on the availability of freshwater [5,6] or its role in an increase of hazardous events [7]. Glacier fluctuations cause massive impacts on glacio-hydrological or geomorphological process systems across various scales. Changes within a cryospheric environment in the form of glacier retreat result, e.g., in local hazard events [8,9], changes in regional water cycle systems [10–12], and sea level rise on a global scale [13,14]. The impacts of glacier retreat in a socio-economic context are far-reaching [15,16], affecting different areas from energy supply [15] to tourism [17,18]. As a consequence, glacier retreat induces hazards due to changing conditions within and different resilience of process systems [8]. Concluding insights from glacier monitoring are able to raise people's awareness to the importance of glaciers for the society [19].

High mountain environments are undergoing major changes due to the impact of the ongoing climate change [2,20]. A large variety of processes—often showing accelerating magnitudes and rates in the last two decades [21]—have been reshaping high mountain environments in recent years [22,23]. Especially areas extensively covered by glaciers—such as the European Alps—show a fast transition from glacially dominated to pro- and paraglacial landscapes since the 1970s [24,25]. Information about, e.g., changes in glacier length, area, and volume are high-confidence indicators of climate change [26,27]. In the European Alps, monitoring the cryosphere, and in particular glaciers, has been performed since the end of the 19th century [28,29]. Since then, there has been an increase in both the number of glaciers observed and the number of measurements per glacier. As a result, countries within the Alps with a long tradition of monitoring systems (such as Austria, France, and Switzerland) have valuable information of glacier change [19,30].

Glaciers are a crucial part of a geomorphological process system and should not be analyzed separately. Different processes occurring with different magnitudes on different spatial and temporal scales are a challenge for establishing a comprehensive monitoring system. Therefore, careful evaluation of every single remote sensing method is essential.

This paper aims to present a quantitative and qualitative evaluation of remote sensing methods for monitoring geomorphic processes in a cryospheric environment. Geomorphic processes under investigation are characterized by different occurrence frequency and different superordinate process types (glacial vs. proglacial processes and valley bottom to rock wall processes). Consequently, particular focus lies on the evaluation of the observation of the dynamics of: (i) the proglacial lake; (ii) icebergs; (iii) the glacier river; (iv) valley-bottom processes; (v) slope processes; and (vi) rock wall processes.

Therefore, this work seeks to provide answers to the following aspects:

- What are the specifications, uncertainties, and constraints of Earth observation techniques in monitoring geomorphic processes in an cryospheric environment?
- How do the Earth observation techniques presented prove to be appropriate for monitoring glacial, and paraglacial processes/landforms of different magnitudes and scales?

This paper provides a review of a variety of Earth observation techniques and demonstrates their applicability in monitoring different geomorphic processes. This extensive overview results in the following breakdown: Section 2 gives a broad overview of the use of Earth observation techniques in the characterization of geomorphic processes in high alpine regions, with special focus on the Alpine region. Section 3 provides an outline of the geomorphic processes and landforms investigated in the study area. Section 4 describes Earth observation methods specifically used for monitoring these processes at the Pasterze Glacier area (in terms of technical specifications, monitoring configurations, and quality assessment). Section 5 presents the quantitative results of Earth observation techniques in monitoring certain geomorphic processes and landforms in the study area. Section 6 discusses the applicability of every single method for characterizing the analysed geomorphic processes. Section 7 provides a classification of Earth observation methods comparing data acquisition specification and processes characteristics. Finally, Section 8 evaluates the applied Earth observation techniques with regard to their suitability for monitoring the processes and landforms, as well as their practical application in the study area.

2. Earth Observation Techniques for Characterizing Processes in Cryospheric Environments

Earth observation techniques provide valuable databases by area wide data acquisition in order to characterize geomorphic processes in a cryospheric environment. Typically, these include high resolution optical data (aerial images and satellite-borne multi-spectral images) as well as data which are subsequently processed to high resolution digital terrain models (DTMs). In Austria, aerial images have been used to, e.g., delineate the extent of glaciers since the 1950s leading to the first so-called Glacier cadaster in 1969 [29] in high mountain applications. Beginning around 2000, airborne (ALS) [31] as well as terrestrial (TLS) [32] laserscanning became a crucial technique to acquire very precise and high resolution surface data. This section provides a short technical description of every Earth observation technique used in this work, and addresses main applications in monitoring geomorphic processes.

2.1. Terrestrial Laserscanning

TLS is a time-of-flight system which allows measurements with a range accuracies of a few centimeters [32]. Laserscanning combines the specifications and hence advantages of laser (directional nature of the rays) and radar (location) [33]. Principles of TLS are summarized in [34,35]. The resulting point cloud is registered with the aid of (reflective) objects such as spheres or cylinders representing known coordinates. Carefully registered point clouds enable precise comparison of multi-temporal measurements. TLS uses the near-infrared section of the spectrum at different wavelength such as approximately 1000 nm for snow and ice applications (e.g., Riegl LPM-i800HA [36] and Riegl VZ6000 [37]), in which the wavelength coincides with the measurement range [38] and approximately 1500 nm for other applications (e.g., Riegl LMS Z620 [24]).

High mountain applications have been a frequent scope in the usage of TLS such as monitoring rock faces [39–42], sediment budgets [43], the evolution of paraglacial areas [44] and subsequent slope instabilities [45], the characterization of rock glaciers [46,47], or snow applications [36]. First long-term measurements on Austrian glaciers were carried out by Bauer et al. [48] at Gössnitzkees (Schober Mountains, Austria), Avian et al. [49] on Pasterze Glacier, and Stötter et al. [50] in Tyrol (e.g., Hintereisferner). Some of these works were the basis for upcoming monitoring networks such as the permanent TLS-observation station 'Im Hinteren Eis' at the Hintereisferner Glacier [51] (using Riegl VZ6000). To assess the glacier mass balance, model input or validation data were discussed by Fischer et al. [29] for small glaciers in Switzerland, Prantl et al. [37] for snowline variations on glaciers, and Gabbud et al. [52] for surface melt rates. Monitoring glacier transition zones (i.e., para- and proglacial areas) caused by glacier retreat were the scope at, e.g., Gepatschferner [53] and Pasterze Glacier [24] in Austria; Aletsch Glacier in Switzerland [45]; the Miage Glacier [54] and Macugnaga Glacier in Italy [55]; or the Brenva Glacier in France [56].

In the last decade, massive glacier retreat, especially of large valley glaciers, has resulted in an increased formation of glacial lakes [57]. Due to natural hazards related to glacial lakes such as glacier lake outburst floods (GLOFs), monitoring these lakes has become a crucial task in cryosphere research [58,59]. Since GLOFs are not among the most frequent hazards in the Alps, monitoring the evolution of glacier lakes using TLS was limited to work on Brenva Glacier [54]. For TLS, methodological consideration discussing monitoring concepts, accuracies and uncertainties were made by Ingensand et al. [60], and in a geomorphological context by Fey and Wichmann [61].

2.2. Radar Satellite Application: Backscatter

SAR instruments generally provide backscatter measurements which are influenced by the terrain structure (surface roughness). High backscatter values are caused by surfaces with higher roughness as incoming radar pulses are scattered in all directions (diffuse reflection) [62]. Contrarily, calm water surfaces show a very smooth surface, reflecting the radar pulse away from the sensor (specular reflection). Therefore, water surfaces typically show lower backscatter values than adjacent surface types and can be measured applying a simple threshold approach. A widely used approach for threshold selection was provided by Otsu [63]. This method, however, relies on a bimodal histogram with a clear minimum, dividing areas of water and land surface. Manual classification was shown applicable to delineate glacial lakes [64] as well as extracting buffered polygons of the lake area in order to obtain a bimodal histogram [65], or recently by level-set segmentation [66].

2.3. Radar Satellite Application: Differential Interferometric Synthetic Aperture Radar (DInSAR)

DInSAR offers a range of approaches to detect small surface deformations with sub-centimeter accuracy. Repeat acquisition of the same constellation can be used to identify small changes in range direction through measured differences in phase. Consequently, it is possible to measure ground deformations in the magnitude of a fraction of the used wavelength [67]. To determine this particular component of the phase (caused by surface displacement), other interfering aspects have to be separated, such as error components due to atmosphere, orbital errors, or phase noise [68]. This can be achieved by using pixels of small phase-noise, which is realized by two reflector types: (i) Permanent or Persistent Scatterers (PS), which consist of a dominating scatterer persisting over time; and (ii) Distributed Scatterers (DS), which feature constant signals caused by different small scatterers [68]. In applying Persistent Scatterer Interferometry (PSI), PS within a time series are used by calculating interferograms in relation to one single master scene [69,70]. Small Baseline Subset (SBAS) is the second multi-temporal InSAR method, which particularly incorporates DS. In contrast to PSI, the retrieval of DS in SBAS becomes increasingly unlikely for interferograms with larger temporal baselines. Therefore, SBAS is not based on one single master scene. Rather, multiple master-slave combinations with small baselines are used in the calculation of the interferograms [71].

In high mountain environments, DInSAR using ERS data was applied to detect slope movements in the Swiss Alps, allowing the implementation of an inventory of mass movement types [72]. However, limitations were identified for steep rock walls and northern and southern facing slopes due to partial illumination by the sensor [72,73]. Slope movements showing varying deformation patterns were further investigated using TerraSAR-X data applying PSI and SBAS [74]. To assess rock glacier movement rates, DInSAR was successfully applied to determine displacement rates [75–77]. Landslide, rock glacier movement rates, and, e.g., surface displacement mapping using DInSAR was determined in order to assess hazards related to GLOFs [78].

DInSAR was also used to measure glacier movement at several study areas [79,80]. However due to snowfall, snowdrift, and melting, which leads to temporal decorrelation, mainly data with short repeat-pass were used [81]. To estimate surface velocities of glaciers, DInSAR using Sentinel-1 data was applied, exhibiting low deformation rates and therefore low decorrelation [82].

2.4. Multi-Spectral Satellite Data

The widespread use of multi-spectral satellite data to monitor and characterize high mountain processes started about 30 years ago. Landsat 5 [83], Landsat ETM+ [84], or ASTER [85,86] were the basis of space-borne glacier mapping mainly using the visible and near-infrared section of the spectrum. Creation of glacier inventories is mainly based on multi-spectral satellite data using automatic procedures [87]. For the Alps, the use of multi-spectral data for monitoring high mountain areas was presented, e.g., for the example of Switzerland in a synoptic view by Huggel et al. [88] as well as the perspectives for a worldwide assessment e.g., using satellite data by Gärtner-Roer et al. [19]. A comprehensive review of global glacier characterization using space-borne sensors was presented by Kääb et al. [89]. For characterizing glacier lake dynamics, Landsat 8 images were used by Li and Sheng [90]. To create a nation-wide glacier lake inventory, Sentinel-2 data were the basis for mapping more than 400 glacier lakes in Norway by Nagy and Andreassen [91] or a classification of glacier lakes by Verma and Ghosh [92].

Sentinel-2 is a two-satellite mission: Sentinel-2A was launched on 23 June 2015 and Sentinel-2B on 07 March 2017 [93]. This constellation yields a revisiting time of five days at the equator showing a substantial improvement in temporal resolution to other missions such as Landsat 8 [94]. Sentinel-2 satellites are equipped with a multi-spectral instrument providing 13 bands from the visible light (VIS), near (NIR), and short wave infrared (SWIR) with up to 10 m spatial resolution [95]. Compared to Landsat 8, the spatial resolution of Sentinel-2 (10 m VIS) is nine times better than Landsat-8 (30 m VIS).

2.5. Structure from Motion—Unmanned Aerial Vehicles

UAV-based images are processed using structure-from-motion (SfM) photogrammetry [96]. SfM photogrammetry uses images captured from different perspectives, automatically assembled to point clouds using image matching techniques. This matching uses the identification of interest points and is based on the Scale-Invariant Feature Transform algorithm [97]. In combination with multi-view stereo (MVS) techniques, SfM photogrammetry allows simultaneously reconstructing dense 3D models, camera positions, and orientations [98,99]. Currently, SfM-MVS photogrammetry is increasingly used for generating ortho-images and DTMs for different applications (e.g., [100–103]. All applications differ in both scaling and flight altitude. Thus, to achieve the necessary image resolution for every particular application, different heights over ground are necessary; e.g., for sediment analysis, low height over ground is required (limit of approximately $H = 12$ m [104]) compared to large scale applications (e.g., $H = 100$ m for mapping river sections [105]).

UAV applications at high mountain environments are challenging due to the following reasons: (1) limited accessibility to unstable surfaces due to hazardous conditions causes constraints in the acquisition of ground data (for registration and accuracy assessment); (2) changing meteorological conditions (temperature, wind, fog); and (3) surface conditions: e.g., proglacial lake water surface covers large parts of the investigation area, which cannot be used in photogrammetry. Analysis using UAV-based SfM-MVS often lack of information about survey design and image measurement and processing [106–108]. In addition, the SfM-MVS approach is somewhat limited, and results such as DTMs and ortho-images can be erroneous [109].

Recently, unmanned aerial vehicles (UAV) have become a well-established platform in acquiring high resolution data, covering various fields of application, e.g., surveying [110–112]. Currently, UAV-based glacio-morphological analyses are increasingly conducted, e.g., the monitoring of glacier surface changes [113–118]. The use of UAVs is also increasing for monitoring fluvial systems [119]. UAV are the basis for data collection for requested bathymetry of river sections (for numerical modeling) and the observation of morphodynamic processes, such as quantifying erosion and accumulation or stream bank failure [105,120]. On a smaller scale, UAV-based SfM analysis are used to determine: (i) the sediment composition and distribution; and (ii) hydraulic parameters, such as grain roughness, etc. [104,121]. For the description of channel characteristics and follow-up hydraulic processes, however, both are necessary: the geometry (DTM) as a basis for hydraulic modeling and the characteristic grain sizes, grain roughness, etc. as its input parameters. For this reason, it is usually required in river engineering to conduct additional flights of characteristic sections at different flight altitudes.

2.6. Automatic Cameras

Due to essential developments in cameras for terrestrial photogrammetry, automatic time-laps cameras have been frequently used in monitoring geomorphological processes in the last decade.

High spatiotemporal resolution, cost efficiency, and independent operating in remote places [122–124] are valuable improvements. In contrast to optical remote sensing techniques, cloud coverage is a by far smaller problem. In the Austrian Alps, a yearly average of 60% of pixels are hidden by clouds [125]. The application of automatic time-lapse cameras as a monitoring tool in mountain regions is versatile: selected examples are glacier terminus position [126], snow cover monitoring [127–131], monitoring mass movements [132], supraglacial lakes drainage events [133], snow melt, and vegetation phenology [134].

3. Study Area: Relevant Pre-Work and Geomorphological Setting

Starting in 1893, Pasterze Glacier was one of the first glaciers constantly monitored in annual measurements, leading to the longest record of measurements of a single glacier [135]. Next to linear and punctual information, the complexity of a glacier system revealed the need for measurements showing more spatial significance such as spatially well distributed measurements. Multi-spectral satellite data were used to quantify changes in glacier extent changes using images from Landsat MSS (1976), Landsat TM (5 during 1984–1992), Landsat ETM+ (2000), and Ikonos (2000) by Hall et al. [136]. Methodological considerations of this work were presented by Hall et al. [137]. DInSAR was only used to determine flow velocity patterns of the Pasterze Glacier by Kaufmann et al. [138] using five ERS–1/2 image pairs between 1995 and 2001. Aerial images were widely used to characterize several processes and impacts: both using multi-temporal ortho-images, Kellerer-Pirklbauer et al. [139] analyzed the influence of supra-glacial debris cover at the Pasterze Glacier tongue and Kaufmann et al. [102] gave a comprehensive analysis of Pasterze Glacier retreat between 2003 and 2009. At Mittlerer Burgstall mountain (MBUG), a first quantification of the large rockfall event in 2007 and possible relations to climate change was given by Kellerer-Pirklbauer et al. [140]. Kaufmann [141] provided a detailed determination using high resolution aerial images. At Pasterze Glacier terminus area, a first assessment of sub-surface ice and glacier lake evolution using TLS and UAV was given by Kellerer-Pirklbauer et al. [142] and Kellerer-Pirklbauer et al. [143].

The study area comprises the catchment area of Pasterze glacier, which includes the maximum extent of Pasterze Glacier of 1851 (Little Ice Age (LIA) glaciation (Figure 1) [102]. Since the LIA-maximum, Pasterze Glacier has undergone a constant retreat, which accelerates since the 1990s [135]. Pasterze Glacier lost 37% of its area (a decrease from 26.5 to 16.6 km^2) and 63% of its volume (a decrease from 3.10 to 1.16 km^3) between 1852 and 2012 [135].

Starting hypsographically at the accumulation area, Pasterze Glacier is characterized by the following sub-areas (Figure 1A(1–7)):

- Accumulation area *Pasterzenboden* (Figure 1A(1)): In 2019, the accumulation area *Pasterzenboden* is a large predominantly flat glacier basin between 3460 m (*Johannisberg* summit) and 2700 m (*Hufeisenbruch*-area) [102].
- Connection between *Pasterzenboden* and Glacier tongue (Figure 1A(2)): The so-called *Hufeisenbruch* summarizes a terrain step from approximately 2700 to 2300 m a.s.l. connecting the latter areas. In 2019, only two small connections are left such as one approximately 50-m-wide serac-area from the so-called *Teufelskampkees* and one larger connection SW from the *MBUG* area.
- Nourishment from tributary glaciers (Figure 1A(3)): The detachment of several tributary glaciers also contribute to the main glacier retreat: In 2019, no tributary glaciers are connected with the main glacier anymore.
- Glacier tongue (Figure 1A(4),C): The tongue of Pasterze Glacier changed its surface characteristics significantly in the last decades. Whereas the glacier tongue showed a predominant clean ice surface, the glacier surface now is increasingly dominated by supraglacial till [139].
- Adjacent slopes at the Pasterze Glacier tongue area: Slope areas at Pasterze Glacier tongue show different characteristics due to predominant geological settings:
  - *Grossglockner* (NE-facing) slope (Figure 1A(5)) consists of a mixture of prasenite and minor mica-schist areas. There is decreasing glaciation with varying extent in every cirque relief (up to 3798 m a.s.l.).

- *Fuscherkarkopf* (FKK) (SW-facing) slope (Figure 1A(6)) is mainly composed of mica-schist and therefore large debris covered. There is minor glaciation at the Burgstall area, relief up to 3331 m a.s.l. Upper slopes are very prone to rock falls, lower slope sections are dominated by consolidation after glacier retreat and linear processes (fluvial erosion). Footslope areas are mainly characterized by dead ice degradation and fluvial erosion.

- Glacier lakes (Figure 1A(7)): Historically, the Pasterze Glacier system includes only the so-called *Sandersee* Lake since 1958 at approximately 2060 m a.s.l., with a maximum extent in 1979 (121,500 km^2). From 2004 onwards, a braided river system evolved into a second lake (*Pasterzensee*), which was established in 2010. A constant decrease of dead-ice islands sections and therefore subsequently prevailing water surface developed since then. Lake Pasterzensee is a main object of interest in this work.

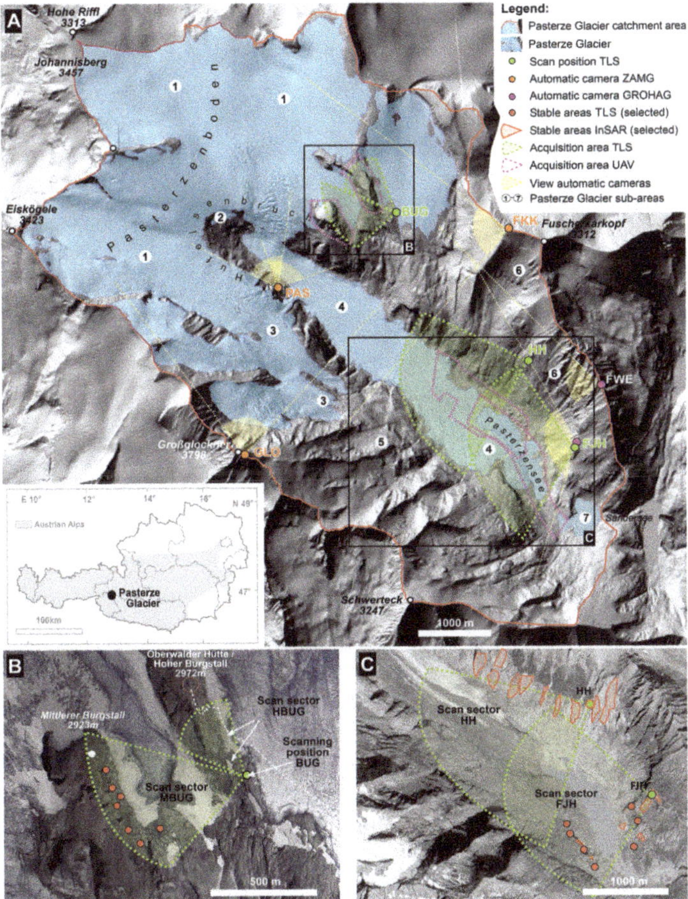

Figure 1. Study area Pasterze Glacier. (**A**) Catchment area Pasterze Glacier area with respective codes for Pasterze Glacier sub-areas (1–7); location of automatic cameras and TLS scanning positions; acquisition areas for TLS and UAV; and view sectors of automatic cameras. (**B**) TLS configuration Burgstall rock fall area (BUG). Red dots indicate stable areas for TLS quality assessment. (**C**) Situation at the Pasterze Glacier terminus and proglacial area. TLS configuration Franz- Josefs-Höhe (FJH) and Hofmanns Hütte (HH). Red areas indicate stable areas for DInSAR quality assessment, red dots stable areas for TLS quality assessment.

4. Application of Earth Observation Techniques at Pasterze Glacier: Data Basis, Spatial and Temporal Variability, Quality Assessment

Based on the technical description presented in Section 2, this section comprises the particular applications of the single monitoring configurations at Pasterze Glacier area indicating: (i) technical specifications of used instruments, and satellite systems; and (ii) monitoring configuration. Furthermore, we present the characteristics of the single datasets and applied approaches (including quality assessment). To ensure interpretability of measurements, monitoring of glacial and proglacial landscape should always be carried out as close as possible to the end of the hydrological year. By definition, the hydrological year ends with the ablation period before the onset of winter (at alpine glaciers mostly in September). Therefore, for the Pasterze Glacier area, at least one measurement is carried out in September designated as annual measurement campaigns.

4.1. Terrestrial Laserscanning

Annual TLS measurements have been carried out since 2001 at the scanning positions, Franz-Josef-Höhe and Hofmanns-Hütte (FJH and HH, Figure 1C). Two devices have been used since 2001: from 2001 to 2009 the Riegl LPM-2k system and from 2009 on the Riegl LMS-Z620 system (Table 1). Technical specifications can be found in [24,49]. A comprehensive overview of previous work including scanning geometries at Pasterze Glacier tongue area is also given in [24].

Table 1. Acquisition dates of TLS data in the respective years of the observation period 2014–2019 [MM-DD].

Year	Date 1	Date 2	Data 3
2014	09-09		
2015	09-12		
2016	08-27		
2017	06-19	08-05	09-22
2018	08-04		
2019	08-03		

In the upper part of the Pasterze Glacier area, a massive rockfall at MBUG occurred in 2007 [140]. Therefore, the TLS monitoring network 'Burgstall rock fall area' was established in 2010, comprising the eastern part of the rock fall area of MBUG and the S-face of the Hohe Burgstall Mountain (HBUG, Figure 1B). The scanning position Burgstall (BUG) is located at the eastern margin of the Wasserfallwinkelkees glacier at a bedrock ridge in 2800.34 m a.s.l. TLS acquisition at Burgstall area was conducted in a distance of 500–750 m at MBUG and a distance of 100–300 m at HBUG (Figure 1B) using six stable, permanent reference points [144]. Thus, the respective ground sampling distances (GSD) at MBUG was 0.25 m (at 650 m distance) and 0.10 m (at 100 m distance) at HBUG. At MBUG, for special analysis such as void size and density for geological interpretation, detail scans with a GSD of 0.15 m were acquired to cover very active rock fall areas.

TLS data processing (e.g., registration) was performed in Riegl RiScan; the rectified point cloud was afterwards exported to Golden Software Surfer 15 to calculate respective DTMs to obtain the area-wide vertical elevation changes and volumetric information in reasonable calculation time. DTMs with different spatial resolutions were calculated: 1 m for the area-wide assessment of vertical surface changes (e.g., for mass balances) and 0.5 m for geomorphological interpretation and process characterization of specific areas of interest [24]. The quality of a TLS measurements is influenced by four factors: instrument calibration, atmospheric conditions, object properties, and scan geometry [145]. As we compare measurements using the same instrument and meteorological conditions are integrated in the processing chain, we focus on the influence of distance measurements and the incidence angle on the accuracy of TLS measurements. Inaccuracies of measurements are assessed by calculating euclidian distances between two point clouds (using CloudCompare) with respect to measurement distance and incidence angle (Figure 2).

Uncertainties of distance measurement show sufficiently small values for incidence angles larger than 50°: at distances of around 1150 m, measurements on rock walls (mean incidence angle 86°) show rather small uncertainties between 0.091 and 0.154 m. The influence of the incidence angle on distance measurements was shown at several stable areas such as 0.042 and 0.049 m (mean incidence angle 50°) and 0.188 and 0.203 m (mean incidence angle 50°) at two adjacent stable areas in a distance of 377 and 390 m. For the single stable areas, uncertainty values were stable over the observation period (Figure 2).

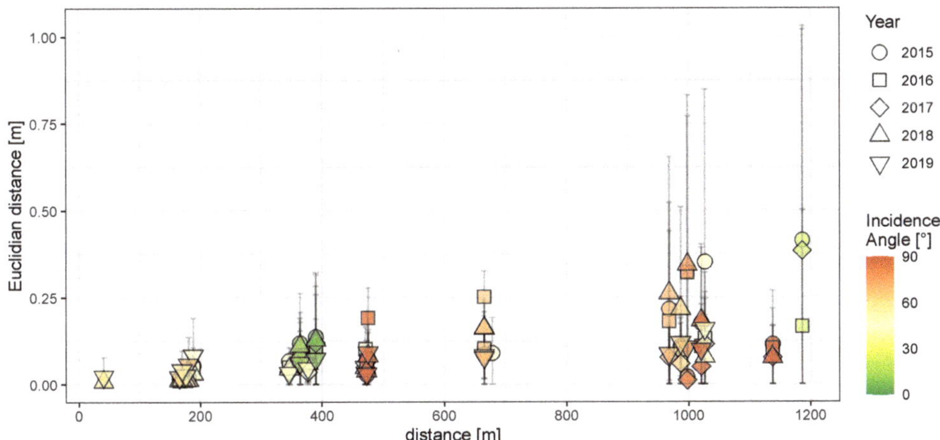

Figure 2. Quality of TLS measurements: Mean euclidian distance between point pairs at stable areas at different measurement distances of the scanning positions FJH and BUG. Color of point pairs indicates incidence angle between laser beam and surface.

4.2. Radar Satellite Application—Backscatter

To quantify the extent of Lake Pasterzensee using backscatter information, Sentinel-1 Single Look Complex (SLC) data were processed using Sentinel Application Platform (SNAP). As precise geolocation is a crucial precondition of comparability, precise orbit files were applied. Pixel values were radiometrically calibrated to derive physical units. Speckle filtering and multi-looking was applied to reduce noise with a subsequent SAR-simulation terrain correction for geometric adjustment. To further reduce noise, the mean of VV and VH data was used to delineate lake extents. Thresholds were set for each scene individually between −14 and −17 dB, and lake extents were calculated per year for every scene taken between June and October. Quality assessment mainly covers co-registration of images on one master image to avoid any existing shift to assure spatial comparability.

Being a mountainous study area, Pasterze Glacier area is widely affected by topographical limitations such as shadowing, foreshortening and layover. which limit a threshold selection based on the image histogram.

4.3. Radar Satellite Application—DInSAR

The DInSAR analysis carried out at Pasterze Glacier is a pilot study for the applicability of Sentinel-1 DInSAR for surface deformation assessment in the entire Grossglockner area. The analysis is based on 149 SLC images of Sentinel-1A and 1B taken in Interferometric Wide swath (IW) between late spring 2017 (2017-06-04) and late fall 2019 (2019-10-28). It covers swaths of 250 km with a spatial resolution of 5 m by 20 m and incidence angles from 29° to 46° with a repeat cycle of six days. Three different DInSAR approaches were investigated and compared. While the underlying data basis is the same in terms of raw input data and temporal resolution, the three methods differ with respect to (aggregated) spatial resolution. Due to decorrelation effects in the winter seasons (snow cover), only summer scenes were used. The area is covered by ascending orbits 44 and 117, and descending orbit 95.

The first approach is based on a SNAP-StaMPS (Stanford Method for Persistent Scatterers)-workflow to perform persistent scatterer interferometry (PSI) using the Stanford Method for Persistent Scatterers [146]. Sentinel-1 pre-processing for PSI was performed utilizing the software SNAP. At first, the optimal master scene was selected for ascending and descending orbits, respectively. The sub-swath covering the region of interest was identified and images were split accordingly. After application of an orbit correction, the images were co-registered and interferograms were computed. Potential PS-points were pre-selected by using the amplitude dispersion, and phase stability was estimated using phase analysis. PS pixels were subsequently filtered and dropped if they were too noisy or if they were influenced by neighboring elements. The wrapped phase was then corrected for spatially-uncorrelated look angle errors followed by phase unwrapping. Eventually, the spatially-correlated look angle error was calculated. Compared to regular grid of aggregated SBAS pixels, the PSI point distribution is more irregular with a point distance of roughly 3 m $\times$ 14 m.

The second approach tested within this study is the P-SBAS (Parallel SBAS) service of European Space Agency (ESA) Geohazards Thematic Exploitation Platform (GEP). This online service provides an unsupervised implementation of the P-SBAS algorithm, which is a parallel computing implementation of the SBAS approach [147,148]. The spatial resolution is specified with 90 m, yet actual point intervals were measured with approximately 60 $\times$ 90 m.

The third approach for assessing surface deformations derived via DInSAR is also based on small baselines and includes the developments in [71,149–151]. For the pre-processing of the Sentinel-1 stack, the joint azimuth shift estimation as in [152] was applied, which is slightly different from the method used in [147]. The analysis was conducted using the Remote Sensing software Graz (RSG [151]) software suite. Results are aggregated to a spatial resolution of 80 $\times$ 80 m.

To assess quality and accuracy of the processed DInSAR results, 24 stable areas were manually identified by delineating areas where no deformations are expected (Figure 1C). Selection of these areas was based on a geomorphological assessment. These areas comprise 19 stable terrain areas (bedrock) and 5 polygons related to artificial structures (buildings and parking lots around TLS scanning position FJH). Areas were chosen considering a trade-off between the size of the polygons and the uniformity with respect to aspect and slope angle. As the area of each single polygon is still comparably small—particularly with respect to the SBAS pixel spacing—deformation results were aggregated the main two categories 'bedrock' and 'infrastructure' in order to provide more robust indication on mean annual (pseudo-)deformation rates. Resulting deformation rates are computed as a linear regression of deformation rates on the date. To quantify variability, we use the standard deviation (SD) of the residuals.

(Pseudo-)deformation rates of the ascending orbit (44) show a higher variability with respect to the (linear) trend across all DInSAR methods and for both the bedrock and the infrastructure cluster (Table 2). This might be explained by the slope exposition, as almost all areas are located on slopes facing southwest. In fact, foreshortening effects are more prominent in ascending orbits than in descending ones. Compared to the bedrock cluster, the standard deviation of residuals is slightly lower for the infrastructure cluster, which is consistent with the expectation that particularly PSI should perform better on PS than on DS [153].

Notably, PSI shows a systematic trend on both orbit directions with a comparably high variability. However, results are consistent across both orbits and both clusters. Arguably, SBAS-based methods seem to be better suited in (high) alpine environments because of their capability to handle DS, too. Both SBAS methods show rather small trends with residual standard deviation of only a few millimeters. Nevertheless one has to keep in mind, that all time series are not optimum as we had to mask out all 'winter-acquisitions' leading to long time periods without data. This has severe negative impacts on, e.g., the removal of atmospheric effects.

Table 2. Annual DInSAR (pseudo-)deformation values in line-of sight (LOS) for stable area clusters. *mean* values refer to the slope of the linear trend within one year (365.25 days), obtained from a linear regression of deformation rates on the date. *SD* denotes the corresponding standard deviation of the residuals. The *infrastructure* cluster comprises polygons of five horizontal artificial areas (building flat roofs and parking areas). The *bedrock* cluster comprises polygons of 17 bedrock areas (Figure 1C). Abbreviations: StaMPS, Stanford Method for Persistent Scatterers; P-SBAS, Parallel Small Baseline Subset; RSG, Remote Sensing Software Graz; GEP, Geohazard Thematic Exploitation Platform.

Stable Area Cluster	DInSAR Method	Orbit 44		Orbit 95	
		Mean [mm]	SD [mm]	Mean [mm]	SD [mm]
Infrastructure	PSI (StaMPS)	−1.89	7.38	1.97	5.01
	P-SBAS (RSG)	0.04	3.20	−0.17	1.35
	P-SBAS (GEP)	–	–	1.76	1.41
Bedrock	PSI (StaMPS)	−1.01	7.92	1.75	6.25
	P-SBAS (RSG)	1.50	5.47	0.44	2.84
	P-SBAS (GEP)	0.39	4.29	3.76	2.04

4.4. Multi-Spectral Satellite Data

Sentinel-2 data were used to enhance inter-annual data availability due to potential high revisiting time. ESA's Sentinel-2 mission consists of two polar-orbiting satellites flying on the same orbit, phased at 180° to each other. Its wide-swath, high-resolution, multi-spectral imaging satellites comprise 13 spectral bands, with four bands at 10 m, six bands at 20 m, and three bands at 60 m spatial resolution covering the Visible and Near Infra-Red (VNIR) and Short Wave Infra-Red (SWIR). The mission has a high revisit time of five days at the equator for both satellites with an orbital swath width of 290 km.

To map glacier lake extension accurately, respective satellite images should be chosen showing the glacier lake's maximum extent. This is ensured when using images without snow and ice cover to avoid misinterpretations. Furthermore, cloud coverage and shadowed areas due to the relief have to be considered. In detail, 27 Sentinel-2 Top of atmosphere Level 1C images (Table 3) were available in the observation period 2015–2019. Due to a misregistration of more than one pixel (>10 m) of multi-temporal Sentinel-2 acquisitions [154], all scenes had to be co-registered to the acquisition of 2016-08-07, which matched satisfactorily with the corresponding TLS dataset. For co-registration the software SNAP was used, which computes the offset between master and slave images by maximizing the cross-correlation within sub-images.

Table 3. Acquisition dates of Sentinel-2 data in the respective years of the observation period 2015–2019 [MM-DD].

Year	Date 1	Date 2	Data 3	Date 4	Date 5	Date 6	Date 7	Date 8
2015	07-04	08-03	09-12					
2016	08-07	08-27	09-26	10-16				
2017	06-13	06-23	08-22	10-11	10-16			
2018	07-18	08-27	08-17	09-21	09-26	10-11	10-16	10-26
2019	06-28	07-23	09-11	09-16	09-21	10-11	10-26	

Subsequently, Sentinel-2 data were classified applying the Normalized Difference Water Index (NDWI), using the green and near-infrared bands [88,155,156]. For this analysis, the respective Bands 3 and 8 of Sentinel-2 images were used:

$$NDWI = \frac{B_{green} - B_{NIR}}{B_{green} + B_{NIR}} = \frac{B_3 - B_8}{B_3 + B_8}$$

For water bodies, the reflectivity of the green light is maximized, while the near infrared reflectivity is typically low and therefore minimized. Water features exhibit positive values, while while soil and vegetation show lower values due to higher reflectivity of near infrared than green light [157]. Thresholding of the resulting NDWI-maps has widely been used [155,158] but the individual values are dependent on the specific application. In delineating glacier lakes, the threshold value must be high enough to distinguish between glacier ice and water but also low enough for the omission of the water pixels [91].

4.5. Structure from Motion

Structure from Motion (SfM) using UAV for monitoring glacier and proglacial area was carried out at the tongue of the Pasterze Glacier, the proglacial area of Pasterze Glacier, the Burgstall rockfall area, and the Oberer Pasterzenboden (Figure 1). Data from Burgstall represent a first flight campaign in order to monitor the entire Burgstall mountains for subsequent geological analysis. Data acquired at the Oberer Pasterzenboden are the basis for glacier mass balance measurements.

The glacial/proglacial transition zone was covered by two UAV surveys in September 2016 and June 2019 (Table 4, VB/GL). Using a fixed-wing UAV (a Quest UAV) in the 2016-11 survey, the consumer grade camera Sony α 6000 was used (E 16 mm F2.8 lens, sensor size is 23.5 × 15.6 mm, resolution 6000 × 4000 px). During the campaign in June 2019, topographic conditions (Figure 1) and platform specifications furthermore necessitated that the southeastern part of the glacier lake was measured using a multi-rotary UAV (DJI Phantom 4. integrated camera (resolution 4000 × 3000 px)). The SfM-MVS photogrammetry processing was based on ground control points (GCPs), which were used for indirect georeferencing of the UAV imagery. GCPs were measured by GNSS solution in real-time kinematic (RTK) mode (EPOSA). The SfM-MVS processing was conducted using the software Agisoft Photoscan (1.2.5 build 2735; 1.3.4 build 5067) with a default key point limit of 40,000 and a maximum of 4000 tie points (sparse point cloud generation). Thereafter, a bundle adjustment and camera self-calibration were conducted followed by a dense point cloud processing.

To independently assess the vertical accuracy of the photogrammetrically processed DTMs, the same geodetic method was used to measure so-called independent check points (ICPs) for validation of geolocation. The mean vertical differences between the DTMs and ICPs was 0.08–0.13 m with a standard deviation of from ±0.12 to ±0.15 m (Table 4). The resulting ortho-images were visually interpreted and the analysis of DTMs focused on the quantification of vertical changes using DTM differencing.

SfM for glacier river characterization comprised the documentation of the channel evolution downstream in autumn 2018 with regard to estimate the future sediment (bed load) input into the reservoir Margaritze [159]. The investigation area (proglacial river, Figure 1C), between the glacier terminus and the delta area of the Lake Pasterzensee (100 m wide and 1000 m long), was mapped with an UAV (Hexacopter model KR615), equipped with compact camera (ILCE-6000; 6000 × 4000 px; Table 4). In the case of the high image resolution for sediment analysis and the depth of the incised river channel, an additional flight for the inaccessible sections was done at surrounding terrain level (around 15 m above water level).

The calculated DTM serves as a basis for the 1D hydraulic model of the proglacial part of the river Möll. The DTM was generated using the program Agisoft [160] and registered using GCPs measured with a GNSS device with RTK provider (APOS). Some of the GCPs (20–30%) serve as ICPs to quantify uncertainties of the dataset. The resulting sparse point cloud was cleaned up by depth filtering and the removing of points with high reprojection error. Finally, the dense point cloud contains around 478 million points (computation time of 14 days, 10 h). The resulting point cloud shows a point density of around 4000 points/m^2, a GSD of 1.59 cm/px and a RMSE (X,Y,Z) of 2.55, 4.38, and 2.44 cm (Table 4). This is a significantly higher accuracy than the one, e.g., presented by Vázquez-Tarrío et al. [121] (GSD = 2 cm/px) and more detailed mapping of the topography is shown.

Table 4. UAV based surveys for the focus of proglacial river (GR) and valley-bottom/glacier lake assessment (VB/GL). Information about the height over ground (HoG), ground sampling distance (GSD), number of images, the root mean square error (RMSE) in both the horizontal plane (xz) and the vertical plane (z), as well as vertical quality of the DTM are indicated.

Focus	Date	HoG [m]	GSD [m]	Number of Images	RMSE XY [m]	RMSE Z [m]	Image Plane [px]	Vertical Quality stdv [m]	Vertical Quality mean [m]
VB/GL	2016-11-03	138	0.032	343	0.06	0.07	0.9	0.12	0.08
VB/GL	2019-06-17	132	0.046	304	0.13	0.16	3.3	0.15	0.13
GR	2018-09-26	34.5	0.0159	1375	0.057	0.024	0.336	–	–

4.6. Automatic Cameras

In the Pasterze Glacier area, six automatic cameras are installed in order to monitor mainly glaciological processes with a very high temporal resolution and, as a side-effect, to validate other methods qualitatively (configuration: Table 5; location: Figure 1). The main focus of cameras operated by ZAMG (Table 5, 1–5) is the assessment of the snow cover extent. Due to complex terrain, different camera locations were selected for a maximum coverage of the glacier. Next to snow cover, individual cameras were also applied for flow velocity or determination of the glacier outline. The panoramic camera at Franz-Josefs-Höhe area (Figure 1 and Table 5, 6) was installed in 2010, which is a valuable source for scientific work and has been used for time-lapse studies of the entire glacier tongue. The five cameras operated by ZAMG are Canon EOS1200D digital single lens reflex cameras. The two cameras at the site Pasterze have a EF-S10-22mm f/3.5-4.5 USM lens. The tree other cameras are equipped with a EF-S 18-55mm III lens.

Table 5. Automatic cameras. Location of cameras are indicated in Figure 1. Cameras 1–5 are operated by ZAMG and Camera 6 is operated by Grossglockner Hochalpenstraße AG (GROHAG).

Nr.	Camera Label	Elevation [m a.s.l.]	Alignment	Recording Rate	Beginning of Recording	Main Topic
1	Burgstall 1 Hufeisen (PAS)	2661	NW	10min	2017-10-27	snow cover
2	Burgstall 2 Vertical (PAS)	2661	NW	10min	2016-06-20	snow cover + velocity
3	Freiwandeck (FWE)	2843	W	10min	2015-10-05	snow cover + glacier outline
4	Fuscherkarkopf (FKK)	3234	W	10min	2015-10-24	snow cover
5	Grossglockner (GLO)	3750	N	10min	2015-10-12	snow cover
6	Franz-Josefs-Höhe (FJH)	2370	panoramic/180° center SW	5min	2010-08-10	surface

5. Results: Monitoring of Processes and Landforms

In the following, the quantitative results of every single Earth observation techniques with respect to particular geomorphic process systems are presented. Geomorphic process systems were distinguished into glacial processes (directly at the glacier) and paraglacial processes (in the vicinity of the glacier since the LIA maximum of 1851).

5.1. Glacial Processes

At Pasterze Glacier, the glacial process group include glacier lake evolution (fluctuations), and the identification of icebergs.

5.1.1. Glacier Lake Evolution

The evolution of Lake Pasterzensee at the terminus area of Pasterze Glacier was analyzed using all three sensor types: TLS, Sentinel-1 for backscatter analysis, and Sentinel-2 (using different bands for calculating the NDWI). TLS delineated the largest lake extent compared to other methods, except for the year 2015, where the value was even slightly lower than for the same month of the preceding year.

Quantitatively, for Sentinel-1, the lake extent nearly doubled in the time from 2015 to 2019. For TLS and Sentinel-2, the increase of extent was larger due to lower values in 2015 and larger values in 2019 compared to Sentinel-1. The lake areas delineated from Sentinel-1 and -2 are similar for similar epochs with some differences in 2016 and in the early season of 2019 (Figure 3). Results exhibit an inter-annual pattern, which is visible in each year: Lake extent increases to a certain level, reaches a peak around early September, and decreases again towards the end of the season. This might be an interesting aspect to be further investigated.

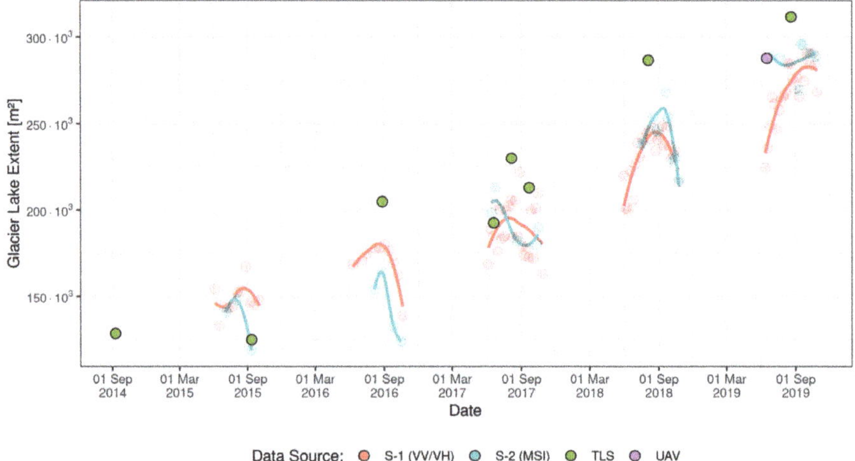

Figure 3. Evolution of the extent of Lake Pasterzensee from 2014-09-01 to 2019-10-01. Abbreviations of data sources: S-1 (VV/VH), Sentinel-1 polarization VV/VH (vertically transmitted and vertically or horizontally received radiation (spatial resolution = 5 × 20 m); S-2 (MSI), Sentinel-2 multispectral instrument (spatial resolution = 10 m); TLS, Terrestrial Laserscanning (spatial resolution = 1 m).

Results of qualitative assessment revealed that TLS data provided very accurate results in delineating the glacier lake extent: rhe mean GSD of TLS raw data is between 0.12 and 0.54 m [24] with a mean error in single point geolocation of approximately 0.15 m (Figure 2). Data gaps due to shadowing are of minor importance at the dead ice zones at the hillslope area, but point density at the transition between the glacier terminus area and the lake is very small. Summing up, uncertainties affecting the delineation of water bodies using TLS data are much smaller than the increase of the extent of Lake Pasterzensee (Figure 4).

For the period 2015–2019, in total, 27 Sentinel-2 scenes (at least three scenes per year) could be used to delineate the extent of Lake Pasterzensee (Table 3). Sentinel-2 images were classified using NDWI, where thresholds of 0.12–0.4 in the different scenes were used to classify water. NDWI values between 0.09 and 0.23 typically indicate ice; water areas are classified with NDWI values < 0.60 [91]. However, lakes with strong turbidity are classified as glacier surface [93], which is proved at Lake Pasterzensee. Glacier lake extent at Lake Pasterzensee show a clear increase of area in the observation period from 136×10^3 ($\pm 12 \times 10^3$) to 288×10^3 ($\pm 8 \times 10^3$) m^2 (Figure 3). To assess uncertainties of automatic glacier lake extent qualitatively, we used: (i) high resolution ortho-images (2018); and (ii) a combination of TLS-based delineation of lake extent and visual interpretation of the automatic camera at FJH (at a 5 min basis) (see Figure 4). Assuming that 0.5 pixels are an average error in image classification, we applied buffers of ±5 m that indicated a mean accuracy of $\pm 34 \times 10^3$ m^2 for water body classification.

The glacial lake delineation based on Sentinel-1 data was conducted using scenes from the same orbit with a revisiting time of 12 days for the years 2015 and 2016 (Sentinel 1A) and 6 days for the years 2017, 2018, and 2019 (Sentinel-1 A and B). For the study area, between 10 and 25 relevant scenes per year could be analyzed during the snow-free period (June to October) between 2015 and 2019. In total, four acquisitions (in June and October) had to be excluded due to partial freezing of the lake. Although mostly weather independent, wind and rain may cause roughening of the water surface leading to a reduction in contrast between water and land surfaces. Furthermore, the noise-like effect of speckle may decrease the accuracy of the delineated lake extent. To reduce this effect, the mean of the available polarizations VV (vertically transmitted and received radiation) and VH (vertically transmitted and horizontally received radiation) was used and overall threshold values of 14–17 dB were applied. To estimate classification accuracies, the change in lake extent was assessed depending on threshold value. A change in threshold of 0.5 dB led to a change of 3–7% depending on contrast conditions between water and terrain.

5.1.2. Proglacial River

In addition to the challenges of using UAV applications in high mountain environments, there are also problems in recording channel geometries. On the one hand, the recording of moving elements, such as the water surface level, provides blanks in recording and no points in the model, and on the other hand the river bed topography cannot be mapped as there is no detection of the bathymetry below the water surface; a general problem of UAV applications for mapping river geometries. Thus, the river bed geometry must be obtained from terrestrial surveying (e.g., GNSS devices) and will finally be combined with the UAV data to get the whole river geometry for high water stages. Thus, it is inevitable to intersect elements with different point density to a final DTM, linear cross-sections with planar information from UAVs. This 'sampling problem' applies to TLS mapping as well.

These results are calculated based on the UAV derived DTM, whose positional accuracy is determined by a point using a GNSS device ($dx = 4$ cm, $dy = 0$ cm). Due to the high mountain environment, just one point was available. Based on the chosen flight altitude, camera and modeling settings, this DTM has a resolution of 1.59 cm/px (Table 4), necessary for the photogrammetric sediment analyses.

5.1.3. Icebergs

Icebergs are a frequent phenomenon at Pasterze Glacier terminus. Contrary to other glaciers, icebergs at Pasterze Glacier seem to develop more from sub-water dead-ice areas than directly from the glacier part itself. Hence, at Pasterze Glacier, we combine analysis of dead-ice areas and icebergs at the glacier lake as dependent processes. The limited lifecycle of icebergs complicate the detection in remote sensing techniques (Figure 5). TLS and multispectral satellite images only detect icebergs randomly and therefore a detection does not support a substantial analysis. The only method providing valuable information is the use of temporal high-frequent automatic cameras. However, since information is provided only qualitatively (occurrence and trajectories), a quantification of iceberg characteristics (e.g., size) cannot be conducted satisfactorily with the available monitoring system at Pasterze Glacier (using automatic cameras).

TLS provides the most accurate delineation of icebergs however in our case at the lowest temporal frequency. In addition, with Sentinel-2 icebergs are clearly distinguishable, although the resolution of 10 m often leads to mixed pixel information for these oftentimes small and highly dissected features. In many cases, clusters of icebergs may seemingly merge into one or be connected to the lake border. Smaller icebergs—clearly detectable with TLS—may also be misclassified. To some extent, icebergs are also distinguishable with Sentinel-1, however the speckled nature of SAR data often hinders a confident delineation. Furthermore, islands, similar to sandbanks, are hardly distinguishable due to their smooth surface and therefore backscatter values similar to water.

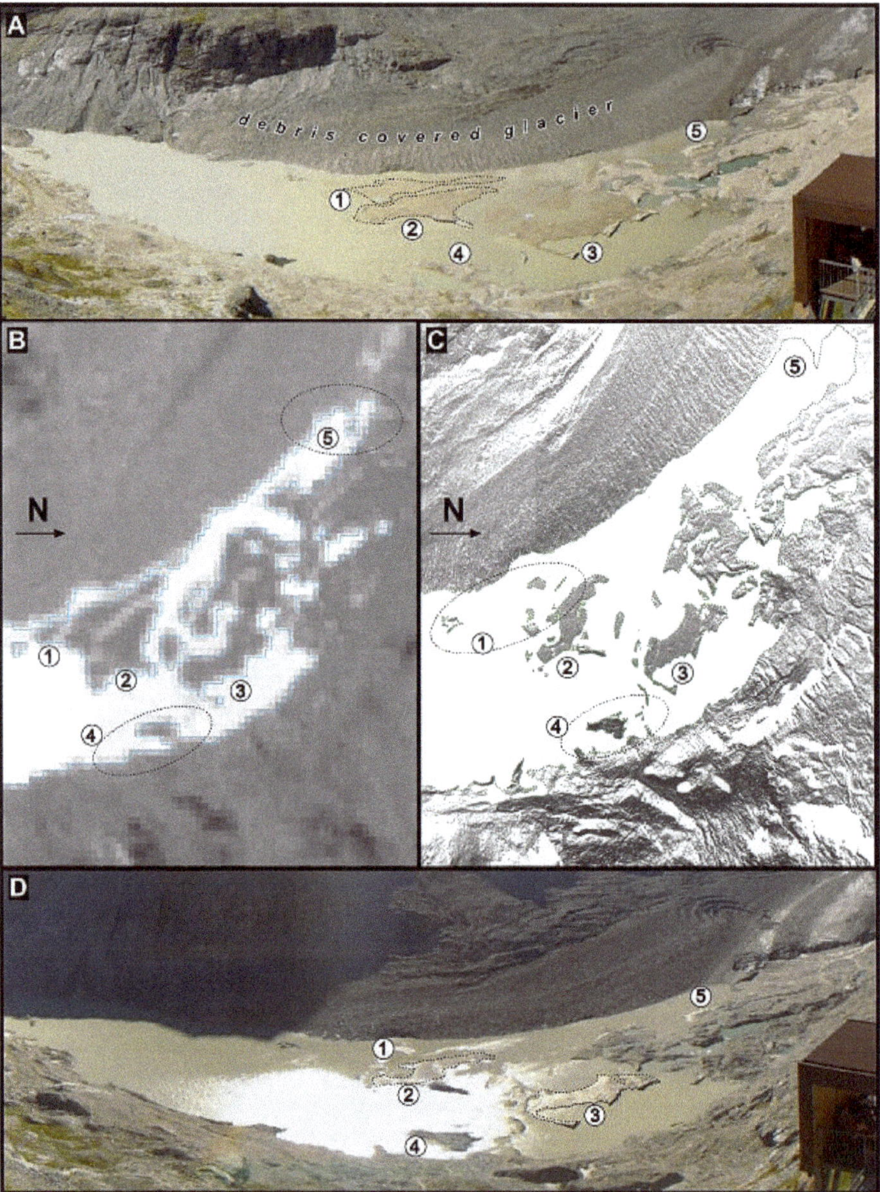

Figure 4. Comparison of glacier lake delineation and different data basis on 2016-08-27: (**A**) automatic Camera at 12:15; (**B**): automatic glacier lake delineation based on Sentinel-2 data at 12:10 using NDWI classification (threshold); (**C**) manuel glacier lake delineation based on point cloud TLS data at 17:00; and (**D**) automatic Camera at 16:25. Codes for areas of interest: 1–3, dead-ice/sander island; 4, dead-ice island/peninsula; 5, glacier terminus area.

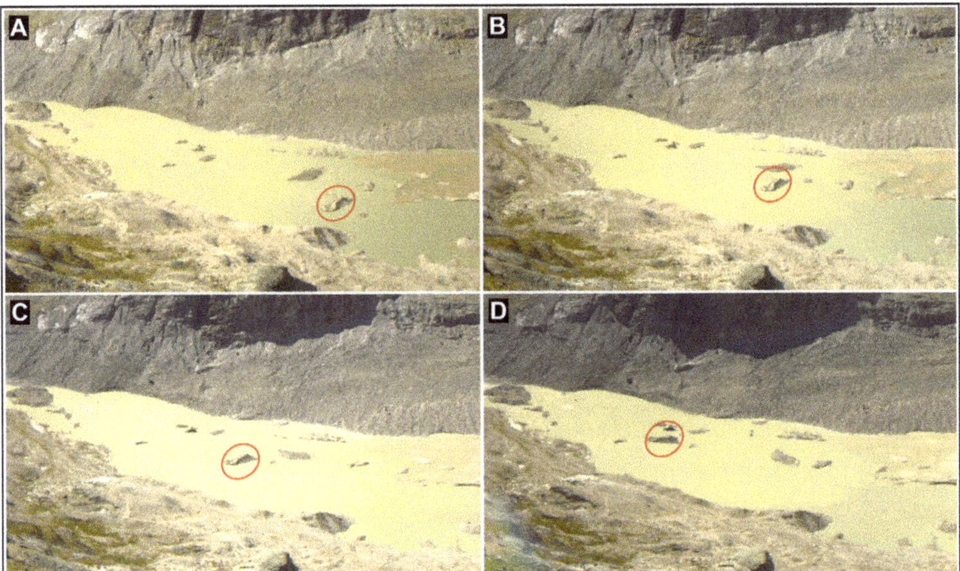

Figure 5. Examples of the high dynamics of icebergs on Lake Pasterzensee (2015-08-28). The iceberg moves about 500 m in 3 h (11:00–14:00 CEST): (**A**) 11:00; (**B**) 12:00; (**C**) 13:00; and (**D**) 14:00. During this process, the iceberg also rotates around its own vertical axis (**C,D**). To automatically track the movement of icebergs, high temporal resolution of the data basis is required. Additionally, due to rotational movements of the iceberg, such automatic monitoring must also be able to process changes in the geometry of the iceberg. Images provided by GROHAG.

5.2. Paraglacial Processes

5.2.1. Valley-Bottom and Slope Processes

At the valley-bottom area of Pasterze Glacier terminus area, TLS was conducted with measurement distances between 300 (dead-ice terraces) and 800 m (debris covered glacier area) to the scanning position. This leads to distance-dependent GSD of 0.12–0.20 m at 300 m and 0.33–0.54 m at 800 m. Depending on the analysis of different process groups, calculated DTMs were resampled to the following spatial resolution: 0.25 m for visual geomorphological interpretation, 0.5 for sectoral, and 1.0 m for area wide calculations [24]. To quantify uncertainties, 13 stable (bedrock) areas beneath the scanning position FJH (Figure 1) were used at distances of 345–1186 m (Figure 2). Calculated uncertainties are 0.083–0.192 m for mean Euclidean distance error and −0.025 to 0.033 m for dz.

TLS based analysis revealed several interesting areas based on the work of [24]. For the period 2018–2019, two areas were exemplarily chosen for further analysis. Figure 6A shows the vertical elevation patterns of a sliding area towards the Lake Pasterzensee. This landform covers an area of 16,750 m^2 with a mean surface subsidence of −0.32 m of the entire area. Within this entire landform, patterns of different vertical surface elevation changes can be detected comprising distinct sub areas such as with maximum dz of −0.92 m Figure 6A(1)), maximum dz of −0.81 m (Figure 6A(2)), and maximum dz of −1.28 m (Figure 6A(3)).

Figure 6B shows typical erosion process with downslope channel structure (Figure 6B(3)) covering an area of 14,500 m^2. In terms of mass balance, erosional (Figure 6B(1)) and deposition processes (Figure 6B(2)) nearly cancel each other out with a calculative dV of 290 m^3 (corresponding to a mean surface subsidence of −0.02 m).

Using Sentinel-1 for DInSAR applications, no significant deformation rates indicating valley bottom or slope processes can be detected in the main study area Pasterze Glacier. However, close to the main study area, DInSAR analysis delivered one particularly interesting process. Ongoing erosional processes are clearly detectable in PSI and SBAS analysis for the landslide Guttal (Figure 7) in the detachment area. We measured deformation values of −6.1 cm/a (away from the satellite) for P-SBAS in orbit 117 and −4.6 cm/a for PSI in orbit 44.

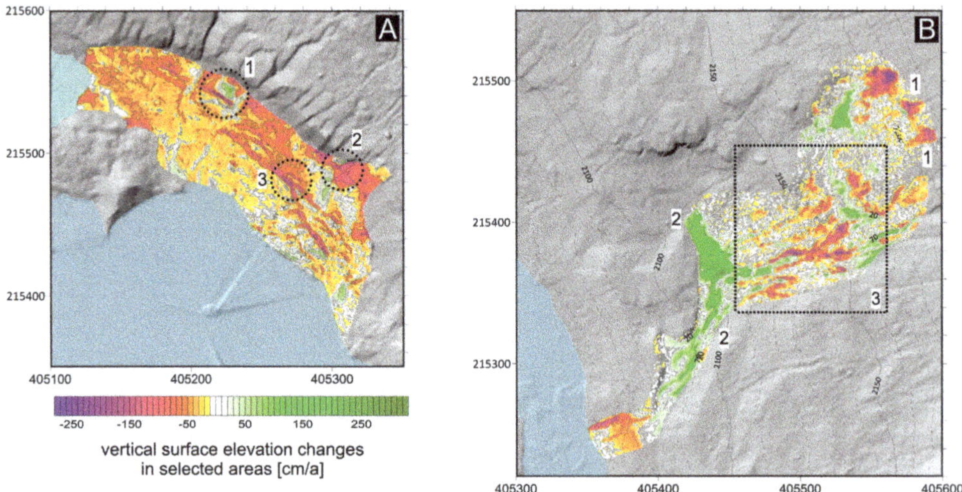

Figure 6. Examples of valley bottom and slope processes at Pasterze glacier terminus area. Calculated vertical surface elevation differences are based on TLS measurements from 2018-08-04 and 2019-08-03. (**A**) Slide at footslope area towards the Lake Pasterzensee, where 1–3 indicate special patterns of vertical surface elevation changes. (**B**) Erosional processes with detectable detachment areas (1), accumulation areas (2), and channel structure. CRS, MGI/Austria GK M31 (EPSG:31258).

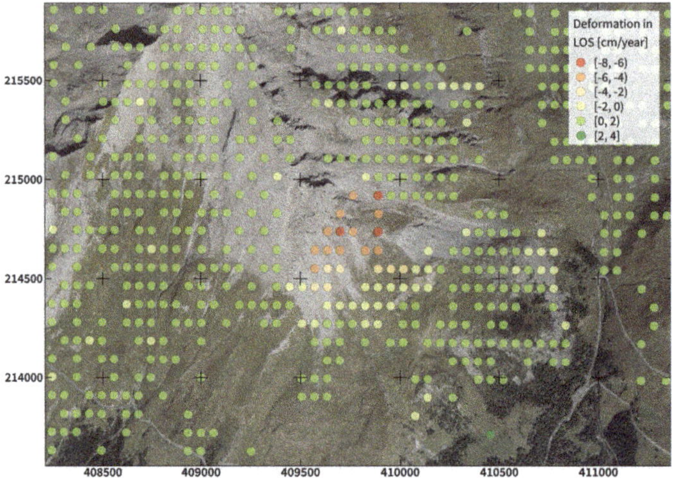

Figure 7. Deformation values of Racherin-Wasserradkopf ridge derived from P-SBAS, relative orbit 117. High deformation values indicate processes within the Guttal landslide. CRS, MGI/Austria GK M31 (EPSG:31258).

5.2.2. Rock Fall Processes

Rock wall processes are monitored at the Burgstall Mountains mainly using TLS. Quality considerations were conducted similar to the glacier terminus area using stable areas Figure 2.

At MBUG, four stable areas were defined showing a good geometric distribution in terms of distance and inclination angle. Calculated mean Euclidean distances vary from 0.045 m (distance 472 m) to 0.166 m (distance 665 m, Figure 2), with calculated dz values from 0.021 m (distance 472 m) to 0.065 m (distance 665 m). At HBUG, six stable areas were defined showing a good geometric distribution in terms of distance and inclination angle. Calculated mean Euclidean distances vary from 0.012 m (distance 166 m) to 0.042 m (distance 188 m, Figure 2), with calculated dz values from 0.009 m (distance 166 m) to 0.037 m (distance 188 m).

6. Discussion

In this chapter, we analyze the applicability and the constraints of single methods used in monitoring particular geomorphological processes at the Pasterze Glacier area.

6.1. Glacial Lakes

TLS was conducted once a year at Pasterze Glacier area, only in 2017 inter-annual TLS data were acquired three times. Measurements at Pasterze Glacier area are limited to three regular measurement campaigns due to logistics and as a consequence costs. However, spatial resolution show remarkable results both in high dynamic (e.g., evolution of Lake Pasterzensee) as well as processes showing small magnitudes (e.g., slope processes) depending on the measurement distances [161]. TLS provides very dense point clouds but lacks in reflectivity especially in close vicinity to the Pasterze glacier terminus.

Water saturated sander and dead-ice bodies as well debris covered glacier do not reflect laser impulses sufficiently and therefore margins and transition zones from water to terrain are not represented accurately. Water at Lake Pasterzensee exhibits high turbidity and diurnal variations of water level which influences signal reflection and interpretability of results. Topographical situations leading to unfavorable scanning geometry (small incidence angle close to the glacier terminus) yield different extents or data gaps, which subsequently have to be interpreted carefully.

The selection of appropriate Sentinel-2 images proved to be difficult due to frequent cloud coverage. Therefore, of the total of about 30 potentially available images taken between June and October, only the images in 2015 and eight images in 2018 are suitable for further analysis. However, in terms of assessing the inter-annual variability, Sentinel-2 images do increase the temporal resolution of data availability, e.g., for delineating glacier lakes. Analysis on a one- to two-monthly basis (in the summer period) can be enabled (Figure 3), which is already sufficient for a single test region with a few lakes and would be a valuable support for area wide quantification of glacier lake extent.

The spatial resolution of 10 m of Sentinel-2 images is a valuable improvement for area-wide assessment compared with available multi-spectral data so far (e.g., Landsat 8: 30 m spatial resolution, panchromatic 15 m). However, in detail, small scale dynamics such as degradation of ice-terraces or the dynamics of sander areas are hardly detectable with Sentinel-2 images. Spatial resolution of Sentinel-2 and reflectivity characteristics of the surface cause misclassifications (Figure 4). Especially shallow water areas (Figure 4 A–D(1,5)) are often classified as terrain and small-scale features are not represented accurately (Figure 4A–D(3)). At Pasterze Glacier, the water level shows distinct diurnal variations of max. 0.5 m especially in August. A comparison of small-scale landforms in shape and size based on TLS and Sentinel-2 data is very problematic due to the rapidly varying water levels (Figure 4A–D(1–3)).

Sentinel-1 provides data with high temporal resolution with a revisiting time of six days (for Sentinel-1 A and B). Since results are hardly affected by cloud cover, this yields clear benefits compared to optical remote sensing for frequent mapping of glacial lakes [64,65]. Because Sentinel-1 is a side-looking sensor, terrain correction with a high resolution DTM is essential in order to achieve correct geolocation in mountainous terrain. The necessity of an appropriate DTM for glacier lake delineation, which is essential in cases of rapidly developing lakes [162,163], was also pointed out by Strozzi et al. [64] and Wangchuk et al. [65]. In this study, a DTM with a spatial resolution of 10 m was used. DTMs with lower resolution, such as SRTM-3 (3 s, i.e. approximately 90 m spatial resolution) would lead to distortions of several tens of meters.

However, compared to other sensors used for glacial lake delineation, SAR data are subject to higher noise levels. Especially the noise-like effect of speckle may lead to classification uncertainties. In addition, special care must be taken regarding acquisitions with lower contrast, when the water surface is roughened due to wind or rain. In this type of situation, the delineation of lake extent or areas which exhibit lower backscatter due to topographic effects may be difficult. Consequently, manual correction of delineation results is necessary. The challenges of speckle and the effects of wind and waves was also pointed out by Strozzi et al. [64], who therefore also preferred a manual classification of glacial lake outlines. Topographic effects may also lead to classification challenges. The slopes facing away from the satellite naturally exhibit lower backscatter values, as the density of scatter points is lower. Very steep slopes may show low values comparable to water surfaces. We therefore chose an orbit with a steep incidence angle in order to minimize these topographic influences. Furthermore, Sentinel-1 provides the same spatial resolution as Sentinel-2 (10×10 m), also leading to mixed pixel information. This mainly affects small island detection in the lake which potentially leads to misclassification. Another source of misclassification are areas of wet sand or wet snow, which exhibit very low backscatter values similar to water [64]. Sander islands (cf. Figure 4A,B) are therefore mostly misclassified due to their low backscatter intensity.

Summarizing glacier lake evolution (at Lake Pasterzensee) is one particular process which benefits from the variety of data availability. TLS and UAV provides information with very high spatial resolution and accuracy in a minimum annual and maximum three times a year temporal resolution. Multi-spectral data Glacier lake extent in late summer season undergoes diurnal variations, which is a crucial indication for subsequent process related data interpretation.

6.2. Icebergs

Because of a short lifecycle, icebergs on Lake Pasterzensee cannot be identified with most remote sensing techniques thus far. Icebergs can be detected using TLS due to the high spatial resolution and accuracy. However, to characterize icebergs, a drawback of using TLS is the low temporal resolution (maximum three times per year). Thus, the detection of icebergs and their general occurrence at the time of measurement and the validity of interpretation of long-term is rather random. The usage of multi-spectral is strongly influenced by turbidity and shallow water areas, which delivers mixed-pixel information. Icebergs can be detected due to different reflectivity but delineation causes inaccuracies due to spatial resolution of the multi-spectral satellite data and the small size of icebergs. However, at Pasterze Glacier, the usage of the automatic camera is a very valuable support for geomorphic interpretation. The automatic camera gives qualitative information about the occurrence and disappearing, the movement (Figure 5), and special dynamics of icebergs as a high dynamic process (e.g., tilting of icebergs [142]). Figure 5 also shows the high dynamics in the movement of several other icebergs within this period of 3 h. Automatic tracking is limited due to sudden rotating and even tilting of the iceberg and accordingly leading to a changing geometry of the iceberg for tracking purposes [143].

To automatically track the movement of this landform, a high temporal resolution is required. In addition, due to the rotational movements of the iceberg in the water, such automatic monitoring must also be able to process changes in the geometry of the iceberg

6.3. River Processes

To determine the potential of erosional processes, the quantification of fluvial processes particularly of the proglacial channel system was based on UAVs and TLS. To describe processes accurately, grain sizes of the sediment have to be determined by using sampling information from several pixels [104]. In detail, roughness coefficients are subsequently deduced from calculated DTMs.

To achieve all requirements from a process point of view, the resulting point density of the point cloud has to be in the range of 15–20 points/m^2. By analyzing statistical parameters (e.g., 2σ) from the vertical elevation differences (dz) of the calculated DTM, correlations between the calculated statistical values and reference samples (from field work) can be observed [121,164]. Consequently characteristic grain sizes can be defined, but no grain size distributions can be determined.

Despite low GSD, the fines fraction is strongly underestimated in both SfM and TLS-based DTMs. Thus, these two applications are limited for characterizing diamictic sediment [121,164]. In [159], the characteristic grain diameter d90 (90% of the particles in the sediment sample are finer than the respective d90 grain size), necessary for hydraulic 1D modeling, could be determined photogrammetrically with the achieved GSD (1.59 cm/px), similar to what Hauer and Pulg [165] obtained during fieldwork. The in-situ determination of any characteristic grain sizes in the proglacial river at the Pasterze, however, is not be possible due to the described characteristics of the torrent.

However, the photogrammetric grain size determination has a major uncertainty: the orientation of each grain may not be determined exactly due to the top view. Thus, the important b-axis for grain size distribution curves probably cannot be measured. Nevertheless, for the determination of large characteristic grain sizes (e.g., d90), photogrammetric evaluation was found to be applicable.

Next to the challenges of using SfM applications in high mountain environments, there are further problems in measuring channel geometries. On the one hand, the recording of, e.g., the water level provides data gaps in the model. To avoid limitations of high water level in assessing river bed topography, low water level situations support area wide determination of bathymetry.

If water level is high and only mono-temporal surveys are possible, the river bed topography must be obtained from terrestrial surveying (e.g., GNSS devices) and will then finally be combined with the SfM-based data to get the entire river geometry. Thus, it is essential to include the information of these linear transverse cross-sections in the 3D information of the SfM-based DTM.

This 'sampling problem' applies to TLS mapping as well. In addition, such high (up to 15 m) and steep slopes, such as the studied canyon-like channel in [159], lead to big 'scan shadows', which, however, can be minimized by changing the position of the TLS device several times. This in turn implies a good accessibility of the channel.

6.4. Valley Bottom and Slope Processes

In high mountain environments, slope processes in the vicinity of ongoing glacier retreat show a large variety of process magnitudes [166]. At Pasterze Glacier slope, processes can be distinguished in areas: (i) reworked by downwasting of dead ice; and (ii) areas mainly stabilized, but partly reworked by gullying and debris faces [24].

For this work, valley bottom and slope processes were analyzed exemplarily comprising changes in dead-ice areas, erosional processes, and small landslides (Figure 6). Due to very fast glacier retreat at Pasterze Glacier these process types show an accentuated temporal succession (e.g., [24]). In the vicinity of the Lake Pasterzensee dead-ice degradation with a vertical subsidence of some decimeters per year are detectable, e.g., using TLS (Figure 6A). High accuracy of data enable the determination of process patterns leading to the fragmentation of landform due to their activity status (Figure 6A(1–3)). Erosional process chains showing lower magnitudes are also clearly detectable indicating areas of erosion and areas of deposition (Figure 6B).

Using multi-spectral images, the information received from quantitative assessment of slope processes is limited. The change of size of dead-ice areas bordering the the glacial lake is the only application which is possible due to significant difference of the spectral information and the spatial resolution of data. Thus, the monitoring of slope processes using multi-spectral surface information is not possible due to the coarse spatial resolution of Sentinel-2 in regard to process magnitudes.

The applicability of DInSAR techniques in characterizing high mountain processes has to be discussed diversely. To measure surface deformations, the approach of using stable areas for testing PSI and SBAS applications provides interesting insights into using different orbits and software applications (StaMPS, GEP, and RSG). However, results do not deliver satisfying results in terms of the required accuracy needed for the characterization of processes at this stage of analysis. Results at slope areas close to the Pasterze Glacier do not show consistent patterns of deformation values, which can be validated with other techniques.

DInSAR analyses show only minor drifts over time. However, deviations of single epochs from the trend line are higher than reported in other publications. This is probably mainly attributable to insufficiently modeled atmospheric turbulences. For the correction of atmospheric phase delays, the methods of Cong et al. [167] are available. However, first experiments with ECMWF (European Centre for Medium-Range Weather Forecasts) ERA5 model parameters confirmed larger inaccuracies of such corrections, especially in the 'turbulent' summer season. Therefore, only annual mean deformation rates were estimated for all DInSAR methods applied in this study. Furthermore, results are based on an assessment in line-of-sight only. This may cause systematic effects in the vertical (up/down) or horizontal (east/west) deformation rate estimation [168].

One particular process, which is clearly detectable with DInSAR techniques, is surface deformation in the upper part of the landslide Guttal, some 4.3 km east of Franz-Josefs-Höhe. This area shows significantly higher deformation values than the surrounding surface in every analysis (cf. Figure 7). Due to the lack of reference data, the validity of the magnitude of the single deformation values cannot be stated in this early stage of using Sentinel-1 data for high mountain applications. There is neither additional information about movement patterns of the Guttal landslide using other techniques, nor recent quantifications in prework [169,170].

6.5. Rockfall Processes

For the characterization of rockfall processes, TLS provides point clouds as a valuable basis for further analysis such as geological interpretation. At both test sites (MBUG and HBUG), analysis delivered information about distinct areas of rock fall, larger block falls, and the changes in the accumulation area of the large rock fall of 2007. Rock fall processes produce small to large blocks, thus the minimum detectable object (MDO) of the methods is a crucial factor. Substantially different measurement distances cause different reasonable GSDs leading to different MDOs. At MBUG, the mean GSD varies from 15 (for special areas) to 40 cm for practical reasons (measurement time and data handling). Consequently, rocks larger than 15 cm are potentially detectable in the accumulation area, as well as, interestingly, changes in the detachment area.

Due to small measurement distances (100–250 m), joints and planes of bedding are detectable in data from HBUG. At MBUG, we adapted the scanning increment the achieve the similar GSD in order to measure special areas of rock falls effectively. In the deposition area of the rock fall in 2007, the positive vertical changes deduce rock fall activity and area-wide negative vertical changes are most likely evidence of block consolidation and/or subsidence due to underlying glacier retreat. In some situations, scanning geometry and topography lead to pseudo-deformations due to misalignments. As geolocation of particular point clouds are not equal, breaklines such as ridges are represented differently.

Qualitative assessment of rock fall activity especially at HBUG east-face was conducted using the automatic camera of Fuscherkarkopf (FKK in Figure 1). Visual inspections of the rock faces were conducted during measurement campaigns to avoid misinterpretations due to inaccurate representation of the surfaces in the models. In the last four years, rock fall activity was not limited to the south-face of HBUG. Frequent rock falls were also reported from the east-face affecting the track to the Oberwalder Hütte, which is one of the main training bases of the Austrian Alpine Club. TLS scanning sectors do not cover this areal thus, for subsequent monitoring of possible rock faces, the automatic camera was a valuable support. In August 2019, a first UAV-based campaign was conducted covering both MBUG and HBUG as reference measurements. The quality assessment of the resulting DTM is still in progress. Annual SfM derived DTMs will be the basis for a comprehensive geological analysis in upcoming years.

7. Synopsis: Processes/Landforms and Data Acquisition

All the consideration made in this section are condensed into a final synopsis of spatial and temporal scales for both methods and processes. To draw conclusions about the applicability of Earth observation techniques for monitoring high alpine environments, both the processes with respect to landforms and the required data are analyzed in spatial and temporal resolution. For this purpose, the processes or consequent landforms are classified with respect to their: (i) persistence of occurrence (lifecycle); and (ii) the extent of occurrence (including the determined morphometric parameter(s)). The data acquisition was quantified with respect to the minimum spatial and temporal resolution in order to monitor these processes and landforms to their changes independently of the sensor specification (Table 6).

The occurrence of Lake Pasterzensee is persistent over a year. The landform itself and the changes are detectable over the parameter *area*. Changes over one year are in the order of >100 m^2 and thus data acquisition with a spatial resolution of 10 m is sufficient to monitor the glacier lake accurately. If the temporal resolution is increased, information about the water body size delivers further information about, e.g., dead-ice degradation, iceberg dynamics, and sediment transport.

Icebergs show pronounced dynamics in lifecycle of days to weeks. These fast changes occurring in size and shape are quantified with the parameter *area* and qualified with movement tracks.

Proglacial rivers are characterized by a high spatiotemporal variability, whereby there may be big differences in the longitudinal direction of supply and sorting of sediments. Near the glacier terminus, lateral channel changes may occur within days to weeks. The already developed and incised part of the channel, however, seems to be stable for months up to a year. Both sections were determined morphometrically over the parameters *area* (=position), *length* (=elongation), and *depth* (*dz*). For the upper section, the short persistence of occurrence leads to minimum observation periods of approximately one day to one month. In terms of sediment budgets, acquisition should be conducted at least monthly close to the glacier terminus with spatial resolution of 1–2 cm to determine grain sizes (b-axis = second longest one). A similar high temporal resolution is necessary for the measurement of large roughness elements in the developed section.

The persistence of occurrence for dead-ice landforms strongly relates to the distance to the glacier. Close to the glacier, they are characterized by a high temporal variation with lifecycles of about weeks. With increasing distance to the glacier, dead-ice landforms persist for an observation year. The determinate morphometric parameters are *area* and vertical elevation changes (*dz*). To detect the area of these landforms, a minimum spatial resolution of 0.5–10 m is necessary (depending on the distance to the glacier). Temporal resolution of one year is sufficient to calculate area-wide characterization of dead-ice areas. Detailed process characterization needs higher temporal resolution due to higher process rates, e.g., lateral melting of dead-ice peninsulas. Qualitative considerations can be done using automatic cameras but they lack quantification so far.

Slope processes were analyzed exemplarily for the processes erosion and landslides. Erosional processes have a lifecycle ranging from a few weeks to a year. They are primarily detectable using the parameter *area* and vertical elevation changes (*dz*). A minimum spatial resolution in the horizontal plane for area of 0.5–1 m is necessary to detect process related landforms and changes. A temporal resolution of at least six months is sufficient for area-wide characterization. An increased temporal resolution provides the differentiation of spontaneous and continuous slope processes (supported by automatic cameras). Landslides show a lifecycle of minimum one year and are detectable by the parameters *area*, *dz*, and *dr*. Rock fall is a spontaneous and extreme rapid process. Rock falls range in size from single stones to large failures, involving several 100,000 m^3 (see Rockfall Burgstall). Due to their infrequent nature, it is almost impossible to monitor the process trigger in detail—except with automatic cameras. Nevertheless, the accumulation of debris over time is persistent and thus the process and its changes is detectable over the parameters *volume* and *dz* with one year.

Overall, the DInSAR results demonstrate the feasibility to monitor deformation even in the challenging alpine environment. Further analysis is required to better understand the effects of DS and PS characteristics, the influence of slope aspect, and exposition and especially best suited atmospheric corrections that can deal with the very local meteorological conditions and large missing data in the analyzed time series.

Table 6. Schematic classification of processes and data acquisition: quantification of lifecycle (persistence), morphometry, spatial resolution, and temporal resolution. Codes: area, area of landform; *dz*, vertical difference; dr, differential range → euclidean distance.

Process/Landform	Process		Data Acquisition	
	Lifecycle	Morphometry	Spatial Resolution	Temporal Resolution
Glacial lake	1 year	area	10 m^2	1 year
Icebergs	1 day to weeks	area	10 m^2	daily
Proglacial river	1 day to 1 month	area	0.5–10 m^2	monthly
		dz	0.01–0.1 m	
		cross section	0.05–0.5 m	
Dead ice	weeks–1 year	area	0.5–10 m^2	monthly
		dz	1–5 m	
Erosion	weeks to 1 year	area	0.5–5 m^2	6 months
		dz	0.5–1 m	
Landslide	weeks to 1 year	area	0.5–10 m^2	monthly
		dz	0.5–1 m	
		dr	0.25–0.40 m	
Rockfall Burgstall	12 months	volume	10 cm^3	1 s to 1 min
		dr	0.25–0.40 m	

8. Conclusions

The characterization of the dynamics of geomorphological processes is always a crucial task in assessing geohazards. The evaluation of the specification, uncertainties, and limitations of monitoring techniques provide valuable information about their applicability. Amongst other aspects, we therefore focused on the comparison of available spatial resolution and feasible, effective temporal resolution.

This article comprises a variety of approaches monitoring geomorphological processes in different magnitudes and temporal as well as spatial scales in the test site Pasterze Glacier area in the observation period 2015–2019. The synoptic usage of data acquired by different remote sensing sensors proves to be a reasonable approach in monitoring geomorphic processes. As different sensors becomes more easily accessible and the usage more frequent in recent years, a closer look at accuracies and effective availability is crucial in terms of interpretability of the geomorphic processes (Table 7).

Table 7. Comparison of applied methods for monitoring geomorphological processes in terms of technical specification and quality characteristics: Codes: coverage, spatial extent of dataset; completion time. time period from the survey to the usable dataset; accuracy. both geolocation and pixel value (elevation for TLS and UAV; reflectivity for backscatter; phase for DInSAR reflectivity for multi-spectral; no additional pixel information for automatic camera); TRL, technology readiness level for the application monitoring geomorphic processes.

Parameter	TLS	UAV	Radar Backscatter	Radar DInSAR	Multi-Spectral	Automatic Camera
Coverage	+	++	+++	+++	+++	++
Spatial resolution	+++	+++	+	+	++	++
Temporal resolution	+	+	+++	+++	+++	+++
Completion time	+	+	++	++	+++	+++
Accuracy	+++	++	++	+	++	+
Survey time	+	++	+++	+++	+++	+++
TRL	+++	+++	+++	++	+++	+++

Nearly all applied techniques were able to quantify glacier lake extent sufficiently. TLS and UAV delivered high resolution datasets with very high accuracy. One drawback is the comparable low temporal resolution due to logistics and costs. Automatic classification of multi-spectral data and radar backscatter analysis provided quantifications with a very high temporal resolution but show some problems in accuracy due to spatial resolution. Summarizing glacier lake extent assessment using this pool of techniques delivered very interesting details in lake evolution.

Conducting sediment analysis such as in the example of river bed characterization, the most efficient method seems to be UAV applications due to: (i) low water depth and high geometric (canyon) and roughness features (boulders); and (ii) limiting errors due to lacking penetration of the water surface. Moreover, terrestrial surveying (e.g., TLS) is limited due to: (i) problems with the scanning geometry (shadowing, GSD); (ii) the dangerous characteristics of torrents (e.g., pronounced riverbed structures and high flow velocities); and (iii) the very steep slopes of torrents.

To quantify valley bottom, slope, and rock wall processes, TLS and UAV also provided precise data with very high spatial resolution. Even slow processes with small magnitudes can be detected over the observation period of one year. UAV could be utilized for large areas consistently, whereas TLS shows drawbacks of small incidence angle or shadowing. To cope with these challenges several scanning positions are necessary, which is not possible at Pasterze Glacier area due to accessibility and time effectiveness. DInSAR using Sentinel-1 data is a rather new application in monitoring geomorphic processes. Spatial patterns of deformation values do not deliver explicit information. We assume that accuracy is significantly dependent on atmospheric influences, which are difficult to correct for, particularly in high alpine environments, although larger deformation values of, e.g., landslide processes are clearly discriminable from adjacent terrain.

Today, automatic cameras become a valuable source of information mostly in a qualitative manner. Especially the availability of public, high quality cameras increasingly improve scientific work. Temporal resolution is a major upgrade in order to better understand processes or to validate processes dynamics. Further development of camera system (e.g., resolution) and available software (registration) will enhance the applicability in upcoming years.

Author Contributions: Conceptualization, M.A. and C.B.; methodology and software, M.A., M.S., B.W., G.W., K.-H.G., M.P., and G.S.; validation, M.A., C.B., M.S., K.-H.G., and M.P.; formal analysis, M.A., M.S., B.W., M.F., A.N., G.W., K.-H.G., M.P., C.H., P.F., G.S., and W.S.; investigation, M.A., M.S., B.W., M.F., A.N., G.W., K.-H.G., M.P., C.H., P.F., G.S., and W.S.; writing—original draft preparation, M.A., C.B., M.S., and M.P.; writing—review and editing, M.A., C.B, and M.S.; visualization, M.A. and M.S.; data curation, MS; and supervision, M.A., C.B. and M.S. All authors have read and agreed to the published version of the manuscript.

Funding: Research initiatives of ZAMG at Pasterze Glacier area are funded diversely: (i) 'Permafrostmonitoring Hohe Tauern' Nationalpark Hohe Tauern, 2016–2018; (ii) Forschungspreis Glockner Ökofonds (GROHAG) 2018; and (iii) 'Permafrostmonitoring Hohe Tauern' Nationalpark Hohe Tauern, 2019–2021. The research activities of Joanneum Research, DIGITAL—Institute for Information and Communication Technologies, are embedded in the SuLaMoSA project running within the Austrian Space Applications Programme (ASAP), and are funded by the Austrian Research Promotion Agency (FFG) grant number 865994. The University of Natural Resources and Life Sciences gratefully acknowledges the financial support by the Austrian Federal Ministry for Digital and Economic Affairs the National Foundation of Research, Technology.

Acknowledgments: We gratefully appreciate the support of the students of University of Graz, Graz University of Technology, and University of Natural Resources and Life Sciences, Vienna. Furthermore, we thank Grossglockner Hochalpenstraßen AG (GROHAG) as a long-term partner for valuable support in terms of logistics, transport and energy supply.

Abbreviations

The following abbreviations are used in this manuscript:

ALS	Airborne Laserscanning
BUG	Burgstall (Scanning position)
DInSAR	Differential Interferometric Synthetic Aperture Radar
DS	Distributed Scatterers
DTM	Digital Terrain Model
ECMWF	European Centre for Medium-Range Weather Forecasts
ERS	European remote sensing satellite
ESA	European Space Agency
FJH	Franz-Josefs-Höhe (Scanning position)
FKK	Fuscherkarkopf (camera position)
FWE	Freiwandeck (camera position)
GCP	Ground control points
GEP	Geohazards Thematic Exploitation Platform

GL	Glacier lake
GLO	Grossglockner (camera position)
GLOF	Glacier lake outburst floods
GROHAG	Grossglockner Hochalpenstraßen AG
GSD	Ground sampling distance
HBUG	Hoher Burgstall
HH	Hofmanns Hütte (Scanning position)
ICP	Independent check point
InSAR	Interferometric synthetic aperture radar
HoG	Height over Ground
IW	Interferometric Wide swath
LIA	Little Ice Age
LOS	Line of sight
MBUG	Mittlerer Burgstall
MDO	Minimum detectable object
MSI	Multi-sprectral instrument
MVS	Multi view stereo
NDWI	Normalized Difference Water Index
NIR	Near infrared
PAS	Burgstall 1, Burgstall 2 (camera position)
PS	Persistent Scatterer
P-SBAS	Parallel Small Baseline Subset
PSI	Persistent Scatterer Interferometry
RMSE	Root-mean-square-error
RTK	Real-time kinematic
S-1	Sentinel-1
S-2	Sentinel-2
SAR	Synthetic aperture radar
SBAS	Small Baseline
SAR	Synthetic aperture radar
SD	Standard deviation
SLC	Single Look Complex
SfM	Structure from motion
SNAP	Sentinel Application Platform
StaMPS	Stanford Method for Persistent Scatterers
SWIR	Short wave infrared
TLS	Terrestrial Laserscanning
TRL	Technology Readiness Level
UAV	Unmanned aerial vehicles
VB	Valley-bottom
VV/VH	Vertically transmitted and vertically or horizontally received radiation
VIS	Visible light
VNIR	Visible and Near Infra-Red
VV	Vertically transmitted and vertically received radiation

References

1. Abram, N.; Gattuso, J.P.; Prakash, A.; Cheng, L.; Chidichimo, M.; Crate, S.; Enomoto, H.; Garschagen, M.; Gruber, N.; Harper, S.; et al. Framing and Context of the Report. In *IPCC Special Report on the Ocean and Cryosphere in a Changing Climate*; Pörtner, H.O., Roberts, D., Masson-Delmotte, V., Zhai, P., Tignor, M., Poloczanska, E., Mintenbeck, K., Alegría, A., Nicolai, M., Okem, A., et al., Eds.; Cambridge University Press: Cambridge, UK; New York, NY, USA, 2019; pp. 73–129.

2. Hock, R.; Rasul, G.; Adler, C.; Cáceres, B.; Gruber, S.; Hirabayashi, Y.; Jackson, M.; Kääb, A.; Kang, S.; Kutuzov, S.; et al. High Mountain Areas. In *IPCC Special Report on the Ocean and Cryosphere in a Changing Climate*; Pörtner, H.O., Roberts, D., Masson-Delmotte, V., Zhai, P., Tignor, M., Poloczanska, E., Mintenbeck, K., Alegría, A., Nicolai, M., Okem, A., et al., Eds.; Cambridge University Press: Cambridge, UK; New York, NY, USA, 2019; pp. 131–202.

3. Collins, M.; Sutherland, M.; Bouwer, L.; Cheong, S.M.; Frölicher, T.; Combes, H.J.D.; Roxy, M.K.; Losada, I.; McInnes, K.; Ratter, B.; et al. Extremes, Abrupt Changes and Managing Risk. In *IPCC Special Report on the Ocean and Cryosphere in a Changing Climate*; Pörtner, H.O., Roberts, D., Masson-Delmotte, V., Zhai, P., Tignor, M., Poloczanska, E., Mintenbeck, K., Alegría, A., Nicolai, M., Okem, A., et al., Eds.; Cambridge University Press: Cambridge, UK; New York, NY, USA, 2019; pp. 598–655.

4.   IPCC. *Climate Change 2013: The Physical Science Basis. Contribution of Working Group I to the Fifth Assessment Report of the Intergovernmental Panel on Climate Change*; Cambridge University Press: Cambridge, UK; New York, NY, USA, 2013; p. 1535. [CrossRef]

5.   Kaser, G.; Grosshauser, M.; Marzeion, B. Contribution potential of glaciers to water availability in different climate regimes. *Proc. Natl. Acad. Sci. USA* **2010**, *107*, 20223–20227. [CrossRef]

6.   Mark, B.G.; French, A.; Baraer, M.; Carey, M.; Bury, J.; Young, K.R.; Polk, M.H.; Wigmore, O.; Lagos, P.; Crumley, R.; et al. Glacier loss and hydro-social risks in the Peruvian Andes. *Glob. Planet Chang.* **2017**, *159*, 61–76. [CrossRef]

7.   Nussbaumer, S.U.; Hoelzle, M.; Hüsler, F.; Huggel, C.; Salzmann, N.; Zemp, M. Glacier Monitoring and Capacity Building: Important Ingredients for Sustainable Mountain Development. *Mt. Res. Dev.* **2017**, *37*, 141–152. [CrossRef]

8.   Kääb, A.; Reynolds, J.M.; Haeberli, W. Glacier and Permafrost Hazards in High Mountains. In *Global Change and Mountain Regions: An Overview of Current Knowledge*; Huber, U.M., Bugmann, H.K.M., Reasoner, M.A., Eds.; Springer: Dordrecht, The Netherlands, 2005; pp. 225–234. [CrossRef]

9.   Vilímek, V.; Zapata, M.L.; Klimeš, J.; Patzelt, Z.; Santillán, N. Influence of glacial retreat on natural hazards of the Palcacocha Lake area, Peru. *Landslides* **2005**, *2*, 107–115. [CrossRef]

10.   Salzmann, N.; Huggel, C.; Rohrer, M.; Stoffel, M. Data and knowledge gaps in glacier, snow and related runoff research—A climate change adaptation perspective. *J. Hydrol.* **2014**, *518*, 225–234. [CrossRef]

11.   Grossi, G.; Caronna, P.; Ranzi, R. Hydrologic vulnerability to climate change of the Mandrone glacier (Adamello-Presanella group, Italian Alps). *Adv. Water Resour.* **2013**, *55*, 190–203. [CrossRef]

12.   Huss, M.; Jouvet, G.; Farinotti, D.; Bauder, A. Future high-mountain hydrology: A new parameterization of glacier retreat. *Hydrol. Earth Syst. Sci.* **2010**, *14*, 815–829. [CrossRef]

13.   Marzeion, B.; Cogley, J.; Richter, K.; Parkes, D. Attribution of global glacier mass loss to anthropogenic and natural causes. *Science* **2014**, *345*, 919–921. [CrossRef]

14.   Zemp, M.; Huss, M.; Eckert, N.; Thibert, E.; Paul, F.; Nussbaumer, S.U.; Gärtner-Roer, I. Brief communication: Ad hoc estimation of glacier contributions to sea-level rise from the latest glaciological observations. *Cryosphere* **2020**, *14*, 1043–1050. [CrossRef]

15.   Schaefli, B.; Manso, P.; Fischer, M.; Huss, M.; Farinotti, D. The role of glacier retreat for Swiss hydropower production. *Renew. Energy* **2019**, *132*, 615–627. [CrossRef]

16.   Vergara, W.; Deeb, A.; Valencia, A.; Bradley, R.; Francou, B.; Zarzar, A.; Grünwaldt, A.; Haeussling, S. Economic impacts of rapid glacier retreat in the Andes. *Eos* **2007**, *88*, 261–264. [CrossRef]

17.   Salim, E.; Mourey, J.; Ravanel, L.; Picco, P.; Gauchon, C. Mountain guides facing the effects of climate change. What perceptions and adaptation strategies at the foot of Mont Blanc? *Rev. Géogr. Alp.* **2019**, *107-4*. [CrossRef]

18.   Mourey, J.; Marcuzzi, M.; Ravanell, L.; Pallendre, F. Effects of climate change on high Alpine mountain environments: Evolution of mountaineering routes in the Mont Blanc massif (Western Alps) over half a century. *Arct. Antarct. Alp. Res.* **2019**, *51*, 176–189. [CrossRef]

19.   Gärtner-Roer, I.; Nussbaumer, S.U.; Hüsler, F.; Zemp, M. Worldwide Assessment of National Glacier Monitoring and Future Perspectives. *Mt. Res. Dev.* **2019**, *39*. [CrossRef]

20.   Gobiet, A.; Kotlarski, S.; Beniston, M.; Heinrich, G.; Rajczak, J.; Stoffel, M. 21st century climate change in the European Alps—A review. *Sci. Total Environ.* **2014**, *493*, 1138–1151. [CrossRef] [PubMed]

21.   Cannone, N.; Diolaiuti, G.; Guglielmin, M.; Smiraglia, C. Accelerating climate change impacts on alpine glacier forefield ecosystems in the European alps. *Ecol. Appl.* **2008**, *18*, 637–648. [CrossRef] [PubMed]

22.   Brighenti, S.; Tolotti, M.; Bruno, M.C.; Wharton, G.; Pusch, M.T.; Bertoldi, W. Ecosystem shifts in Alpine streams under glacier retreat and rock glacier thaw: A review. *Sci. Total Environ.* **2019**, *675*, 542–559. [CrossRef] [PubMed]

23.   Carey, M.; Molden, O.C.; Rasmussen, M.B.; Jackson, M.; Nolin, A.W.; Mark, B.G. Impacts of Glacier Recession and Declining Meltwater on Mountain Societies. *Ann. Am. Assoc. Geogr.* **2017**, *107*, 350–359. [CrossRef]

24.   Avian, M.; Kellerer-Pirklbauer, A.; Lieb, G. Geomorphic consequences of rapid deglaciation at Pasterze Glacier, Hohe Tauern Range, Austria, between 2010 and 2013 based on repeated terrestrial laser scanning data. *Geomorphology* **2018**, *310*, 1–14. [CrossRef]

25.   Zemp, M.; Frey, H.; Gärtner-Roer, I.; Nussbaumer, S.; Hoelzle, M.; Paul, F.; Haeberli, W.; Denzinger, F.; Ahlstrøm, A.; Anderson, B.; et al. Historically unprecedented global glacier decline in the early 21st century. *J. Glaciol.* **2015**, *61*, 745–762. [CrossRef]

26. Bojinski, S.; Verstraete, M.; Peterson, T.; Richter, C.; Simmons, A.; Zemp, M. The concept of Essential Climate Variables in support of climate research, applications, and policy. *Bull. Am. Meteorol. Soc.* **2014**, *95*, 1431–1443. [CrossRef]

27. Haeberli, W.; Hoelzle, M.; Paul, F.; Zemp, M. Integrated monitoring of mountain glaciers as key indicators of global climate change: The European Alps. *Ann. Glaciol.* **2007**, *46*, 150–160. [CrossRef]

28. GLAMOS. The Swiss Glaciers 1880-2016/17. In *Yearbooks of the Cryospheric Commission of the Swiss Academy of Sciences (SCNAT)*; Glaciological Reports No 1-138; VAW/ETH Zurich: Zurich, Switzerland, 2018. [CrossRef]

29. Fischer, M.; Huss, M.; Kummert, M.; Hoelze, M. Application and validation of long-range terrestrial laser scanning to monitor the mass balance of very small glaciers in the Swiss Alps. *Cryosphere* **2016**, *10*, 1279–1295. [CrossRef]

30. Zemp, M.; Thibert, E.; Huss, M.; Stumm, D.; Rolstad Denby, C.; Nuth, C.; Nussbaumer, S.U.; Moholdt, G.; Mercer, A.; Mayer, C.; et al. Reanalysing glacier mass balance measurement series. *Cryosphere* **2013**, *7*, 1227–1245. [CrossRef]

31. Baltsavias, E. Airborne laser scanning: Basic relations and formulas. *ISPRS J. Photogramm.* **1999**, *54*, 199–214. [CrossRef]

32. Slob, S.; Hack, R. 3D Terrestrial Laser Scanning as a New Field Measurement and Monitoring Technique. In *Engineering Geology for Infrastructure Planning in Europe. A European Perspective*; Hack, R., Azzam, R., Charlier, R., Eds.; Springer: Berlin/Heidelberg, Germany, 2004; pp. 179–190. [CrossRef]

33. Ravanel, L.; Bodin, X.; Deline, P. Using Terrestrial Laser Scanning for the Recognition and Promotion of High-Alpine Geomorphosites. *Geoheritage* **2014**, *6*, 129–140. [CrossRef]

34. Shan, J.; Toth, C. *Topographic Laser Ranging and Scanning: Principles and Processing*, 2nd ed.; CRC Press: New York, NY, USA, 2009; p. 590.

35. Heritage, G.L.; Large, A.R.G., Eds. *Laser Scanning for the Environmental Sciences*; Wiley-Blackwell: Hoboken, NJ, USA, 2009. [CrossRef]

36. Prokop, A.; Schirmer, M.; Rub, M.; Lehning, M.; Stocker, M. A comparison of measurement methods: Terrestrial laser scanning, tachymetry and snow probing for the determination of the spatial snow-depth distribution on slope. *Ann. Glaciol.* **2008**, *49*, 210–216. [CrossRef]

37. Prantl, H.; Nicholson, L.; Sailer, R.; Hanzer, F.; Juen, I.; Rastner, P. Glacier Snowline Determination from Terrestrial Laser Scanning Intensity Data. *Geosciences* **2017**, *7*, 30. [CrossRef]

38. Jaboyedoff, M.; Oppikofer, T.; Abellán, A.; Derron, M.H.; Loye, A.; Metzger, R.; Pedrazzini, A. Use of LIDAR in landslide investigations: A review. *Nat. Hazards* **2010**, *61*, 5–28. [CrossRef]

39. Abellán, A.; Vilaplana, J.M.; Calvet, J.; García-Sellés, D.; Asensio, E. Rockfall monitoring by Terrestrial Laser Scanning—Case study of the basaltic rock face at Castellfollit de la Roca (Catalonia, Spain). *Nat. Hazards Earth Syst. Sci.* **2011**, *11*, 829–841. [CrossRef]

40. Kenner, R.; Phillips, M.; Danioth, C.; Denier, C.; Thee, P.; Zgraggen, A. Investigation of rock and ice loss in a recently deglaciated mountain rock wall using terrestrial laser scanning: Gemsstock, Swiss Alps. *Cold Reg. Sci. Technol.* **2011**, *67*, 157–164. [CrossRef]

41. Draebing, D.; McColl, S.; Jacobs, B. Identification of periglacial processes and their link to rockfalls. In Proceedings of the 5th European Conference on Permafrost, Chamonix-Mont Blanc, France, 23 June–1 July 2018; pp. 505–506.

42. Zangerl, C.; Fey, C.; Prager, C. Deformation characteristics and multi-slab formation of a deep-seated rock slide in a high alpine environment (Bliggspitze, Austria). *Bull. Eng. Geol. Environ.* **2019**, *78*, 6111–6130. [CrossRef]

43. Bremer, M.; Sass, O. Combining airborne and terrestrial laser scanning for quantifying erosion and deposition by a debris flow event. *Geomorphology* **2012**, *138*, 49–60. [CrossRef]

44. Carrivick, J.L.; Geilhausen, M.; Warburton, J.; Dickson, N.E.; Carver, S.J.; Evans, A.J.; Brown, L.E. Contemporary geomorphological activity throughout the proglacial area of an alpine catchment. *Geomorphology* **2013**, *188*, 83–95. [CrossRef]

45. Kos, A.; Amann, F.; Strozzi, T.; Delaloye, R.; von Ruette, J.; Springman, S. Contemporary glacier retreat triggers a rapid landslide response, Great Aletsch Glacier, Switzerland. *Geophys. Res. Lett.* **2016**, *43*, 12466–12474. [CrossRef]

46. Bodin, X.; Schoeneich, P.; Jaillet, S. High-resolution DEM extraction from terrestrial LiDAR topometry and surface kinematics of the creeping alpine permafrost: The Laurichard rock glacier case study (southern French Alps). In Proceedings of the 9th International Conference on Permafrost, Fairbanks, AK, USA, 29 June–3 July 2008.

47. Avian, M.; Kellerer-Pirklbauer, A.; Bauer, A. LiDAR for monitoring mass movements in permafrost environments at the cirque Hinteres Langtal, Austria, between 2000 and 2008. *Nat. Hazards Earth Syst. Sci.* **2009**, *9*, 1087–1094. [CrossRef]

48. Bauer, A.; Kaufmann, V.; Kellerer-Pirklbauer, A.; Avian, M.; Paar, G. Terrestrial Laser Scanning for Glacier Monitoring: A Comparison to Standard Geodetic and Photogrammetric Methods, and Documentation of the Glacier Retreat of Goessnitzkees (Schober Group, Austria) between 2000 and 2005. In Proceedings of the 8th International Symposium of High Mountain Remote Sensing Cartography, La Paz, Bolivia, 21–27 March 2005.

49. Avian, M.; Lieb, G.K.; Kellerer-Pirklbauer, A.; Bauer, A. Variations of Pasterze Glacier (Austria) between 1994 and 2006—Combination of different data sets for spatial analysis. In Proceedings of the 9th International Symposium on High Mountain Remote Sensing Cartography (HMRSC-IX), Graz, Austria, 14–15 September 2006; Volume 43, pp. 79–88.

50. Stötter, J.; Bremer, M.; Mayr, A.; Rutzinger, M.; Sailer, R.; Zieher, T. *Laserscanning am Institut für Geographie —Ein Überblick*; Technical Report; Universität Innsbruck: Innsbruck, Austria, 2017.

51. Sailer, R.; Kaser, G.; Stötter, J.; Strasser, U. A permanent Terrestrial Laser Scanning Station a Hintereisferner, Rofental, Ötztal Alps. In *Workshop Long-Term Research in Mountain Areas. Book of Abstracts*; Challhart, N., Erschbamer, B., Beguin, D., Prunier, P., Eds.; Eigenverlag, Universität Innsbruck: Innsbruck, Austria, 2015; p. 26.

52. Gabbud, C.; Micheletti, N.; Lane, S. Lidar measurement of surface melt for a temperate Alpine glacier at the seasonal and hourly scales. *J. Glaciol.* **2019**, *61*, 963–974. [CrossRef]

53. Baewert, H.; Morche, D. Coarse sediment dynamics in a proglacial fluvial system (Fagge River, Tyrol). *Geomorphology* **2014**, *218*, 88–97. [CrossRef]

54. Deline, P.; Grange, C.; Jaillet, S.; Tamburini, A. Sept ans de suivi de la dynamique de la falaise de glace du lac du Miage (massif du Mont Blanc) par scanner laser terrestre. *Collect. EDYTEM* **2011**, *12*, 95–106. [CrossRef]

55. Godone, D.; Godone, F. The Support of Geomatics in Glacier Monitoring: The Contribution of Terrestrial Laser Scanner. In *Laser Scanner Technology*; InTech: London, UK, 2012; doi:10.5772/33463. [CrossRef]

56. Conforti, D.; Deline, P.; Mortara, G.; Tamburini, A. Terrestrial scanning LiDAR technology applied to study the evolution of the ice-contact Miage lake (Mont Blanc, Italy). In Proceedings of the 9th Alpine Glaciological Meeting, Milan, Italy, 24–25 February 2005.

57. Otto, J.C. Proglacial Lakes in High Mountain Environments: Landform and Sediment Dynamics in Recently Deglaciated Alpine Landscapes. In *Geomorphology of Proglacial Systems*; Heckmann, T., Morche, D., Eds.; Springer: Berlin/Heidelberg, Germany, 2019; pp. 231–247._14. [CrossRef]

58. Ravanel, L.; Lambiel, C.; Oppikofer, T.; Mazotti, B.; Jaboyedoff, M. Evolution of a highly vulnerable ice-cored moraine: Col des Gentianes, Swiss Alps. *Geophys. Res. Abstr.* **2012**, *14*, EGU2012-11574.

59. Fan, J.; Wang, Q.; Liu, G.; Zhang, L.; Guo, Z.; Tong, L.; Peng, J.; Yuan, W.; Zhou, W.; Yan, J.; et al. Monitoring and Analyzing Mountain Glacier Surface Movement Using SAR Data and a Terrestrial Laser Scanner: A Case Study of the Himalayas North Slope Glacier Area. *Remote Sens.* **2019**, *11*, 625. [CrossRef]

60. Ingensand, H.; Ryf, A.; Schulz, T. Performances and Experiences in Terrestrial Laserscanning. In Proceedings of the 6th Conference on Optical 3D Measurement Techniques, Zurich, Switzerland, 22–25 September 2003; pp. 505–506.

61. Fey, C.; Wichmann, V. Long-range terrestrial laser scanning for geomorphological change detection in alpine terrain—Handling uncertainties. *Earth Surf. Process. Landf.* **2016**, *42*, 789–802. [CrossRef]

62. Ulaby, F.T.; Moore, R.K.; Fung, A. *Microwave Remote Sensing—Active and Passive*; Artech House: Norwood, MA, USA, 1982.

63. Otsu, N. A Threshold Selection Method from Gray-Level Histograms. *IEEE Trans. Syst. Man Cybern.* **1979**, *9*, 62–66. [CrossRef]

64. Strozzi, T.; Wiesmann, A.; Kääb, A.; Joshi, S.; Mool, P. Glacial lake mapping with very high resolution satellite SAR data. *Nat. Hazards Earth Syst. Sci.* **2012**, *12*, 2487–2498. [CrossRef]

65. Wangchuk, S.; Bolch, T.; Zawadzki, J. Towards automated mapping and monitoring of potentially dangerous glacial lakes in Bhutan Himalaya using Sentinel-1 Synthetic Aperture Radar data. *Int. J. Remote Sens.* **2019**, *40*, 4642–4667. [CrossRef]

66. Zhang, M.; Chen, F.; Tian, B.; Liang, D.; Yang, A. High-Frequency Glacial Lake Mapping Using Time Series of Sentinel-1A/1B SAR Imagery: An Assessment for the Southeastern Tibetan Plateau. *Int. J. Environ. Res. Public Health* **2020**, *17*, 1072. [CrossRef]

67. Woodhouse, I. *Introduction to Microwave Remote Sensing*; CRC Press: Boca Raton, FL, USA, 2006.

68. Crosetto, M.; Monserrat, O.; Cuevas-González, M.; Devanthéry, N.; Crippa, B. Persistent scatterer interferometry: A review. *ISPRS J. Photogramm.* **2016**, *115*, 78–89. [CrossRef]

69. Ferretti, A.; Prati, C.; Rocca, F. Permanent scatterers in SAR interferometry. *IEEE Trans. Geosci. Remote Sens.* **2001**, *39*, 8–20. [CrossRef]

70. Hooper, A.; Zebker, H.; Segall, P.; Kampes, B. A new method for measuring deformation on Volcanoes and other natural terrains using InSAR Persistent Scatterers. *Geophys. Res. Lett.* **2004**, *31*, 1–5. [CrossRef]

71. Berardino, P.; Fornaro, G.; Lanari, R.; Sansosti, E. A new algorithm for surface deformation monitoring based on small baseline differential SAR interferograms. *IEEE Trans. Geosci. Remote Sens.* **2002**, *40*, 2375–2383. [CrossRef]

72. Delaloye, R.; Lambiel, C.; Lugon, R.; Raetzo, H.; Strozzi, T. ERS InSAR for Detecting Slope Movement in a Periglacial Mountain Environment (Western Valais Alps, Switzerland). *Grazer Schriften Der Geogr. Und Raumforsch.* **2007**, *43*, 113–120.

73. Barboux, C.; Delaloye, R.; Lambiel, C. Inventorying slope movements in an Alpine environment using DInSAR. *Earth Surf. Process. Landf.* **2014**, *39*, 2087–2099. [CrossRef]

74. Barboux, C.; Strozzi, T.; Delaloye, R.; Wegmüller, U.; Collet, C. Mapping slope movements in Alpine environments using TerraSAR-X interferometric methods. *ISPRS J. Photogramm.* **2015**, *109*, 178–192. [CrossRef]

75. Kenyi, L.; Kaufmann, V. Estimation of rock glacier surface deformation using sar interferometry data. *IEEE Trans. Geosci. Remote Sens.* **2003**, *41*, 1512–1515. [CrossRef]

76. Lambiel, C.; Delaloye, R.; Raetzo, H.; Lugon, R.; Strozzi, T. ERS InSAR for Assessing Rock Glacier Activity. In Proceedings of the 9th International Conference on Permafrost, Fairbanks, AK, USA, 29 June–3 July 2008.

77. Villarroel, C.; Beliveau, G.T.; Forte, A.; Monserrat, O.; Morvillo, M. DInSAR for a Regional Inventory of Active Rock Glaciers in the Dry Andes Mountains of Argentina and Chile with Sentinel-1 Data. *Remote Sens.* **2018**, *10*, 1588. [CrossRef]

78. Scapozza, C.; Ambrosi, C.; Cannata, M.; Strozzi, T. Glacial lake outburst flood hazard assessment by satellite Earth observation in the Himalayas (Chomolhari area, Bhutan). *Geogr. Helv.* **2019**, *74*, 125–139. [CrossRef]

79. Schneevoigt, N.J.; Sund, M.; Bogren, W.; Kääb, A.; Weydahl, D.J. Glacier displacement on Comfortlessbreen, Svalbard, using 2-pass differential SAR interferometry (DInSAR) with a digital elevation model. *Pol. Rec.* **2011**, *48*, 17–25. [CrossRef]

80. Goldstein, R.M.; Engelhardt, H.; Kamb, B.; Frolich, R.M. Satellite Radar Interferometry for Monitoring Ice Sheet Motion: Application to an Antarctic Ice Stream. *Science* **1993**, *262*, 1525–1530. [CrossRef]

81. Rott, H. Advances in interferometric synthetic aperture radar (InSAR) in earth system science. *Prog. Phys. Geogr.* **2009**, *33*, 769–791. [CrossRef]

82. Sánchez-Gámez, P.; Navarro, F.J. Glacier Surface Velocity Retrieval Using D-InSAR and Offset Tracking Techniques Applied to Ascending and Descending Passes of Sentinel-1 Data for Southern Ellesmere Ice Caps, Canadian Arctic. *Remote Sens.* **2017**, *9*, 442. [CrossRef]

83. Hall, D.K.; Chang, A.T.; Siddalingaiah, H. Reflectances of glaciers as calculated using Landsat-5 thematic mapper data. *Remote Sens. Environ.* **1988**, *25*, 311–321. [CrossRef]

84. Kääb, A.; Paul, F.; Maisch, M.; Hoelzle, M.; Haeberli, W. The new remote sensing derived Swiss glacier inventory: II. First results. *Ann. Glaciol.* **2002**, *34*, 362–366. [CrossRef]

85. Kääb, A. Combination of SRTM3 and repeat ASTER data for deriving alpine glacier flow velocities in the Bhutan Himalaya. *Remote Sens. Environ.* **2005**, *94*, 463–474. [CrossRef]

86. Kääb, A.; Huggel, C.; Fischer, L.; Guex, S.; Paul, F.; Roer, I.; Salzmann, N.; Schlaefli, S.; Schmutz, K.; Schneider, D.; Strozzi, T.; Weidmann, Y. Remote sensing of glacier- and permafrost-related hazards in high mountains: An overview. *Nat. Hazard Earth Syst.* **2005**, *5*, 527–554. [CrossRef]

87. Paul, F.; Kääb, A.; Maisch, M.; Kellenberger, T.; Haeberli, W. The new remote-sensing- derived Swiss glacier inventory: I. Methods. *Ann. Glaciol.* **2002**, *34*, 355–361. [CrossRef]

88. Huggel, C.; Kääb, A.; Haeberli, W.; Teysseire, P.; Paul, F. Remote sensing based assessment of hazards from glacier lake outbursts: A case study in the Swiss Alps. *Can. Geotech. J.* **2002**, *39*, 316–330. [CrossRef]

89. Kääb, A.; Bolch, T.; Casey, K.; Heid, T.; Kargel, J.S.; Leonard, G.J.; Paul, F.; Raup, B.H. Glacier Mapping and Monitoring Using Multispectral Data. In *Global Land Ice Measurements from Space*; Springer: Berlin/Heidelberg, Germany, 2014; pp. 75–112. [CrossRef]

90. Li, J.; Sheng, Y. An automated scheme for glacial lake dynamics mapping using Landsat imagery and digital elevation models: A case study in the Himalayas. *Int. J. Remote Sens.* **2012**, *33*, 5194–5213. [CrossRef]

91. Nagy, T.; Andreassen, L. *Glacier Lake Mapping with Sentinel-2 Imagery in Norway*; NVE Rapport 40-2019; Norwegian Water Resources and Energy Directorate (NVE): Oslo, Norway, 2019, p. 54.

92. Verma, P.; Ghosh, S.K. Classification of glacial lakes using integrated approach of DFPS technique and gradient analysis using Sentinel 2A data. *Geocarto Int.* **2018**, *34*, 1075–1088. [CrossRef]

93. Paul, F.; Winsvold, S.; Kääb, A.; Nagler, T.; Schwaizer, G. Glacier Remote Sensing Using Sentinel-2. Part II: Mapping Glacier Extents and Surface Facies, and Comparison to Landsat 8. *Remote Sens.* **2016**, *8*, 575. [CrossRef]

94. Li, J.; Roy, D.P. A Global Analysis of Sentinel-2A, Sentinel-2B and Landsat-8 Data Revisit Intervals and Implications for Terrestrial Monitoring. *Remote Sens.* **2017**, *9*, 902. [CrossRef]

95. Zhang, H.K.; Roy, D.P.; Yan, L.; Li, Z.; Huang, H.; Vermote, E.; Skakun, S.; Roger, J.C. Characterization of Sentinel-2A and Landsat-8 top of atmosphere, surface, and nadir BRDF adjusted reflectance and NDVI differences. *Remote Sens. Environ.* **2018**, *215*, 482–494. [CrossRef]

96. Ullman, S. The interpretation of structure from motion. *Proc. R. Soc. Lond. Ser. B Biol. Sci.* **1979**, *203*, 405–426. [CrossRef]

97. Lowe, M. Distinctive Image Features from Scale-Invariant Keypoints. *Int. J. Comput. Vis.* **2004**, *60*, 91–110. [CrossRef]

98. Brown, M.; Lowe, D.G. Unsupervised 3D object recognition and reconstruction in unordered datasets. In Proceedings of the 5th International Conference on 3D Digital Imaging and Modeling, Ottawa, ON, Canada, 13–16 June 2005.

99. Snavely, N.; Seitz, S.M.; Szeliski, R. Modeling the World from Internet Photo Collections. *Int. J. Comput. Vis.* **2007**, *80*, 189–210. [CrossRef]

100. Westoby, M.; Brasington, J.; Glasser, N.; Hambrey, M.; Reynolds, J. 'Structure-from-Motion' photogrammetry: A low-cost, effective tool for geoscience applications. *Geomorphology* **2012**, *179*, 300–314. [CrossRef]

101. Fonstad, M.A.; Dietrich, J.T.; Courville, B.C.; Jensen, J.L.; Carbonneau, P.E. Topographic structure from motion: A new development in photogrammetric measurement. *Earth Surf. Process. Landf.* **2013**, *38*, 421–430. [CrossRef]

102. Kaufmann, V.; Kellerer-Pirklbauer, A.; Lieb, G.K.; Slupetzky, H.; Avian, M. Glaciological Studies at Pasterze Glacier (Austria) Based on Aerial Photographs. In *Monitoring and Modeling of Global Changes: A Geomatics Perspective*; Li, J., Yang, X., Eds.; Springer: Dordrecht, The Netherlands, 2015; pp. 173–198._9. [CrossRef]

103. Micheletti, N.; Chandler, J.H.; Lane, S.N. Structure from Motion (Sfm) Photogrammetry. In *Geomorphological Techniques*; Cook, S.J., Clarke, L.E., Nield, J.M., Eds.; British Society for Geomorphology: London, UK, 2015; pp. 1–12.

104. Langhammer, J.; Lendzioch, T.; Miřijovský, J.; Hartvich, F. UAV-Based Optical Granulometry as Tool for Detecting Changes in Structure of Flood Depositions. *Remote Sens.* **2017**, *9*, 240. [CrossRef]

105. Hemmelder, S.; Marra, W.; Markies, H.; Jong, S.M.D. Monitoring river morphology & bank erosion using UAV imagery—A case study of the river Buëch, Hautes-Alpes, France. *Int. J. Appl. Earth Obs.* **2018**, *73*, 428–437. [CrossRef]

106. James, M.R.; Chandler, J.H.; Eltner, A.; Fraser, C.; Miller, P.E.; Mills, J.P.; Noble, T.; Robson, S.; Lane, S.N. Guidelines on the use of structure-from-motion photogrammetry in geomorphic research. *Earth Surf. Process. Landf.* **2019**, *44*, 2081–2084. [CrossRef]

107. O'Connor, J.; Smith, M.J.; James, M.R. Cameras and settings for aerial surveys in the geosciences. *Prog. Phys. Geogr.* **2017**, *41*, 325–344. [CrossRef]

108. James, M.R.; Robson, S.; Smith, M.W. 3-D uncertainty-based topographic change detection with structure-from-motion photogrammetry: Precision maps for ground control and directly georeferenced surveys. *Earth Surf. Process. Landf.* **2017**, *42*, 1769–1788. [CrossRef]

109. James, M.R.; Robson, S. Straightforward reconstruction of 3D surfaces and topography with a camera: Accuracy and geoscience application. *J. Geophys. Res. Earth Surf.* **2012**, *117*. [CrossRef]

110. Remondino, F.; Barazzetti, L.; Nex, F.; Scaioni, M.; Sarazzi, D. UAV Photogrammetry for mapping and 3D modeling—Current status and future perspectives. *ISPRS Int. Arch. Photogramm. Remote. Sens. Spat. Inf. Sci.* **2012**, *XXXVIII-1/C22*, 25–31. [CrossRef]

111. Colomina, I.; Molina, P. Unmanned aerial systems for photogrammetry and remote sensing: A review. *ISPRS J. Photogramm. Remote Sens.* **2014**, *92*, 79–97. [CrossRef]

112. González-Jorge, H.; Martínez-Sánchez, J.; Bueno, M.; .; Arias, P. Unmanned Aerial Systems for Civil Applications: A Review. *Drones* **2017**, *1*, 2. [CrossRef]

113. Immerzeel, W.; Kraaijenbrink, P.; Shea, J.; Shrestha, A.; Pellicciotti, F.; Bierkens, M.; de Jong, S. High-resolution monitoring of Himalayan glacier dynamics using unmanned aerial vehicles. *Remote Sens. Environ.* **2014**, *150*, 93–103. [CrossRef]

114. Bhardwaj, A.; Sam, L.; Akanksha.; Martín-Torres, F.J.; Kumar, R. UAVs as remote sensing platform in glaciology: Present applications and future prospects. *Remote Sens. Environ.* **2016**, *175*, 196–204. [CrossRef]

115. Seier, G.; Kellerer-Pirklbauer, A.; Wecht, M.; Hirschmann, S.; Kaufmann, V.; Lieb, G.K.; Sulzer, W. UAS-Based Change Detection of the Glacial and Proglacial Transition Zone at Pasterze Glacier, Austria. *Remote Sens.* **2017**, *9*, 549. [CrossRef]

116. Seier, G.; Sulzer, W.; Lindbichler, P.; Gspurning, J.; Hermann, S.; Konrad, H.M.; Irlinger, G.; Adelwöhrer, R. Contribution of UAS to the monitoring at the Lärchberg-Galgenwald landslide (Austria). *Int. J. Remote Sens.* **2018**, *39*, 5522–5549. [CrossRef]

117. Bash, E.; Moorman, B.; Gunther, A. Detecting Short-Term Surface Melt on an Arctic Glacier Using UAV Surveys. *Remote Sens.* **2018**, *10*, 1547. [CrossRef]

118. Fugazza, D.; Scaioni, M.; Corti, M.; D'Agata, C.; Azzoni, R.S.; Cernuschi, M.; Smiraglia, C.; Diolaiuti, G.A. Combination of UAV and terrestrial photogrammetry to assess rapid glacier evolution and map glacier hazards. *Nat. Hazards Earth Syst. Sci.* **2018**, *18*, 1055–1071. [CrossRef]

119. Carrivick, J.L.; Smith, M.W. Fluvial and aquatic applications of Structure from Motion photogrammetry and unmanned aerial vehicle/drone technology. *Wiley Interdiscip. Rev. Water* **2018**, *6*, e1328. [CrossRef]

120. Heritage, G.; Entwistle, N. Drone Based Quantification of Channel Response to an Extreme Flood for a Piedmont Stream. *Remote Sens.* **2019**, *11*, 2031. [CrossRef]

121. Vázquez-Tarrío, D.; Borgniet, L.; Liébault, F.; Recking, A. Using UAS optical imagery and SfM photogrammetry to characterize the surface grain size of gravel bars in a braided river (Vénéon River, French Alps). *Geomorphology* **2017**, *285*, 94–105. [CrossRef]

122. How, P.; Hulton, N.; Buie, L.; Benn, D. PyTrx: A Python-Based Monoscopic Terrestrial Photogrammetry Toolset for Glaciology. *Front. Earth Sci.* **2020**, *8*. [CrossRef]

123. Felbauer, L.; Mergili, M.; Neureiter, A.; Hynek, B. Automated processing of terrestrial photos for glacier monitoring at Hoher Sonnblick (Austria). *Geophys. Res. Abstr.* **2019**, *21*, EGU2019-11133.

124. Härer, S.; Bernhardt, M.; Corripio, J.; Schulz, K. PRACTISE—Photo Rectification And ClassificaTIon SoftwarE (V.1.0). *Geosci. Model Dev.* **2013**, *6*, 837–848. [CrossRef]

125. Dumont, M.; Gascoin, S. Optical Remote Sensing of Snow Cover. In *Land Surface Remote Sensing in Continental Hydrology*; Elsevier: Amsterdam, The Netherlands, 2016; pp. 115–137. [CrossRef]

126. Kick, W. Long-Term Glacier Variations Measured by Photogrammetry. A Re-Survey of Tunsbergdalsbreen after 24 Years. *J. Glaciol.* **1966**, *6*, 3–18. [CrossRef]

127. Portenier, C.; Hüsler, F.; Härer, S.; Wunderle, S. Towards a webcam-based snow cover monitoring network: Methodology and evaluation. *The Cryosphere Discuss* **2019**. [CrossRef]

128. Huss, M.; Sold, L.; Hoelzle, M.; Stokvis, M.; Salzmann, N.; Farinotti, D.; Zemp, M. Towards remote monitoring of sub-seasonal glacier mass balance. *Ann. Glaciol.* **2013**, *54*, 75–83. [CrossRef]

129. Parajka, J.; Haas, P.; Kirnbauer, R.; Jansa, J.; Blöschl, G. Potential of time-lapse photography of snow for hydrological purposes at the small catchment scale: Potential of time-lapse photography of snow for hydrological purposes. *Hydrol. Process.* **2012**, *26*, 3327–3337. [CrossRef]

130. Salvatori, R.; Plini, P.; Giusto, M.; Valt, M.; Salzano, R.; Montagnoli, M.; Cagnati, A.; Crepaz, G.; Sigismondi, D. Snow cover monitoring with images from digital camera systems. *Ital. J. Remote Sens.* **2011**, *137*, 137–145. [CrossRef]

131. Farinotti, D.; Magnusson, J.; Huss, M.; Bauder, A. Snow accumulation distribution inferred from time-lapse photography and simple modelling. *Hydrol. Process.* **2010**, *24*, 2087–2097. [CrossRef]

132. Kenner, R.; Phillips, M.; Limpach, P.; Beutel, J.; Hiller, M. Monitoring mass movements using georeferenced time-lapse photography: Ritigraben rock glacier, western Swiss Alps. *Cold Reg. Sci. Technol.* **2018**, *145*, 127–134. [CrossRef]

133. Danielson, B.; Sharp, M. Development and application of a time-lapse photograph analysis method to investigate the link between tidewater glacier flow variations and supraglacial lake drainage events. *J. Glaciol.* **2013**, *59*, 287–302. [CrossRef]

134. Ide, R.; Oguma, H. A cost-effective monitoring method using digital time-lapse cameras for detecting temporal and spatial variations of snowmelt and vegetation phenology in alpine ecosystems. *Ecol. Inform.* **2013**, *16*, 25–34. [CrossRef]

135. Lieb, G.; Kellerer-Pirklbauer, A. Die Pasterze, Österreichs größter Gletscher, und seine lange Messreihe in einer Ära massiven Gletscherschwundes. In *Gletscher im Wandel. 125 Jahre Gletschermessdienst des Alpenvereins*; Fischer, A., Patzelt, G., Achrainer, M., Groß, G., Lieb, G., Kellerer-Pirklbauer, A., Bendler, G., Eds.; Springer: Berlin/Heidelberg, Geramny, 2018; pp. 31–51.

136. Hall, D.K.; Bayr, K.J.; Bindschadler, R.A.; Schöner, W. Changes in the Pasterze Glacier, Austria, as Measured from the Ground and Space. In Proceedings of the 58th Eastern Snow Conference, Ottawa, ON, Canada, 17–19 May 2001; pp. 1–7.

137. Hall, D.K.; Bayr, K.J.; Schöner, W.; Bindschadler, R.A.; Chien, J.Y. Consideration of the errors inherent in mapping historical glacier positions in Austria from the ground and space (1893–2001). *Remote Sens. Environ.* **2003**, *86*, 566–577. [CrossRef]

138. Kaufmann, V.; Kellerer-Pirklbauer, A.; Kenyi, L. Satellite-based measurement of the surface displacement of the largest glacier in Austria. In Proceedings of the 4th Symposium of the Hohe Tauern National Park for Research in Protected Areas, Kaprun, Austria, 17–19 September 2009; pp. 145–149.

139. Kellerer-Pirklbauer, A.; Lieb, G.; Gspurning, J. The response of partially debris-covered valley glaciers to climate change: The Example of the Pasterze Glacier (Austria) in the period 1964 to 2006. *Geogr. Ann. Ser. A Phys. Geogr.* **2008**, *90*, 1–17. [CrossRef]

140. Kellerer-Pirklbauer, A.; Lieb, G.; Avian, M.; Carrivick, J. Climate change and rock fall events in high mountain areas: Numerous and extensive rock falls in 2007 at Mittlerer Burgstall, central Austria. *Geogr. Ann. Ser. A Phys. Geogr.* **2012**, *94*, 59–78. [CrossRef]

141. Kaufmann, V. Geomorphometrische Dokumentation des Felssturzes (2007) am Mittleren Burgstall, Glocknergruppe) des Felssturzes (2007). In Proceedings of the 6. Treffen des Arbeitskreises Permafrost, Salzburg, Austria, 24–26 October 2013.

142. Kellerer-Pirklbauer, A.; Avian, M.; Götz, J.; Hirschmann, S.; Sulzer, W.; Seier, G.; Douglas, I.; Lieb, G.; Wecht, M.; Ziesler, C. Rapid supraglacial-lake to proglacial-lake transition in a sediment-rich environment (Pasterze Glacier, Austria). *Geophys. Res. Abstr.* **2019**, *21*, EGU2019–17765.

143. Kellerer-Pirklbauer, A.; Avian, M.; Bernsteiner, F.; Gahleitner, H.; Götz, J.; Lieb, G.; Ziesler, C. Quantification of ice-breakup events and iceberg dynamics in a highly dynamical proglacial lake in Austria (Pasterze Glacier). *Geophys. Res. Abstr.* **2020**, *22*, EGU2020-13760. [CrossRef]

144. Wujanz, D.; Avian, M.; Krueger, D.; Neitzel, F. Identification of stable areas in unreferenced laser scans for automated geomorphometric monitoring. *Earth Surf. Dynam.* **2018**, *6*, 303–317. [CrossRef]

145. Soudarissanane, S.; Lindenbergh, R.; Menenti, M.; Teunissen, P. Incidence angle influence on the quality of terrestrial laser scanning points. In Proceedings of the ISPRS Workshop Laserscanning 2009, Paris, France, 1–2 September 2009; pp. 183–188.

146. Hooper, A.; Segall, P.; Zebker, H. Persistent scatterer interferomtric synthetic aperture radar for crustal deformation analysis, with application to Volcán Alcedo, Galápagos, Galapagos. *J. Geophys. Res.-Solid* **2007**, *112*. [CrossRef]

147. Casu, F.; Elefante, S.; Imperatore, P.; Zinno, I.; Manunta, M.; De Luca, C.; Lanari, R. SBAS-DInSAR Parallel Processing for Deformation Time-Series Computation. *IEEE J. STARS* **2014**, *7*, 3285–3296. [CrossRef]

148. Luca, C.D.; Cuccu, R.; Elefante, S.; Zinno, I.; Manunta, M.; Casola, V.; Rivolta, G.; Lanari, R.; Casu, F. An On-Demand Web Tool for the Unsupervised Retrieval of Earth's Surface Deformation from SAR Data: The P-SBAS Service within the ESA G-POD Environment. *Remote Sens.* **2015**, *7*, 15630–15650. [CrossRef]

149. Pepe, A.; Lanari, R. On the Extension of the Minimum Cost Flow Algorithm for Phase Unwrapping of Multitemporal Differential SAR Interferograms. *IEEE Trans. Geosci. Remote Sens.* **2006**, *44*, 2374–2383. [CrossRef]

150. Pepe, A.; Yang, Y.and Manzo, M.; Lanari, R. Improved EMCF-SBAS Processing Chain Based on Advanced Techniques for the Noise-Filtering and Selection of Small Baseline Multi-Look DInSAR Interferograms. *IEEE Trans. Geosci. Remote Sens.* **2015**, *53*, 4394–4417. [CrossRef]

151. Esch, C.; Köhler, J.; Gutjahr, K.; Schuh, W. On the Analysis of the Phase Unwrapping Process in a D-InSAR Stack with Special Focus on the Estimation of a Motion Model. *Remote Sens.* **2019**, *11*, 2295. [CrossRef]

152. Yague-Martinez, N.; Zan, F.; Prats-Iraola, P. Coregistration of Interferometric Stacks of Sentinel-1 TOPS Data. *IEEE Geosci. Remote Sens. Lett.* **2017**, *14*, 1002–1006. [CrossRef]

153. Pasquali, P.; Cantone, A.; Riccardi, P.; Defilippi, M.; Ogushi, F.; Gagliano, S.; Tamura, M. Mapping of Ground Deformations with Interferometric Stacking Techniques. In *Land Applications of Radar Remote Sensing*; InTech: London, UK, 2014. [CrossRef]

154. Yan, L.; Roy, D.P.; Li, Z.; Zhang, H.; Huang, H. Sentinel-2A multi-temporal misregistration characterization and an orbit-based sub-pixel registration methodology. *Remote Sens. Environ.* **2018**, *215*, 495–506. [CrossRef]

155. Watson, C.S.; King, O.; Miles, E.S.; Quincey, D.J. Optimising NDWI supraglacial pond classification on Himalayan debris-covered glaciers. *Remote Sens. Environ.* **2018**, *217*, 414–425. [CrossRef]

156. Du, Y.; Zhang, Y.; Ling, F.; Wang, Q.; Li, W.; Li, X. Water Bodies' Mapping from Sentinel-2 Imagery with Modified Normalized Difference Water Index at 10-m Spatial Resolution Produced by Sharpening the SWIR Band. *Remote Sens.* **2016**, *8*, 354. [CrossRef]

157. McFeeters, S.K. The use of the Normalized Difference Water Index (NDWI) in the delineation of open water features. *Int. J. Remote Sens.* **1996**, *17*, 1425–1432. [CrossRef]

158. Miles, K.E.; Willis, I.C.; Benedek, C.L.; Williamson, A.G.; Tedesco, M. Toward Monitoring Surface and Subsurface Lakes on the Greenland Ice Sheet Using Sentinel-1 SAR and Landsat-8 OLI Imagery. *Front. Earth Sci.* **2017**, *5*. [CrossRef]

159. Paster, M.; Flödl, P.; Pulg, U.; Habersack, H.; Skoglund, R.; Neureiter, A.; Weyss, G.; Hynek, B.; Hauer, C. Channel evolution processes in a diamictic glacier foreland: Implications on downstream sediment supply and hydropower use: Case study Pasterze/Austria. *Cold Reg. Sci. Technol.* **2020**, submitted.

160. Agisoft. *Agisoft PhotoScan User Manual*; Professional Edition, Version 1.4; Agisoft, LLC.: Saint Petersburg, Russia, 2018.

161. Wiesenegger, H.; Kum, G.; Slupetzky, H. 'Unterer Eisbodensee'—A good example for the future evolution of glacial lakes in Austria? In Proceedings of the 6th Symposium of the Hohe Tauern National Park for Research in Protected Areas, Mittersill, Austria, 2–3 November 2017; pp. 721–725.

162. Narama, C.; Duishonakunov, M.; Kääb, A. Daiyrov, M.; Abdrakhmatov, K. The 24 July 2008 outburst flood at the western Zyndan glacier lake and recent regional changes in glacier lakes of the Teskey Ala-Too range, Tien Shan, Kyrgyzstan. *Nat. Hazards Earth Syst. Sci.* **2010**, *10*, 647–659. [CrossRef]

163. Rounce, D.R.; Byers, A.C.; Byers, E.A.; McKinney, D.C. Brief communication: Observations of a glacier outburst flood from Lhotse Glacier, Everest area, Nepal. *Cryosphere* **2017**, *11*, 443–449. [CrossRef]

164. Heritage, G.L.; Milan, D.J. Terrestrial Laser Scanning of grain roughness in a gravel-bed river. *Geomorphology* **2009**, *113*, 4–11. [CrossRef]

165. Hauer, C.; Pulg, U. The non-fluvial nature of Western Norwegian rivers and the implications for channel patterns and sediment composition. *CATENA* **2018**, *171*, 83–98. [CrossRef]

166. Geilhausen, M.; Otto, J.C.; Schrott, L. Spatial distribution of sediment storage types in two glacier landsystems (Pasterze & Obersulzbachkees, Hohe Tauern, Austria). *J. Maps* **2012**, *8*, 242–259. [CrossRef]

167. Cong, X.; Balss, U.; Gonzalez, F.R.; Eineder, M. Mitigation of Tropospheric Delay in SAR and InSAR Using NWP Data: Its Validation and Application Examples. *Remote Sens.* **2018**, *10*, 1515. [CrossRef]

168. Yin, X.; Busch, W. Nutzung der Sentinel-1 Aufnahmekonfigurationen zur Ableitung von Bodenbewegungskomponenten im Rahmen eines radarinterferometrischen Bodenbewegungsmonitorings. In *Tagungsband Geomonitoring*; Busch, W., Ed.; 2018; TU Clausthal: Clausthal-Zellerfeld, Germany, pp. 119–138.

169. Horninger, G.; Weiss, E. Engineering Geology in mountainous regions. *Abh. Der Geol. Bundesanst.* **1980**, *34*, 257–286.

170. Weidner, S.; Moser, M. Der Bewegungsablauf tiefgreifender Hangdeformationen. In *Geoforum Umhausen*; Huber, U.M., Bugmann, H.K.M., Reasoner, M.A., Eds.; Verein Geoforum Tirol: Umhausen, Austria, 2001; Volume 2, pp. 14–25.

 *remote sensing*

Article

Estimating Daily Inundation Probability Using Remote Sensing, Riverine Flood, and Storm Surge Models: A Case of Hurricane Harvey

Jiayong Liang [1] and Desheng Liu [2],*

[1] Cloud to Street, Brooklyn, NY 11217, USA; jiayong@cloudtostreet.info
[2] Department of Geography, The Ohio State University, Columbus, OH 43210, USA
* Correspondence: liu.738@osu.edu

Received: 29 February 2020; Accepted: 5 May 2020; Published: 8 May 2020

Abstract: Heavy precipitation and storm surges often co-occur and compound together to form sudden and severe flooding events. However, we lack comprehensive observational tools with high temporal and spatial resolution to capture these fast-evolving hazards. Remotely sensed images provide extensive spatial coverage, but they may be limited by adverse weather conditions or platform revisiting schedule. River gauges could provide frequent water height measurement but they are sparsely distributed. Riverine flood and storm surge models, depending on input data quality and calibration process, have various uncertainties. These lead to inevitable temporal and spatial gaps in monitoring inundation dynamics. To fill in the observation gaps, this paper proposes a probabilistic method to estimate daily inundation probability by combining the information from multiple sources, including satellite remote sensing, riverine flood depth, storm surge height, and land cover. Each data source is regarded as a spatial evidence layer, and the weight of evidence is calculated by assessing the association between the evidence presence and inundation occurrence. Within a Bayesian model, the fusion results are daily inundation probability whenever at least one data source is available. The proposed method is applied to estimate daily inundation in Harris, Texas, impacted by Hurricane Harvey. The results agree with the reference water extent, high water mark, and extracted tweet locations. This method could help to further understand flooding as an evolving time-space process and support response and mitigation decisions.

Keywords: fusion; inundation probability; remote sensing; Hurricane Harvey; NDWI; ADCIRC

1. Introduction

In the past 50 years, over 650 Atlantic cyclones have lead to life loss, property damage, and psychological consequences [1,2]. On average, in the U.S., two to three tropical cyclones cause about 50 deaths per year. Nearly 90% of fatalities are water-related, such as those caused by drowning, among which the storm surge is responsible for roughly half of the total deaths. The deadliest storm from 1963 to 2012 was Katrina, costing nearly 40% of the fatalities [3]. In 2017, the hurricane season in the Atlantic got the most attention due to its abnormal intensity and enormous damage, notably from Harvey, Irma, and Maria. For example, Harvey brought record-setting rainfall and caused devastating flooding. During Hurricane Harvey, about half of the casualties were found outside the FEMA 500-year floodplain [4]. The majority of deaths (80%) occurred within the first week after the hurricane landfall. To facilitate early warning and prevent damage, risk maps need to be well prepared and communicated with emergency agencies and the public as frequently as possible.

As greenhouse gases continue to increase in concentration, the resulting rising sea level and intensifying storm activity will increase flood risk in the coming decades [5–8]. The heat content in the Gulf of Mexico was highest in the summer of 2017, and decreased dramatically after the dismissal of Hurricane Harvey, in the form of unprecedented rainfall. Some locations observed more than 1500 mm precipitation. It is estimated that climate change could increase rainfall by 35% [9]. The annual probability of such precipitation has increased six-fold (to 6%) since the late twentieth century. Flood characteristics with a high spatial and temporal resolution are scarce, such as information of the return period, height, and inundation extent [10].

Remote sensing is the most widely used and effective technique to map large-scale flood extents. Open water can be distinguished from dry land based on its spectral characteristics, which have low reflectance in visible and infrared bands, or based on its low emitted radiation and backscattering in microwave bands. On the one hand, studies using optical sensors can employ more spectral information and benefit from a long archive of consistent observations [11,12]. On the other hand, active and passive microwave signals can penetrate clouds, and thus sensors using the microwave are able to monitor inundation regardless of adverse illumination and weather. There is increasing availability of microwave sensors, and their applications in flood mapping have been published [13–18]. Remote sensing has advantages in both long-time-period and near-real-time applications. However, due to the current design tradeoff between revisit time and spatial resolution in public satellite data, continuous observations with both high spatial and temporal resolution are absent. No single sensor can provide reliable daily observation for rapid flood response.

Flood inundation models are effective for risk mapping, damage assessment, forecast and engineering [19–21]. Based on hydrological simulations, these models require streamflow measurements and forecasts for flooding predictions and early-warning [22]. The river monitoring system is far more sophisticated in developed regions than in the developing countries where flood risk is increasing. With high-quality data acquisition and processing, empirical methods are straightforward and accurate. Other state-of-the-art flood modeling approaches are hydrodynamic models and conceptual models. The hydrodynamic models describe water motion by assuming conversion of mass and momentum, yet the conceptual models are non-physically based and fastest in computation for large scale applications [19]. Assessing the hazard of storm surge flooding demands an understanding of storm activity, local storm surge, and sea level. While there are numerous models addressing those topics, various uncertainties persist [5]. Model uncertainties have various sources, ranging from model inputs (like precipitation, streamflow, topography, etc.), underlying stochastic processes, to incomplete understanding of the underlying hydrologic mechanisms [19].

Numerous attempts have been made to facilitate flood observation and rapid response. However, one single source of information could not provide comprehensive coverage with high spatial and temporal resolution. To fill the gaps of lacking consistent flood maps, this study proposes a probabilistic method to combine multi-source data, including remote sensing data, estimation from riverine flooding and storm surge models, and underlying surface features, into a daily inundation risk map. The proposed method can estimate inundation probability during an emergency, through an open and flexible framework. It is able to deal with different data availability for different days. When at least one data source is accessible, the fusion method could estimate inundation probability. Data acquired by observation or model, of various qualities, are combined by the Bayesian framework. To our knowledge, this is the first method to estimate daily inundation probability using discrete data sources with different mechanisms, aiming to capture fast-changing hurricane impact. The following sections detail the proposed method (Section 2), data collection and processing (Section 3), results (Section 4), discussion (Section 5), and conclusions (Section 6).

2. Method

The proposed method fuses different data sources into a series of probabilistic flood maps, indicating the inundation probability for each pixel. Each data source could contain information for specific time points, and here the smallest unit of the time period is a day. The estimation of inundation probability is conducted on a daily basis, whenever data are available either from satellite imagery, flood models using gauge input, and/or storm surge models. Each data source is regarded as an evidence layer, and multiple pieces of evidence of one pixel are combined into a probabilistic value to show inundation occurrence. The evidence is weighted by a Bayesian framework, and more details are given in the following subsections. Section 2.1 describes the weight of evidence theory, along with the validation and assessment scheme given in Section 2.2.

2.1. Weight of Evidence

Weight of evidence is a model to combine different information based on the Bayesian conditional probability framework [23,24]. Suppose that there are $N(M)$ pixels in the area of interest, among which event A occurs in $N(A)$ pixels. In other words, there are $N(A)$ inundated pixels. The prior probability P_{prior} could be estimated as $N(A)/N(M)$ when a pixel is selected randomly showing the event A. Furthermore, there are $N(B)$ pixels that evidence B are present, and $N(\overline{B})$ pixels that B is absent or not observed. With such information, the probability of A conditioning on B could be estimated. Given the event A, a positive (W^+) and a negative (W^-) weight could be estimated for the spatial evidence B, using the formulas

$$W^+ = \ln \frac{P(B|A)}{P(B|\overline{A})} \tag{1}$$

$$W^- = \ln \frac{P(\overline{B}|A)}{P(\overline{B}|\overline{A})} \tag{2}$$

where the conditional probabilities $P(B|A)$, $P(B|\overline{A})$, $P(\overline{B}|A)$, and $P(\overline{B}|\overline{A})$ are calculated by the occurrences or absences of the event and spatial evidence as likelihood ratios,

$$P(B|A) = \frac{N(B \cap A)}{N(A)} \tag{3}$$

$$P(B|\overline{A}) = \frac{N(B \cap \overline{A})}{N(\overline{A})} \tag{4}$$

$$P(\overline{B}|A) = \frac{N(\overline{B} \cap A)}{N(A)} \tag{5}$$

$$P(\overline{B}|\overline{A}) = \frac{N(\overline{B} \cap \overline{A})}{N(\overline{A})} \tag{6}$$

where $N(B \cap A)$ is the amount of pixels that both inundation (event A) and evidence B are observed; $N(B \cap \overline{A})$ means that evidence B is present but there is no inundation, similarly for $P(\overline{B}|A)$ and $P(\overline{B}|\overline{A})$. A large sample size is required for statistically significant weights where the significance of weights could be estimated by

$$S^2 W^+ = \frac{1}{N(B \cap A)} + \frac{1}{N(B \cap \overline{A})} \tag{7}$$

$$S^2 W^- = \frac{1}{N(\overline{B} \cap A)} + \frac{1}{N(\overline{B} \cap \overline{A})} \tag{8}$$

Based on the positive and negative weights, the contrast of weights is defined as their difference

$$C = W^+ - W^- \tag{9}$$

This contrast of weights C measures the spatial association between the event A and the spatial evidence B, in which $C > 0$ suggests a positive spatial association and $C < 0$ a negative one. When C

approaches zero, there is no obvious spatial association. The statistical significance of spatial association could be calculated by studentizing C,

$$S(C) = \sqrt{S^2W^+ + S^2W^-} \tag{10}$$

$S(C)$ can help to combine the multiclass evidence into predictor maps. When a confidence level is provided, it is compared to $S(C)$ to finalize the weight, either positive W^+ or negative W^-. In this study, the evidence layers are assumed to be positively associated with water presence. When $S(C)$ is larger than a predefined confidence level, the corresponding positive weight of the subset of the evidence layer would be used.

The predictor map is obtained by combining the presence and absence of spatial evidence. For a unique condition, the posterior probability is estimated by prior probability and the weights of spatial evidence

$$P_{post} = \frac{e^{\sum_{j=1}^{n} W_j + \ln O(A)}}{1 + e^{\sum_{j=1}^{n} W_j + \ln O(A)}} \tag{11}$$

where

$$O(A) = P_{prior}/\left(1 - P_{prior}\right) \tag{12}$$

W_j denotes the positive weight (W^+_j) of the spatial evidence B_j in this study, and $O(A)$ is the prior odd of inundation event A. The variance of P_{post} could be estimated by the variance of weights as

$$s^2\left(P_{post}\right) = \left[\frac{1}{N(A)} + \sum_{j=1}^{n} s^2\left(W_j\right)\right] \times P_{post}^{\,2} \tag{13}$$

2.2. Validation and Assessment

The proposed method estimates inundation probability, which is difficult to be directly evaluated. The available ground reference data are either numeric or binary, including optical images, high water marks, published flood maps, and posted flood locations. In order to evaluate the proposed method, the inundation probability is compared to a referenced water extent derived from an optical image, which is acquired on the same day of prediction, as well as a published flood map. Given the validated water extent, a quantile–quantile plot could be generated as a reliable diagram [15,25]. The reliable diagram shows the actual occurrence ratios and the estimated probability using discrete intervals. For an example of statistically reliable estimation, among the pixels with 80–90% inundation probability, around 85% of these pixels could be actually flooded according to the validated water extent. Ideally, the reliable diagram is a dot plot aligning a 1:1 line. The difference between the ideal 1:1 line and the estimation could be measured by their weighted root-mean-square difference (WRMS) ε

$$\varepsilon = \sqrt{\frac{\sum_{i=1}^{N} \left(\widetilde{p}_i - \hat{p}_l\right)^2 n_i}{\sum_{i=1}^{N} n_i}} \tag{14}$$

where N is the number of intervals that the range of probability values $[0, 1]$ split into for ε calculation, and n_i is the pixel number of the ith interval; $\widetilde{p}_i$, and $\hat{p}_i$ are the estimated and validated inundation probability.

Additionally, daily inundation probability is estimated over an extended period of time and collected. Over that time period, the maximum of the daily inundation probability for each pixel is calculated and compared to the high water mark and flood location. The optical images are collected from two commercial platforms (Planet and DMCii). The high water mark dataset is collected and distributed by FEMA. This dataset provides information on qualified flood location and its associated flood depth, which is the height of flood water above ground level. The flood location is extracted from Twitter using the Location Extractor API [26]. The published flood map is generated by FEMA, for which extensive remote sensing data have been used.

3. Data

The data used as spatial evidence layer include SAR backscatter, optical water index, riverine flood depth, storm surge simulated water height, and land cover. The description and preprocessing of each data source are provided in the following sub-sections, along with data collected for the study case in Harris, Texas, during Hurricane Harvey.

3.1. SAR Backscatter Intensity

The SAR could provide all-sky observations by the received signal of backscatter intensity, which is sensitive to surface roughness. Water, as a smooth surface, usually shows a rather low backscatter value. As a part of the Copernicus initiative, the Sentinel-1 program focuses on land and ocean monitoring and emergency response. It consists of two polar-orbiting satellites, among which Sentinel-1A was launched in April 2014 and Sentinel-1B in April 2016. They both carry a C-band (5.4 GHz) imaging SAR [27]. With an aim to build a long-term data archive, the Sentinel-1 inherits SAR data from European Remote Sensing satellite, ENVISAT, and the Canadian satellite RADARSAT, and continues their observations. The imagery products are free of charge to all users and fast delivered, at best within hours for emergency response. The Sentinel-1 consists of four imaging modes with different resolution and coverage, single and dual-polarization, and improved revisit time. The repeat cycle for one satellite is 12 days at the equator and 6 days for the constellation.

Following standard instruction, the SAR preprocessing in this study is conducted by using the Sentinel Application Platform (SNAP) from the European Space Agency, including radiometric calibration and terrain correction. The ground range detected (GRD) product contains information needed to convert digital pixel values into calibrated backscatter intensity. This backscatter intensity of each pixel is then transformed into the decibel unit as spatial evidence. In the process of geometric correction, the elevation data are downloaded from the National Elevation Dataset (NED) with a spacing of 1/3 arc-second. The SNAP also provides automatically downloaded digital elevation models (DEMs) from several products. Speckle has been reduced to some extent for the GRD products, so additional filtering except the default setting in SNAP is skipped to preserve the fine pixel resolution. This study uses backscatter intensity to generate spatial layers and calculate weights. For urban flooding in densely developed areas, interferometry coherence could be useful in future studies.

3.2. Optical Water Index

Surface water extent is obvious in the optical imagery so water classification is rather straightforward using an optical image. The spectral characteristics of water are special, and the Normalized Difference Water Index (NDWI, [28]) is one of the most widely used spectral indices to delineate water.

$$NDWI = (green - NIR)/(green + NIR) \tag{15}$$

When an optical image is available, its spectral bands of green and near-infrared (NIR) are used to calculate the NDWI and used as an evidence component. In this study, two optical images were collected from the two constellations of satellites, namely Planet and DMCii. Provided by PlanetScope, the Planet image includes RGB and NIR bands, with a spatial resolution of 3 m. The DMCii image also has three spectral bands (R, G, NIR) but a coarse spatial resolution of 22 m. Both radiometric calibration and registration are conducted for the collected satellite images from PlanetScope and DMCii. From the two optical images, the PlanetScope one is used to estimate the weight, and the DMCii one is used as a reference to assess the result.

3.3. Riverine Flood Depth

The riverine flood depth is calculated by the river flow model and gauge readings, which is processed by FEMA. It requires a high-resolution (usually 5 m) digital elevation model (DEM) from USGS. Stream or river gauges with reference to mean sea level are collected from the National Weather Service.

When the water level in the river is rising, the calculated flood extent may not reflect reality. This calculation needs to be continued and regularly updated as the flooding evolves. On a daily basis, gauges report their water stage readings when the river reaches its crest. If some location cannot observe crests during the time frame of interest, the highest record would be reported and used. All available gauge readings are collected and converted into the same datum (NAVD88). The updated water stage readings, crest or highest records are used to provide water surface elevation as the sum of readings and referenced sea level. The result is called the standardized water surface elevation.

Gauges are spatially sparse so that information between existing gauges needs to be estimated by interpolation. The interpolation processing relies on the understanding of the hydraulic situation—how water flows on a terrain surface. To facilitate this interpolation, the National Hydrography Data flow lines are used to create reinforcement points. A flow line depicts the path of a stream or river, and the gauges are located along a river. While the flow lines link the gauges, points on the flow line intersections and between gauges are used as reinforcement points. Water levels of the reinforcement points are estimated based on the gauge's readings and the distance to the gauges located on the upstream and downstream.

Once the water level elevations are available for the gauges and reinforcement points, a triangulated irregular network (TIN) is generated to represent the continuous water surface. This water surface is then converted to a raster image, with the same geo-reference and grid alignment as the ground DEM data. Overlaying the water surface with the DEM, flood depth can be calculated as the difference between the water surface elevation and ground elevation (similar to [29]). Flood extent is identified as regions where the flood depth is positive. For isolated flood areas, if they are not adjacent to gauges and reinforcement points, they are regarded as disconnected false flooding areas and are removed. The flooded region with depth information is the final product and denoted as the riverine flood depth. The flood depth is categorized into a multi-class set and is used as one spatial evidence to estimate the presence probability of flooding.

3.4. Storm Surge Simulated Water Height

The Advanced CIRCulation (ADCIRC) model is a finite element model to forecast or estimate water height caused by storms [30,31]. Millions of elements are used in the domain of interest where the element sizes are various, ranging from ten meters along the coast to ten kilometers in the deep ocean. For each element, wind and pressure fields are required and obtained from multiple hydraulic data sources. The ADCIRC model outputs include water surface elevation and velocity over the domain continuously. This study makes use of the estimated water surface elevation to interpolate water height caused by storm surge for the study region.

The storm surge model nodes calculate water height continuously, and the daily maximum records are used to generate the storm surge surface. The dense point set is transformed into a continuous surface through Thiessen polygon creation. Although the point set is relatively dense, the discrete points could not be directly combined with the other raster data. Thiessen polygons, also known as Voronoi diagram, are generated from the point set, in which the polygons share the same value, here water height, as the center point. In a space X, a Thiessen polygon R_k surrounding a point P_k could be expressed as

$$R_k = \left\{ x \in X \middle| d(x,\ P_k) \le d\left(x, P_j\right) \text{ for all } j \ne k \right\} \tag{16}$$

where the distance function $d(\ )$ used in this study is the Euclidean distance. After the Thiessen polygons are created and their values of water height are assigned, they are converted into rasters to match other data sets, regarding georeferenced and grid properties. Then the height is categorized as the riverine flood depth data using multiple quantiles of the total data.

3.5. Land Cover

The land cover information is obtained from the National Land Cover Database (NLCD) at a resolution of 30 m [32,33]. According to the modified Anderson Level II classification system, there are 16 classes of land use and cover in the NLCD product. The Level I classification is used in this study to aggregate the land cover types, including water, developed, barren, forest, shrubland, herbaceous, planted/cultivated, and wetlands. Each land cover type is used as layers of spatial evidence.

3.6. Data Collection for Hurricane Harvey

In August 2017, Hurricane Harvey swept Southeast Texas as the wettest and costliest tropical cyclone. More than 1000 mm of rain was received in many regions over four days, with a peak of over 1500 mm rain accumulations. The catastrophic flooding caused the death of over 100 people, displacement of more than 40,000 people and damage worth over 125 billion US dollars, according to the National Oceanic and Atmospheric Administration (NOAA). Harvey was formed on 17 August 2017, moved towards the northwest and made multiple landfalls. With an intensification phase of a Category 4 hurricane, Harvey made landfall on 25 August at San Jose Island and Holiday Beach, after leaving Barbados and Saint Vincent. After that, a torrential downpour of rain led to unprecedented flooding. Oil refineries were shut down, declining energy production, and some chemical plants suffered from explosions due to a power outage [10]. On 2 September, Harvey weakened and dissipated.

In Texas, the Department of Public Safety estimated that about 300,000 structures and 500,000 vehicles had been destroyed or damaged [4]. Houston metropolitan areas observed over 990 mm of precipitation during August, the wettest record since 1892. On 27 August, the National Weather Service office in Houston measured 408 mm of daily rainfall accumulations. Emergencies of flash flooding were issued several times from 26 August onwards. Harris County Flood Control District estimated that 25–30% of the county was submerged, around 450 square miles. The regions in Figure 1 are investigated, including the Addicks and Barker Reservoirs in the left part. These reservoirs were built by the US Army Corps of Engineers with aims to prevent flood damages to the city of Houston.

Harris County is the site of interest in this study, shown in Figure 1. According to the US Census Bureau, Harris is the most populous county in Texas and ranked third in population across the country, home to 4.5 million people. The county seat, Houston, is the fourth largest city in the U.S. In Figure 2, developed areas with different intensities are depicted in various shades of red (Land Cover Class 21–24), making up a large portion of the county area.

Located in the Gulf Coastal Plain, the region of interest is mainly made up of clay-based soils and it is low-lying. The digital elevation model and the derived slope are shown in Figures 3 and 4. The northwest part is higher in elevation than the southeast part where most rivers and streams flow. The slope is usually small, except along the mountain ridges and the river banks. This area is a vast floodplain depending on San Jacinto River and Buffalo Bayou to carry away flood and stormwater.

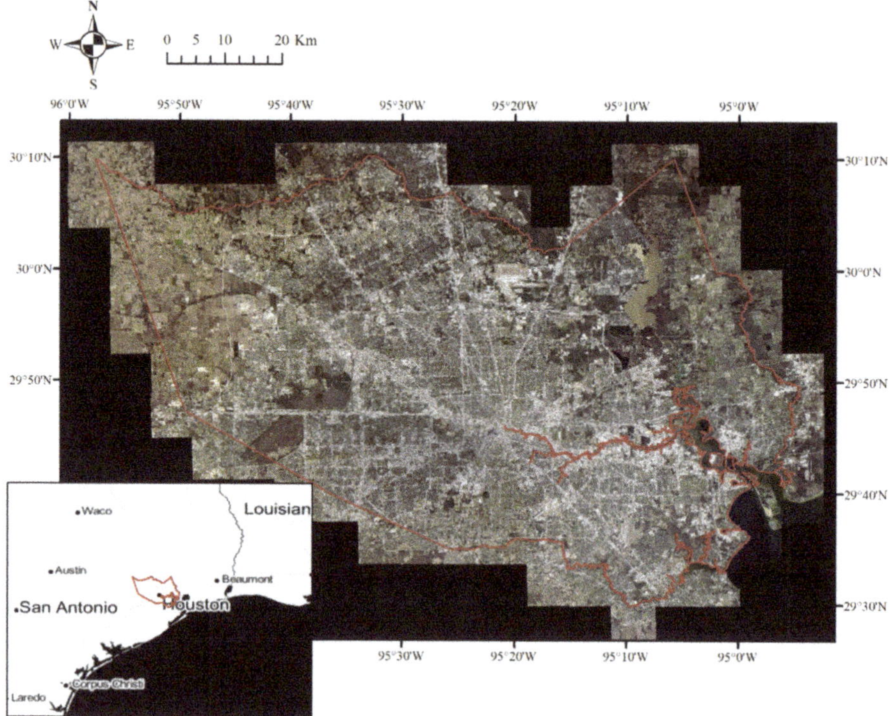

Figure 1. The study site Harris County, Texas, aerial image composited by RGB bands.

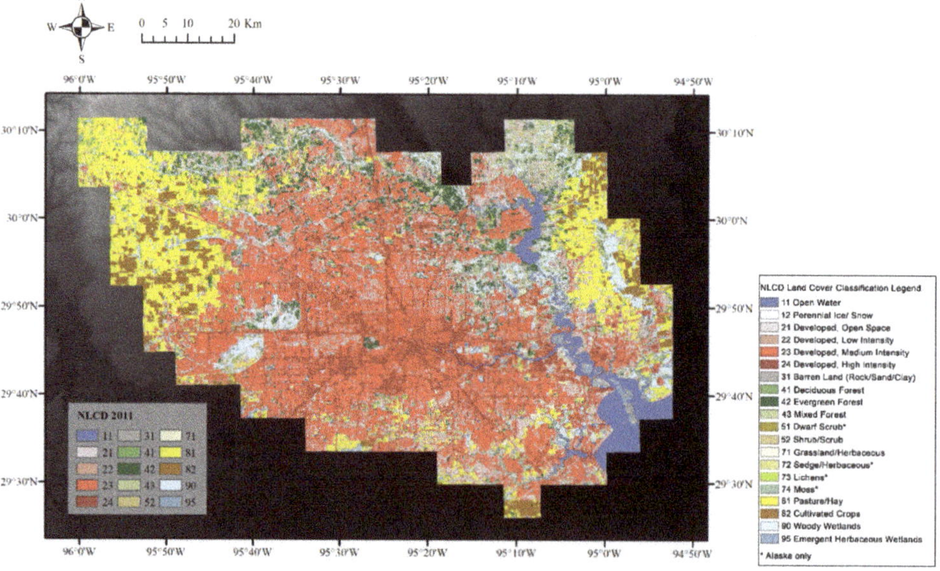

Figure 2. Land use and land cover from NLCD 2011.

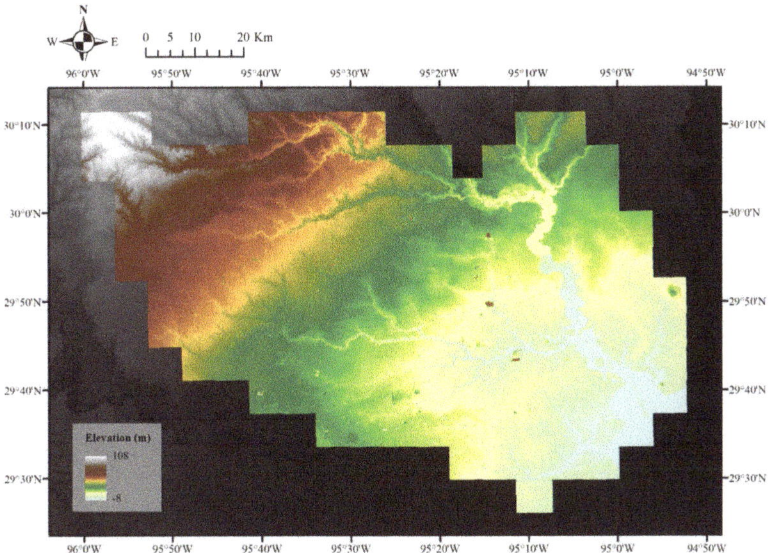

Figure 3. Elevation of Harris, Texas.

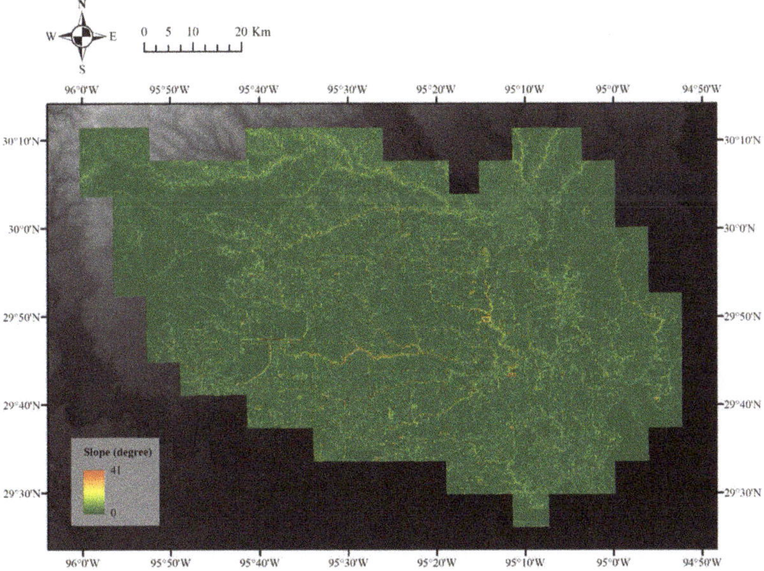

Figure 4. Slope of Harris, Texas.

The digital elevation model (DEM) was collected from the National Map 3D Elevation Program, and the source data are products of LIDAR point clouds, with a spatial resolution or ground spacing of 10 m. Based on the DEM, the slope is determined by the change rate of the surface in vertical and horizontal directions. An increase in slope suggests that the terrain is changing from a relatively flat state to a steeper state. As water flows from regions of high elevation to low elevation and usually is trapped in flat areas, the slope is a reasonable indicator of inundation. However, for this study site, a large portion of the area is of a very small slope, so this may not add much useful information as a spatial evidence layer.

The riverine floodplain and depth were calculated using gauges and flow lines shown in Figure 5. The nodes to simulate storm surge by the ADCIRC model are shown in Figure 6, with an enlarged frame for the dense points. Notice that the storm surge simulation could not provide full coverage of the study site and it is available for the coastal regions.

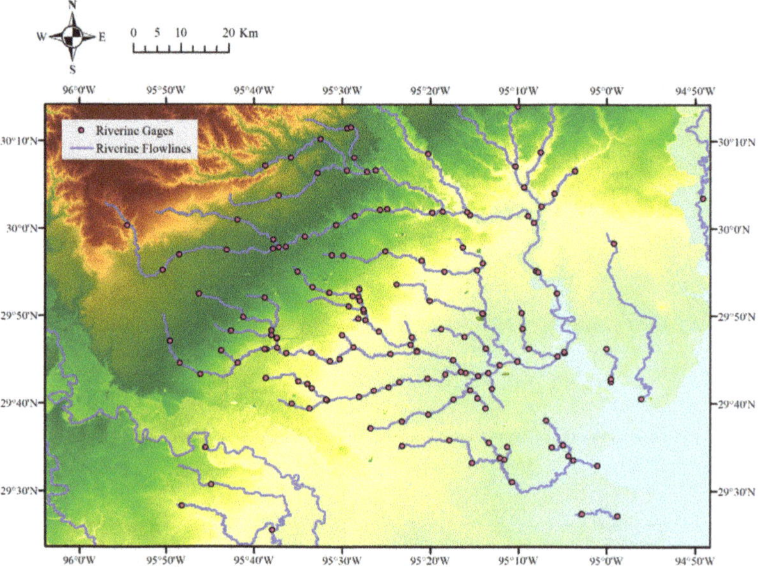

Figure 5. Riverine gauges and flow lines used for FEMA flood depth calculation.

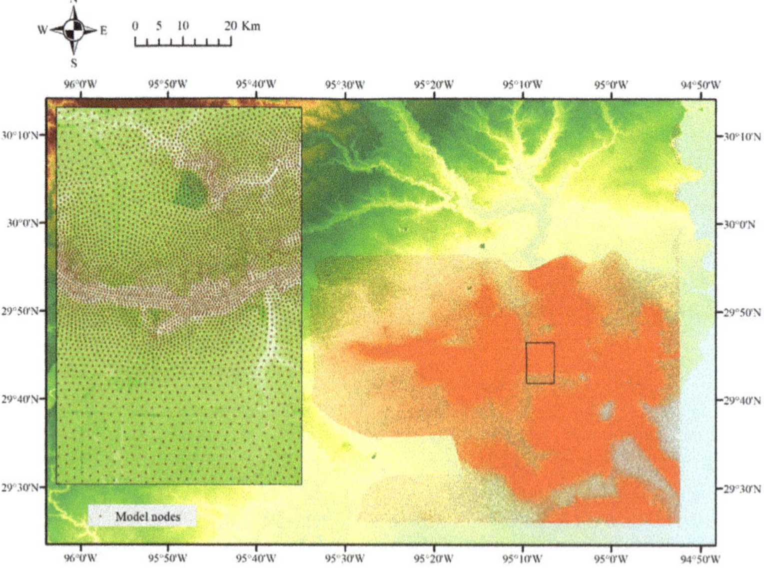

Figure 6. Storm surge model simulation nodes.

The riverine flood depth was collected and processed on 27–30 August, and 1 September (list in Table 1). The ADCIRC model outputs, simulated water heights, were collected and processed as the previous section from 27 to 30 August. As shown in Figure 7, the SAR images were acquired from Sentinel-1 on 18, 24, 30 August, and 5 September; the Planet and DMCii images were used to calculated NDWI on 31 August and 1 September. The Planet image could not cover the whole study region, missing the eastern part. The inundation areas were detected using the satellite images (Planet and Sentinel-1) and support vector machine classification with manually selected training samples. Then the inundation extent was used to estimate the weights of different spatial evidence based on the Bayesian conditional probability framework (Section 2.1).

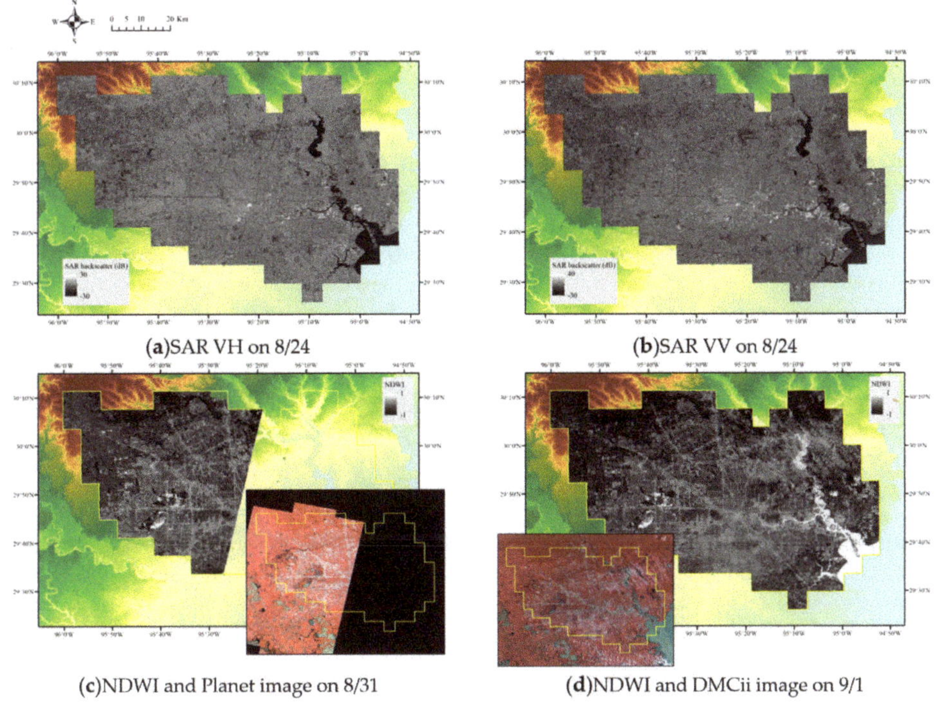

(**a**)SAR VH on 8/24 (**b**)SAR VV on 8/24

(**c**)NDWI and Planet image on 8/31 (**d**)NDWI and DMCii image on 9/1

Figure 7. Remotely sensed data from SAR (**a**,**b**) and optical (**c**,**d**) sensors.

Table 1. Available dates for different information sources.

Sources	8/18	8/24	8/27	8/28	8/29	8/30	8/31	9/1	9/5
River stream gauge			√	√	√	√		√	
Storm surge model			√	√	√	√			
Planet							√		
DMCii								√	
Sentinel-1	√	√				√			√

To assess the estimated inundation, satellite images and flood maps from multiple data sources were used as the ground reference, including supervised classification on the high-resolution optical image from DMCii and flood maps released by FEMA.

4. Results

In this study, the water extent derived from the satellite image acquired on August 30 (Sentinel-1) and 31 (Planet) was used to estimate the weights of various spatial evidence, and the resulting weight of evidence is listed in Table 2. Each evidence layer is divided into multiple subsets, according to quantiles or land cover types. As the riverine flood depth and the storm surge water height can be available daily, the depth and height of the day that the remotely sensed image was acquired are used to generate the multi-class evidence layers. The ranges of different variable classes are also included in Table 2, which correspond to quantiles of the spatial evidence or land cover types. Regions of low SAR backscatter intensity (for both polarization), high NDWI, flood depth or storm surge height have larger positive weights W^+ and contrasts C, indicating possibilities of positive spatial association. The areas with SAR backscatter intensity below -18 dB for polarization VH (vertical transmit, horizontal receive) and below -25 dB for VV (vertical transmit and receive) are possibly open water. In addition, open water (Land Cover Class 11) and bare soil (31) are also highly related to inundation, having positive weights of 4.45 and 1.42. Grassland (71) and cropland (82) are less relevant (weights lower than 0.25). The standard deviations of weights and contrast are small, thanks to the large sample size (amount of image pixels). The studentized contrast $C/S(C)$ is used to determine the final weights of spatial evidence, with a bold value in Table 2, suggesting the final weight selected to calculate inundation probability. Since the five types of spatial evidence are assumed to be positively related to inundation, only positive weights are used for the presence of spatial evidence. As mentioned in Section 2, the variance of the posterior probability P_{post} could be estimated by the variance of weights by Equation (13). The variance of weights, S^2W^+ listed in Table 2, are small, ranging from 8×10^{-7} to 1×10^{-4}. Based on Equation (13) and the calculated variance of weights, the variance of inundation probability P_{post} is small.

The predictor map, the posterior probability of inundation, is obtained by combining the presence of different forms of spatial evidence and their weights. The prior odd of inundation is estimated as the ratio of open water to the total area of interest. In Figures 8 and 9, the posterior probabilities of inundation are shown for Harris, Texas, on 18, 24, 27–30 August, and 1 and 5 September 2017. During this period, the water extent increased and submerged areas could be found along the rivers and streams, and around the reservoirs (Addicks and Barker Reservoir). From 27 August (Figure 8c), flood water could be found near the center of Houston, along the streams. From 18 August to 1 September (Figures 8d–g and 9), inundation probability around the two reservoirs was increased. However, the inundation evolution is not smooth. The water area near the reservoirs expands in Figure 8d–f, with relatively low inundation probability. In Figure 8g, the water area decreases but the inundation probability increases. This inconsistency may be due to data availability. River stream gauge data are available from 28 to 30 August and 1 September, the storm surge model 28 to 10 August, and satellite data from 30 August to 1 September. Except for 28 and 29 August, the daily data availability is different, even for two consecutive days. Such a difference in data availability and data quality could lead to the inconsistency of estimated probability.

Table 2. Calculation of initial weights (W^+, W^-), contrast (C), variances of weights and contrast $\left(S^2W^+,\ S^2W^-, S(C)\right)$, studentized contrast ($C/S(C)$), and final weight $\left(W_{final}\right)$ selected for different evidence layers. Each row is a spatial layer extracted from one spatial evidence that its value falls between the lower and upper boundaries.

Layer	Lower	Upper	W^+	W^-	C	$S^2(W^+)$	$S^2(W^-)$	$S(C)$	$C/S(C)$	W_{final}
SAR(VH)	−29.321	−18.268	1.85379	−0.0432	1.89697	7.15×10^{-6}	2.91×10^{-7}	0.00273	695.543	1.85379
SAR(VV)	−37.414	−25.61	3.4625	−0.1255	3.58796	6×10^{-6}	3.13×10^{-7}	0.00251	1427.72	3.4625
NDWI	0.12139	0.15152	0.16514	−0.0036	0.16879	1.19×10^{-5}	2.83×10^{-7}	0.00349	48.3063	0.16514
	0.15152	0.19437	0.91802	−0.0292	0.94721	6.36×10^{-6}	2.9×10^{-7}	0.00258	367.425	0.91802
	0.19437	0.30909	2.4979	−0.1535	2.65136	2.87×10^{-6}	3.24×10^{-7}	0.00179	1482.94	2.4979
	0.30909	0.51724	6.87541	−0.4697	7.34514	3.84×10^{-5}	4.34×10^{-7}	0.00623	1178.79	6.87541
	0.51724	0.78641	12.1489	−0.479	12.6279	7.25×10^{-3}	4.38×10^{-7}	0.08513	148.333	12.1489
Land cover	10	19	4.44941	−1.1515	5.6009	2.20×10^{-6}	8.51×10^{-7}	0.00175	3205.45	4.44941
	30	39	1.4161	−0.023	1.43915	1.08×10^{-5}	2.85×10^{-7}	0.00333	432.416	1.4161
	70	79	0.24489	−0.0073	0.2522	5.07×10^{-6}	1.67×10^{-7}	0.00229	110.168	0.24489
	80	89	0.13477	−0.0314	0.16614	8.11×10^{-7}	2.02×10^{-7}	0.00101	165.116	0.13477
	90	99	0.04837	−0.0043	0.05269	1.93×10^{-6}	1.76×10^{-7}	0.00145	36.2931	0.04837
River flood	5.95694	8.14822	0.27213	−0.006	0.27816	1.14×10^{-5}	2.84×10^{-7}	0.00341	81.4998	0.27213
	8.14822	10.3568	1.82436	−0.0959	1.92023	3.29×10^{-6}	3.08×10^{-7}	0.0019	1011.98	1.82436
	10.3568	13.1327	2.22213	−0.1375	2.35964	2.8×10^{-6}	3.2×10^{-7}	0.00177	1336.4	2.22213
	13.1327	17.879	1.4284	−0.0544	1.4828	4.72×10^{-6}	2.96×10^{-7}	0.00224	662.283	1.4284
	17.879	61.2138	1.53856	−0.0616	1.60014	4.41×10^{-6}	2.98×10^{-7}	0.00217	737.28	1.53856
Storm surge	−2.0662	0.00566	3.99019	−0.0564	4.04662	1.88×10^{-5}	2.92×10^{-7}	0.00437	925.317	3.99019
	0.00566	0.02091	6.35328	−0.0747	6.42796	1.2×10^{-4}	2.97×10^{-7}	0.01097	586.072	6.35328
	0.02091	0.06425	5.22234	−0.07	5.29238	4.37×10^{-5}	2.96×10^{-7}	0.00663	798.172	5.22234
	0.06425	0.08073	6.13127	−0.0742	6.20543	9.75×10^{-5}	2.97×10^{-7}	0.00989	627.651	6.13127
	0.08073	0.08867	6.082	−0.074	6.15601	9.31×10^{-5}	2.97×10^{-7}	0.00966	636.949	6.082
	0.08867	0.09885	6.83344	−0.0757	6.90912	1.9×10^{-4}	2.97×10^{-7}	0.01377	501.644	6.83344
	0.09885	0.11356	6.16258	−0.0743	6.23688	1×10^{-4}	2.97×10^{-7}	0.01003	621.95	6.16258
	0.11356	0.34259	3.52539	−0.0484	3.57383	1.56×10^{-5}	2.9×10^{-7}	0.00398	896.943	3.52539
	0.34259	1.667	3.23446	−0.0428	3.27726	1.45×10^{-5}	2.89×10^{-7}	0.00385	851.103	3.23446
	1.667	11.182	1.21562	−0.0083	1.22395	2.65×10^{-5}	2.8×10^{-7}	0.00517	236.549	1.21562

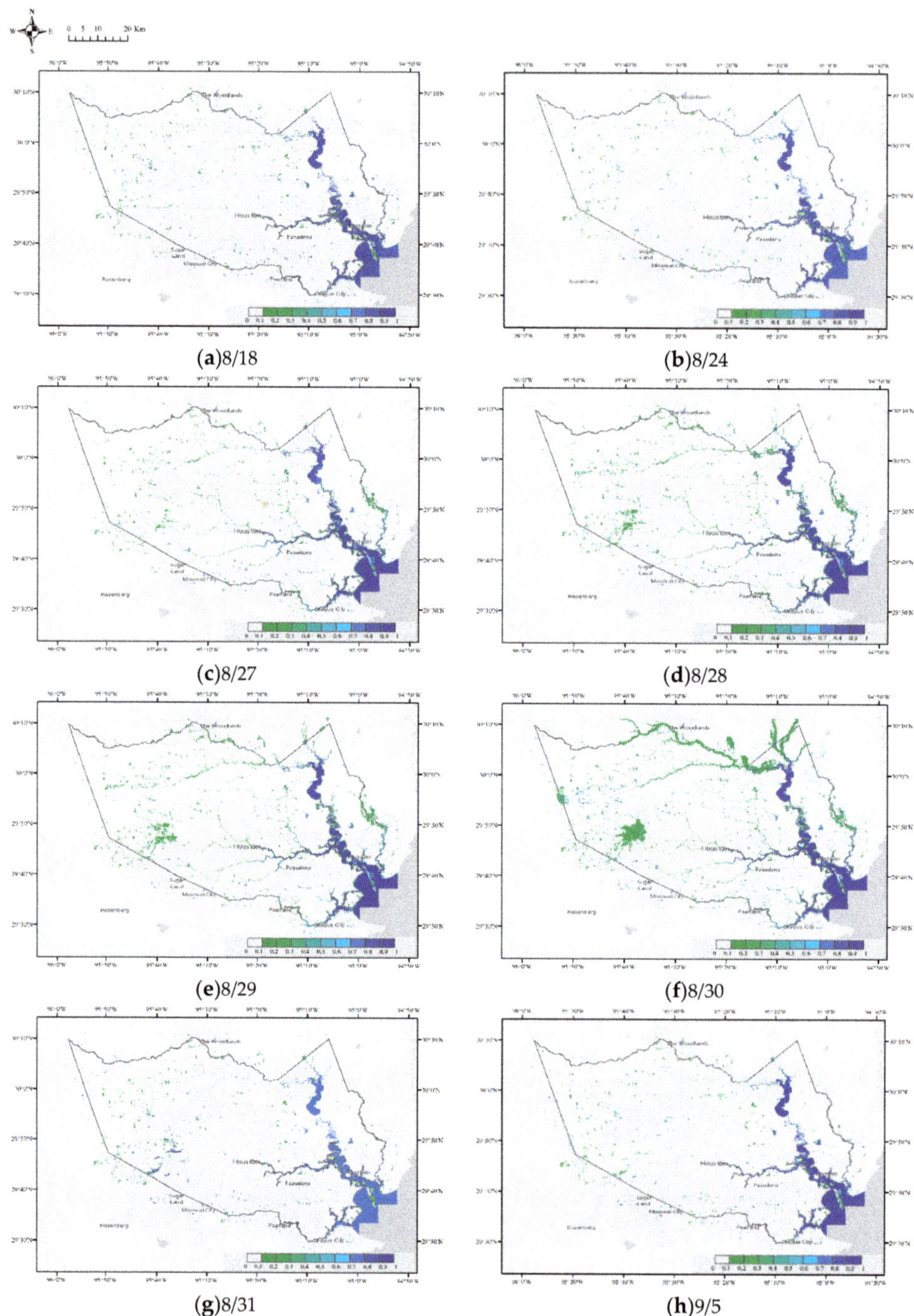

Figure 8. (**a–h**) The posterior probability of inundation from 18 August to 5 September 2017.

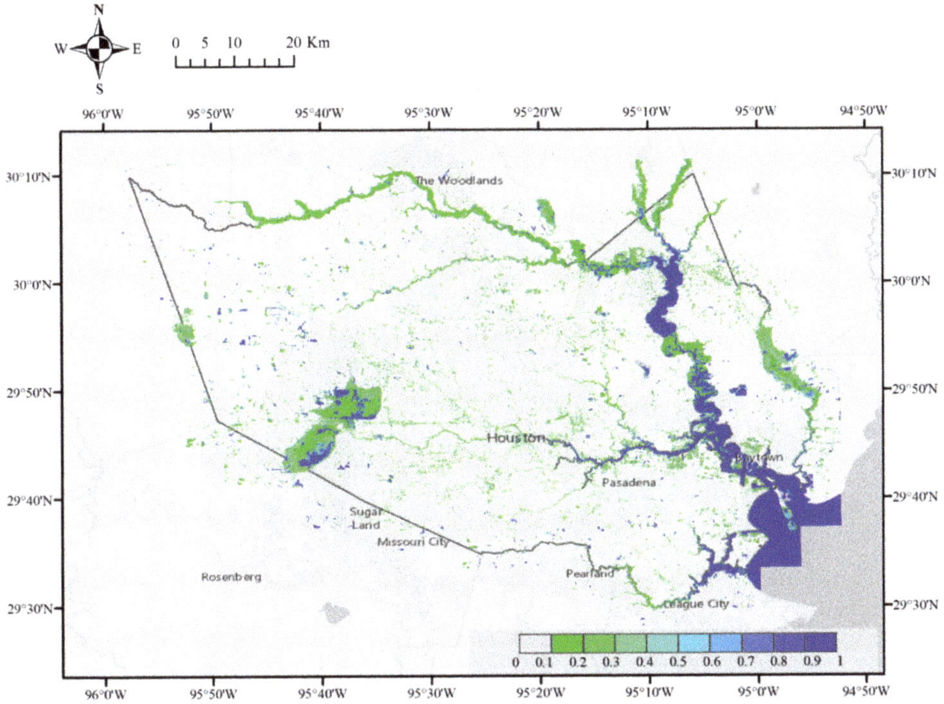

Figure 9. The posterior probability of inundation on 1 September 2017.

There is no available storm surge simulation on 1 September, so the inundation probability map (Figure 9) is generated by remote sensing data, riverine flood depth, and land cover data. An optical image is used to assess the effectiveness of the method. This image was collected by DMCii with a multi-spectral sensor and a spatial resolution of 22 m. Six locations in Harris are selected to show detailed water distribution, as in the yellow boxes in Figure 10.

As shown in Figures 11–16, the validated water extent (blue areas in (b)) agrees well with the regions of inundation probability larger than 0.9. The estimation generated from the proposed method is more accurate than the FEMA flood map, which tends to overestimate the water extent.

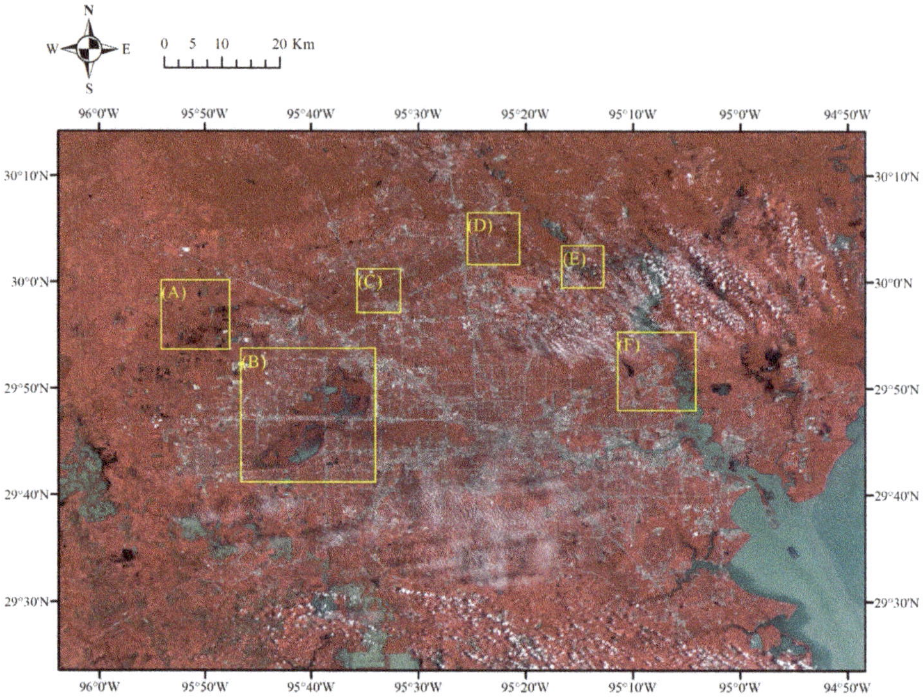

Figure 10. The DMCii image acquired on 1 September, along with boxes frames indicating flooded areas to be investigated.

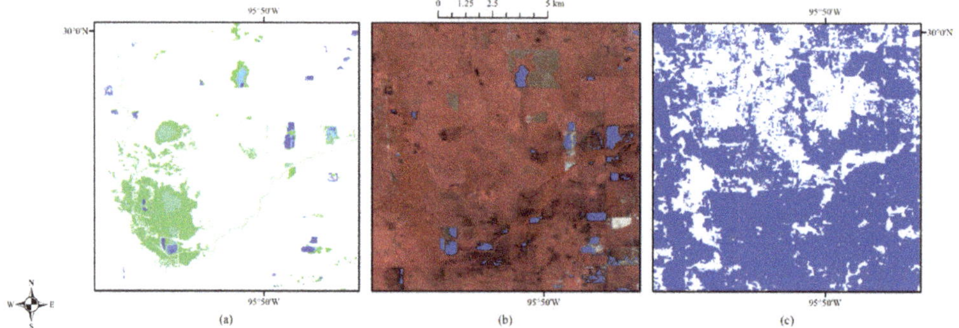

Figure 11. Enlarged site (A): inundation probability (**a**), the corresponding optical image on September 1 (**b**), and (**c**) FEMA flood map, with water shown in blue.

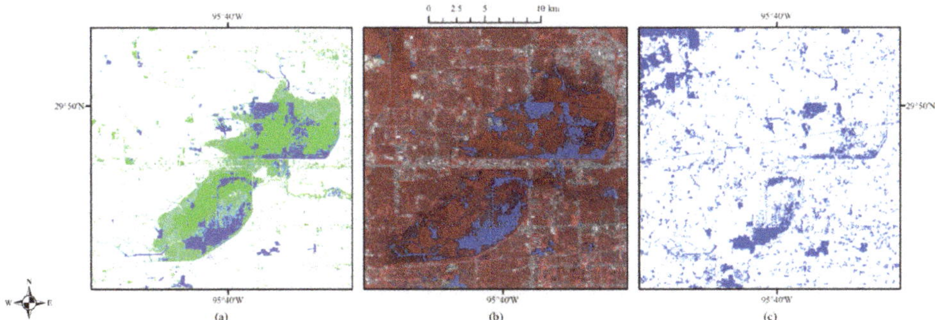

Figure 12. Enlarged site (B): inundation probability (**a**), the corresponding optical image on 1 September (**b**), and (**c**) FEMA flood map, with water shown in blue.

Figure 12 shows the Addicks Reservoir and the Barker Reservoir, which is a 'dry' reservoir covered by vegetation usually. From these two reservoirs, water was released with control towards the Buffalo Bayou after 28 August. Despite the original attempts to protect the neighboring area, the Addicks Reservoir began to spill out after reaching its capacity. The recorded water level during Hurricane Harvey was the highest since the construction of the reservoirs.

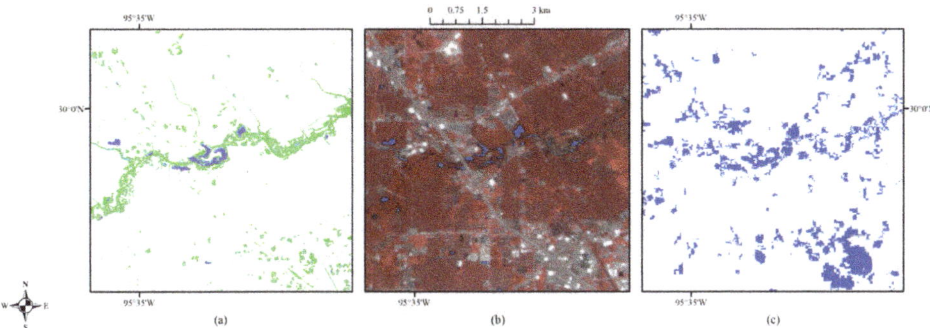

Figure 13. Enlarged site (C): inundation probability (**a**), the corresponding optical image on 1 September (**b**), and (**c**) FEMA flood map, with water shown in blue.

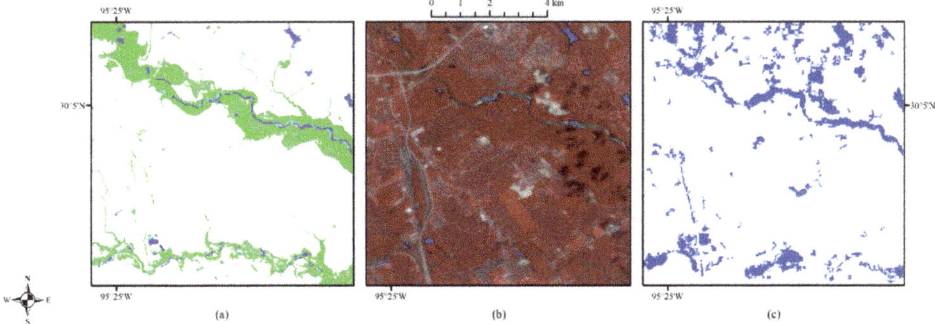

Figure 14. Enlarged site (D): inundation probability (**a**), the corresponding optical image on 1 September (**b**), and (**c**) FEMA flood map, with water shown in blue.

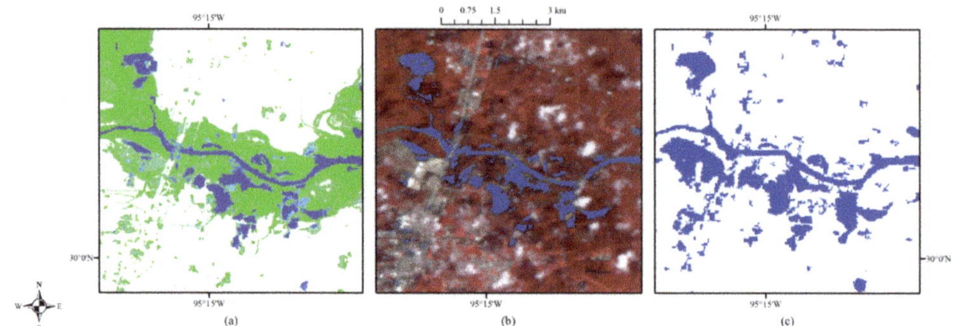

Figure 15. Enlarged site (E): inundation probability (**a**), the corresponding optical image on 1 September (**b**), and (**c**) FEMA flood map, with water shown in blue.

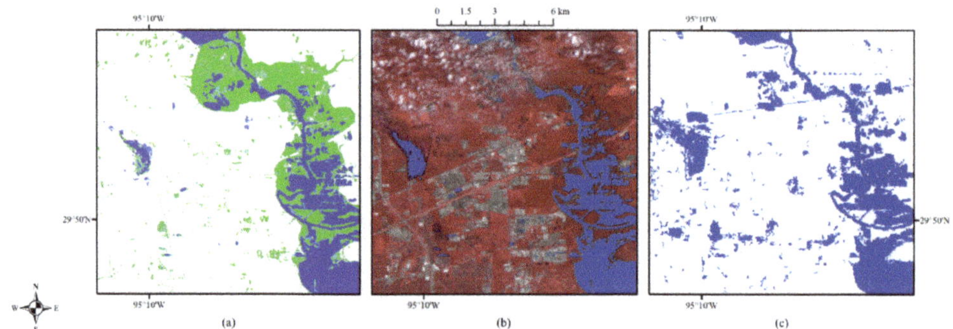

Figure 16. Enlarged site (F): inundation probability (**a**), the corresponding optical image on 1 September (**b**), and (**c**) FEMA flood map, with water shown in blue.

Based on the time series maps from 18 August to 5 September (Figures 8 and 9), the maximum estimated inundation probability for each pixel along the time series is produced and shown in Figure 17, along with extracted tweets and high water marks. The observed points are usually inside or near the regions of high inundation probability, especially around the streams or large water bodies. For the areas with low inundation probability, if those regions are at high flood risk, for example, places with a short distance to the river and with low elevation, inundation could be possible. However, the low inundation probability in potential flooded urban areas implies a major limitation of the proposed fusion method.

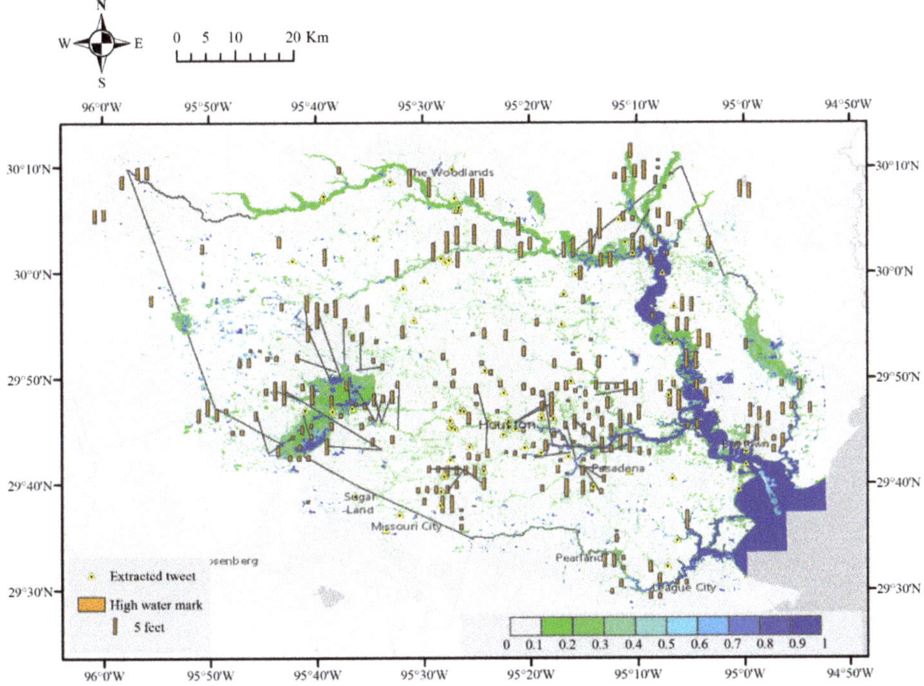

Figure 17. Extracted tweets and high water marks on top of the max-inundation probability from 14 August to 5 September when data are available.

Similarly to the calculation of maximum inundation probability, the mean and standard deviation are also calculated and shown in Figure 18. High mean inundation probability suggests normal or permanent water, and areas of high standard deviation are generally flooded regions where surface water is uncommon. Using the maps of mean and standard deviation, flooded regions with severe impact could be identified, such as Addicks Reservoir, Barker Reservoir, and the riverside near downtown Huston.

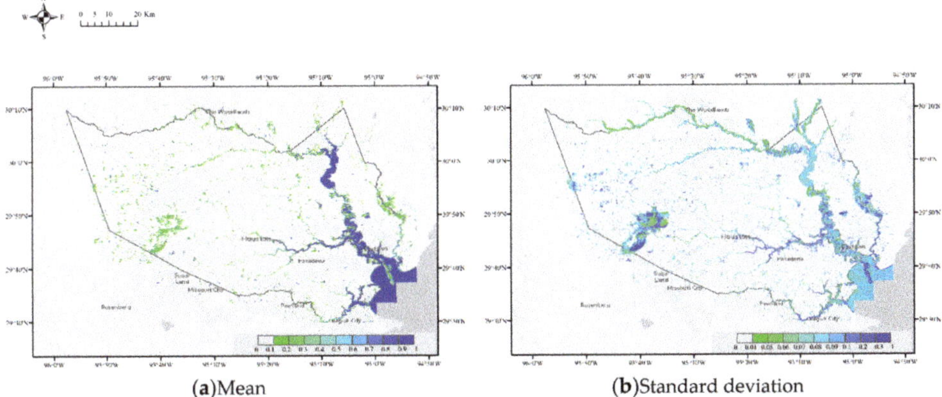

(a) Mean **(b)**Standard deviation

Figure 18. (**a,b**) The mean and standard deviation of inundation probability from 17 August to 5 September when data are available.

A reliable diagram of 1 September (Figure 19) is used to quantitatively assess the prediction. The probability range [0, 1] is split into ten intervals [0, 0.1], (0.1, 0.2], ..., (0.9, 1]. Pixels of estimated probability within each interval form a subset, and in such a subset, the validated water extent ratio is used as the validated probability. Generally, the prediction is close to the validation, especially for the regions with relatively small (<0.5) or large (around 0.7 and 0.9) inundation probability. Though the prediction reliabilities of intervals (0.5, 0.6] and (0.7, 0.9) are lower than other intervals, the pixel amount of these intervals is rather small, as listed in Table 3. In order to consider the pixel number, the weighted root-mean-square difference ε was calculated, with a value of 0.0686.

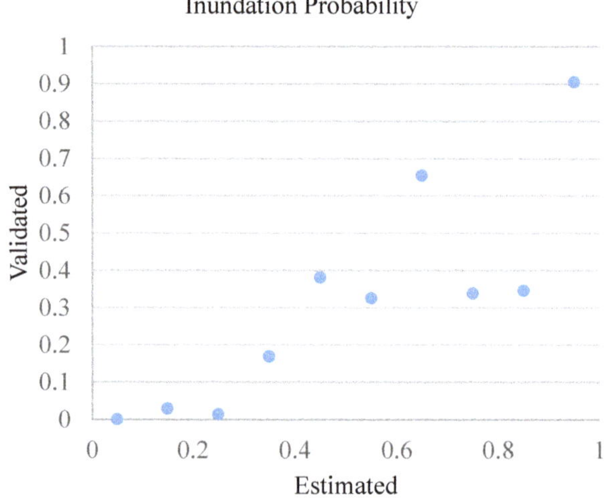

Figure 19. Estimated and validated probability.

Table 3. The number of pixels for different predicted probability intervals.

Predicted Probability	Number of Pixels
[0, 0.1]	101,463,434
(0.1, 0.2]	3,457,280
(0.2, 0.3]	1,687,411
(0.3, 0.4]	1,438,002
(0.4, 0.5]	119,149
(0.5, 0.6]	37,904
(0.6, 0.7]	69,981
(0.7, 0.8]	269,509
(0.8, 0.9]	151,728
(0.9, 1]	3,422,574

5. Discussion

In this paper, we proposed a framework to fuse multi-source information to generate daily maps of inundation probability. We incorporated remote sensing data, riverine flood depth, storm surge simulation, and land use and land cover. The fusion framework makes use of available data of a given day, because the multi-source data could not be acquired with the same frequency. In line with previous studies [11–18], the remotely sensed inundation extent is clearly useful to observe flood events, but such a technique could be sensitive to the timing of image acquisition. The SAR images used in this study are theoretically accessible every six days, but observation gaps still exist. The optical images, though more informative, are limited due to heavy cloud coverage, which is often present during a flood event. When the optimal data sources are limited, we utilized other information sources and provided a series of consistent probability maps. We found that each data source has some limitations and uncertainty, but the fused result is more accurate than each component in terms of identifying the inundation extent. The estimated inundation probabilities using fused components are more realistic than the ones using a single data source.

Each data source has limitations in mapping inundation. In Figure 20, the light blue regions indicate areas of high uncertainty. The land use and land cover data are static during the flood evolution, so the probabilities estimated from the land cover are also constant for different days. Areas with high probability in NDWI and SAR backscatter are correlated with the high probability areas in the fusion result. In this study case, NDWI estimates an extensive area of rather low inundation probability. SAR backscatter estimates less water extent but with higher inundation probability than that from NDWI. Although the probability maps estimated from NDWI and SAR backscatter are for different days, the difference in probability could be dramatic for the same event.

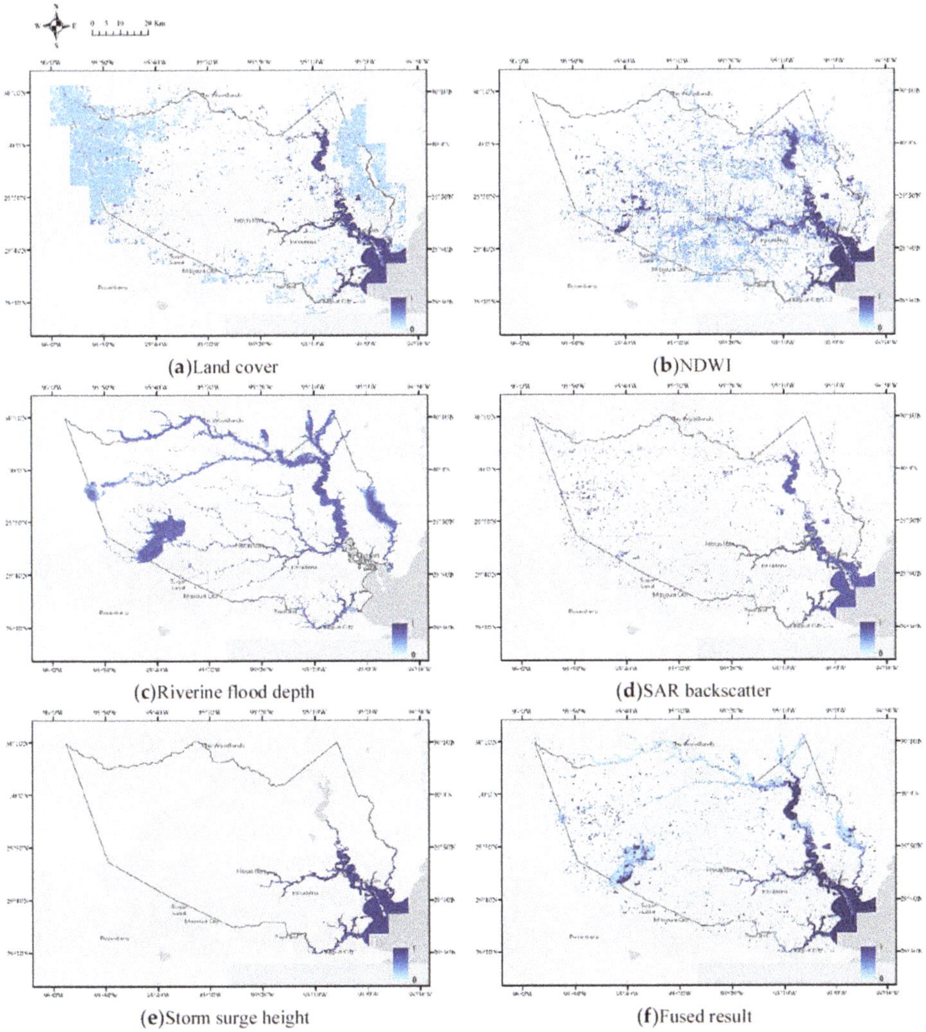

(**a**)Land cover   (**b**)NDWI

(**c**)Riverine flood depth   (**d**)SAR backscatter

(**e**)Storm surge height   (**f**)Fused result

Figure 20. (**a–f**) Estimated inundation probability using single and fused components.

The riverine flood depth relies on the estimation of water surface elevation, which could inevitably include some inaccuracies or even errors. First, gauge readings are usually fairly accurate, but on the one hand, not every record is uploaded and usable. On the other hand, the daily highest water level may not correspond to the water level of the flood crest. Second, interpolation based on gauges and reinforcement points may not capture the reality among areas without reliable observation. Third, flood depth calculation using DEM would suffer from the propagated error of the input data. Lastly, the removal of isolated flooded regions could ignore some actually flooded areas.

Similarly, the water heights simulated by the storm surge model have their uncertainties from input data qualities and modeling reliability. Furthermore, the transformation from storm surge nodes (points) to surface (raster) via Thiessen polygons might have some local issues. For example, the boundaries of river banks or coasts show artifacts with abrupt transitions. More complex interpolation and post-processing considering topography could be helpful. One possible solution would be using the initial storm surge model mesh to interpolate water heights, but given the fine resolution in the coastal mesh, the improvement may not be significant. Nevertheless, the Thiessen polygon generation would be more efficient for near-real-time or real-time applications, because one set of polygons could be applied to the flood event for a few days or even months, without repeating interpolation processes.

The fused result (Figure 20f) is a statistical combination of the different components (Figure 20a–e). In the Results section, we compared the estimated inundation probability with the remotely sensed water extent and observation points (flood Tweets and high water mark). Areas with high probability tend to be flooded (Figures 11–16). However, observation points can be found near places of relatively low probability, especially in the city center. This is a major limitation of the proposed method. Although the observation points are not evenly distributed and of different scales (much higher resolution than the spatial evidence layers), the points are reliable data showing water presence. The overlaid map of points and probability shows that regions with low probability can be flooded when those regions are at high flood risk. Future studies can consider flood risks, such as distance to rivers or low-elevation places, modeled flood return period, and historic flood frequency. In addition, additional data in urban areas can be beneficial, including sewer systems and SAR interferometry coherence.

We also tried to include elevation and slope information but later found that the elevation-based information could not improve the results. The weight of evidence modeling assumes conditional independence among different predictors. Since the elevation data were used to calculate riverine flood depth and slope, the elevation and slope do not provide additional information as other independent spatial evidence.

The spatial evidence layers used in this study are assumed to be independent as they are collected from independent measurements. These layers are combined through a probabilistic approach so that a higher probability value suggests a lower level of uncertainty. However, the uncertainty over each day of the time period varies for different places, because the availability of spatial evidence layers is different most of the days. Another issue that has not been addressed is the propagation of uncertainty in calculating inundation probability. Although the measurements of remote sensors and river gauges are assumed to be accurate enough, the weights of spatial evidence are calculated for non-overlapped subsets of each layer. For example, the final weight W_{final} of river flood depth between 5.96 and 8.15 feet is 0.27, and W_{final} of depth between 8.15 and 10.36 feet is 1.82. Within a subset (e.g., 5.96–8.15), the uncertainty is assumed to be even. Areas of flood depth 8.14 and 8.16 feet have quite different final weights (0.27 and 1.82). Consequently, uncertainty may propagate from layer preprocessing, so layer value could end up falling in a less reliable weight.

In Harris County, Texas, only 15% of homes have flood insurance, according to the National Flood Insurance Program (NFIP). Meanwhile, flooding has been common since the founding of Houston; for example, the previous floods in May 2015 and April 2016. Houston had been submerged by three 500-year floods in three consecutive years before Harvey. Due to the lack of planning and zoning restrictions, Houston's urban sprawl is astonishing, even within floodplains. Around the reservoirs in Harris, the population increased from 400,000 in the 1940s to six million in 2017. To address the increasing challenge from environment and society [34], mitigation and adaptation measures, including seawalls, levees, and building codes, need to be employed. All these measures need to be informed by timely and continuous inundation maps. As disasters like Harvey will happen with increasing frequency and intensity, mitigation and adaptation are required based on an understanding of event evolution. By providing the inundation probability, this study could be readily implemented.

6. Conclusions

Extreme precipitation and sea level rise in a warming climate are aggravating flooding—a billion-dollar disaster, as called by the NOAA. Flood response, mitigation and adaptation require a comprehensive understanding of inundation evolution, yet our flood observations are limited regarding spatial and temporal coverage. This study tried to improve flood detection and observation in coastal areas by combining current satellite remote sensing-based techniques, in situ information from gauge measures, and storm surge model outputs. We proposed a fusion framework to integrate available information based on Bayesian conditional probability. The fusion product is a daily inundation probability map utilizing all accessible data. The input data could be different for each day, but the output is consistent within the probability range [0, 1]. Probabilistic flood maps could incorporate uncertainty propagated from the input data. The maps are useful for rapid response when important decisions such as collecting emergency aid or relocating residents need to be made.

Although this proposed methodological framework could provide useful results for the case studies, there are several limitations regarding input data uncertainty and error propagation. The accuracy of the proposed flood mapping methods varies in different regions of an image, but the error models between geographic characters have not been investigated. Lacking such information about the error, the generalizability of the proposed algorithms would be limited. The fusion model was trained locally for one study site, so additional consideration would be necessary when transferring the model to new sites.

Author Contributions: Conceptualization, J.L. and D.L.; methodology and validation, J.L. and D.L.; writing—original draft preparation, J.L.; writing—review and editing, D.L.; funding acquisition, D.L. All authors have read and agreed to the published version of the manuscript.

Funding: This research was funded by National Science Foundation Award, grant number 1520870 "Hazards SEES: Social and Physical Sensing Enabled Decision Support for Disaster Management and Response".

Acknowledgments: We would like to thank Ethan Kubatko, Srinivasan Parthasarathy, and Steven Quiring for their comments in developing this study.

References

1. Shultz, J.M.; Galea, S. Mitigating the mental and physical health consequences of hurricane harvey. *JAMA* **2017**, *318*, 1437–1438. [CrossRef] [PubMed]
2. Rappaport, E.N. Fatalities in the United States from Atlantic tropical cyclones: New data and interpretation. *Bull. Am. Meteorol. Soc.* **2014**, *95*, 341–346. [CrossRef]
3. Jonkman, S.N.; Maaskant, B.; Boyd, E.; Levitan, M.L. Loss of life caused by the flooding of new orleans after hurricane katrina: Analysis of the relationship between flood characteristics and mortality. *Risk Anal.* **2009**, *29*, 676–698. [CrossRef] [PubMed]
4. Jonkman, S.N.; Godfroy, M.; Sebastian, A.; Kolen, B. Brief communication: Loss of life due to Hurricane Harvey. *Nat. Hazards Earth Syst. Sci.* **2018**, *18*, 1073–1078. [CrossRef]
5. Rahmstorf, S. Rising hazard of storm-surge flooding. *Proc. Natl. Acad. Sci. USA* **2017**, *114*, 11806–11808. [CrossRef]
6. Adhikari, P.; Hong, Y.; Douglas, K.R.; Kirschbaum, D.B.; Gourley, J.; Adler, R.; Robert Brakenridge, G. A digitized global flood inventory (1998–2008): Compilation and preliminary results. *Nat. Hazards* **2010**, *55*, 405–422. [CrossRef]

7. Garner, A.J.; Mann, M.E.; Emanuel, K.A.; Kopp, R.E.; Lin, N.; Alley, R.B.; Horton, B.P.; DeConto, R.M.; Donnelly, J.P.; Pollard, D. Impact of climate change on New York City's coastal flood hazard: Increasing flood heights from the preindustrial to 2300 CE. *Proc. Natl. Acad. Sci. USA* **2017**, *114*, 11861–11866. [CrossRef]

8. Trenberth, K.E.; Cheng, L.; Jacobs, P.; Zhang, Y.; Fasullo, J. Hurricane harvey links to ocean heat content and climate change adaptation. *Earths Future* **2018**, *6*, 730–744. [CrossRef]

9. Risser, M.D.; Wehner, M.F. Attributable human-induced changes in the likelihood and magnitude of the observed extreme precipitation during hurricane harvey. *Geophys. Res. Lett.* **2017**, *44*, 12457–12464. [CrossRef]

10. Khakzad, N.; Van Gelder, P. Vulnerability of industrial plants to flood-induced natechs: A Bayesian network approach. *Reliab. Eng. Syst. Saf.* **2018**, *169*, 403–411. [CrossRef]

11. Feyisa, G.L.; Meilby, H.; Fensholt, R.; Proud, S.R. Automated Water Extraction Index: A new technique for surface water mapping using Landsat imagery. *Remote Sens. Environ.* **2014**, *140*, 23–35. [CrossRef]

12. Pekel, J.-F.; Cottam, A.; Gorelick, N.; Belward, A.S. High-resolution mapping of global surface water and its long-term changes. *Nature* **2016**, *540*, 418–422. [CrossRef] [PubMed]

13. Pekel, J.F.; Vancutsem, C.; Bastin, L.; Clerici, M.; Vanbogaert, E.; Bartholomé, E.; Defourny, P. A near real-time water surface detection method based on HSV transformation of MODIS multi-spectral time series data. *Remote Sens. Environ.* **2014**, *140*, 704–716. [CrossRef]

14. Schumann, G.J.P.; Moller, D.K. Microwave remote sensing of flood inundation. *Phys. Chem. Earth* **2015**, *83–84*, 84–95. [CrossRef]

15. Giustarini, L.; Hostache, R.; Kavetski, D.; Chini, M.; Corato, G.; Schlaffer, S.; Matgen, P. Probabilistic flood mapping using synthetic aperture radar data. *IEEE Trans. Geosci. Remote Sens.* **2016**, *54*, 6958–6969. [CrossRef]

16. Chini, M.; Hostache, R.; Giustarini, L.; Matgen, P. A Hierarchical Split-Based Approach for Parametric Thresholding of SAR Images: Flood Inundation as a Test Case. *IEEE Trans. Geosci. Remote Sens.* **2017**, *55*, 6975–6988. [CrossRef]

17. Shen, X.; Anagnostou, E.N.; Allen, G.H.; Brakenridge, G.R.; Kettner, A.J. Near-real-time non-obstructed flood inundation mapping using synthetic aperture radar. *Remote Sens. Environ.* **2018**, *221*, 302–315. [CrossRef]

18. Chini, M.; Pelich, R.; Pulvirenti, L.; Pierdicca, N.; Hostache, R.; Matgen, P. Sentinel-1 InSAR coherence to detect floodwater in urban areas: Houston and Hurricane Harvey as a test case. *Remote Sens.* **2019**, *11*, 107. [CrossRef]

19. Teng, J.; Jakeman, A.J.; Vaze, J.; Croke, B.F.W.; Dutta, D.; Kim, S. Flood inundation modelling: A review of methods, recent advances and uncertainty analysis. *Environ. Model. Softw.* **2017**, *90*, 201–216. [CrossRef]

20. Noh, S.J.; Lee, J.H.; Lee, S.; Seo, D.J. Retrospective dynamic inundation mapping of hurricane harvey flooding in the houston metropolitan area using high-resolution modeling and high-performance computing. *Water* **2019**, *11*, 597. [CrossRef]

21. Wing, O.E.; Sampson, C.C.; Bates, P.D.; Quinn, N.; Smith, A.M.; Neal, J.C. A flood inundation forecast of Hurricane Harvey using a continental-scale 2D hydrodynamic model. *J. Hydrol. X* **2019**, *4*, 100039. [CrossRef]

22. Alfieri, L.; Burek, P.; Dutra, E.; Krzeminski, B.; Muraro, D.; Thielen, J.; Pappenberger, F. GloFAS—Global ensemble streamflow forecasting and flood early warning. *Hydrol. Earth Syst. Sci.* **2013**, *17*, 1161–1175. [CrossRef]

23. Carranza, E.J.M. Weights of evidence modeling of mineral potential: A Case study using small number of prospects, abra, philippines. *Nat. Resour. Res.* **2004**, *14*, 173–187. [CrossRef]

24. Rosser, J.F.; Leibovici, D.G.; Jackson, M.J. Rapid flood inundation mapping using social media, remote sensing and topographic data. *Nat. Hazards* **2017**, *87*, 103–120. [CrossRef]

25. Horritt, M.S.; Mason, D.C.; Luckman, A.J. Flood boundary delineation from Synthetic Aperture Radar imagery using a statistical active contour model. *Int. J. Remote Sens.* **2001**, *22*, 2489–2507. [CrossRef]

26. Hussein, S. Al-olimat, steven gustafson, jason mackay, krishnaprasad thirunarayan, and amit sheth. A practical incremental learning framework for sparse entity extraction. In Proceedings of the 27th International Conference on Computational Linguistics (COLING 2018), Santa Fe, NM, USA, 20–26 August 2018.

27. Torres, R.; Snoeij, P.; Geudtner, D.; Bibby, D.; Davidson, M.; Attema, E.; Potin, P.; Rommen, B.; Floury, N.; Brown, M.; et al. GMES Sentinel-1 mission. *Remote Sens. Environ.* **2012**, *120*, 9–24. [CrossRef]

28. McFeeters, S.K. The use of the Normalized Difference Water Index (NDWI) in the delineation of open water features. *Int. J. Remote Sens.* **2007**, *17*, 1425–1432. [CrossRef]

29. Nobre, A.D.; Cuartas, L.A.; Hodnett, M.; Rennó, C.D.; Rodrigues, G.; Silveira, A.; Saleska, S. Height above the nearest drainage—A hydrologically relevant new terrain model. *J. Hydrol.* **2011**, *404*, 13–29. [CrossRef]

30. Dawson, C.; Kubatko, E.J.; Westerink, J.J.; Trahan, C.; Mirabito, C.; Michoski, C.; Panda, N. Discontinuous Galerkin methods for modeling hurricane storm surge. *Adv. Water Resour.* **2011**, *34*, 1165–1176. [CrossRef]

31. Dietrich, J.C.; Dawson, C.N.; Proft, J.M.; Howard, M.T.; Wells, G.; Fleming, J.G.; Luettich, R.A.; Westerink, J.J.; Cobell, Z.; Vitse, M.; et al. Real-Time Forecasting and Visualization of Hurricane Waves and Storm Surge Using SWAN+ADCIRC and FigureGen. In *Computational Challenges in the Geosciences*; The IMA Volumes in Mathematics and Its Applications; Springer: New York, NY, USA, 2013; Volume 156, pp. 49–70.

32. Homer, C.; Dewitz, J.; Yang, L.; Jin, S.; Danielson, P.; Xian, G.; Coulston, J.; Herold, N.; Wickham, J.; Megown, K. Completion of the 2011 National Land Cover Database for the conterminous United States–representing a decade of land cover change information. *Photogramm. Eng. Remote Sens.* **2015**, *8*, 345–354.

33. Yang, L.; Jin, S.; Danielson, P.; Homer, C.; Gass, L.; Bender, S.M.; Case, A.; Costello, C.; Dewitz, J.; Fry, J.; et al. A new generation of the United States National Land Cover Database: Requirements, research priorities, design, and implementation strategies. *ISPRS J. Photogramm. Remote Sens.* **2018**, *146*, 108–123. [CrossRef]

34. Sebastian, A.; Gori, A.; Blessing, R.B.; van der Wiel, K.; Bass, B. Disentangling the impacts of human and environmental change on catchment response during Hurricane Harvey. *Environ. Res. Lett.* **2019**, *14*, 124023. [CrossRef]

MDPI

St. Alban-Anlage 66

4052 Basel

Switzerland

Tel. +41 61 683 77 34

Fax +41 61 302 89 18

www.mdpi.com

Remote Sensing Editorial Office

E-mail: remotesensing@mdpi.com

www.mdpi.com/journal/remotesensing